Weed Science
Cannabis Controversies and Challenges

Weed Science

Cannabis Controversies and Challenges

Godfrey Pearlson
Yale University, New Haven, CT, USA

ACADEMIC PRESS
An imprint of Elsevier

ELSEVIER

Academic Press is an imprint of Elsevier
125 London Wall, London EC2Y 5AS, United Kingdom
525 B Street, Suite 1650, San Diego, CA 92101, United States
50 Hampshire Street, 5th Floor, Cambridge, MA 02139, United States
The Boulevard, Langford Lane, Kidlington, Oxford OX5 1GB, United Kingdom

Notices
Knowledge and best practice in this field are constantly changing. As new research and experience broaden our understanding, changes in research methods, professional practices, or medical
treatment may become necessary.

Practitioners and researchers must always rely on their own experience and knowledge in evaluating and using any information, methods, compounds, or experiments described herein. In using
such information or methods they should be mindful of their own safety and the safety of others,
including parties for whom they have a professional responsibility.

To the fullest extent of the law, neither the Publisher nor the authors, contributors, or editors,
assume any liability for any injury and/or damage to persons or property as a matter of products
liability, negligence or otherwise, or from any use or operation of any methods, products, instructions, or ideas contained in the material herein.

Library of Congress Cataloging-in-Publication Data
A catalog record for this book is available from the Library of Congress

British Library Cataloguing-in-Publication Data
A catalogue record for this book is available from the British Library

ISBN: 978-0-12-818174-4

For information on all Academic Press publications
visit our website at https://www.elsevier.com/books-and-journals

Publisher: Nikki Levy
Acquisitions Editor: Joslyn Chaiprasert-Paguio
Editorial Project Manager: Tracy Tufaga
Production Project Manager: Paul Prasad Chandramohan
Designer: Miles Hitchen

Typeset by Thomson Digital

Working together
to grow libraries in
developing countries

www.elsevier.com • www.bookaid.org

*To my wife Judy and my daughters
Becky and Lishy, who were
cheerleaders for this project.*

Contents

7. Epidemiology

8. Toxicology

12. Summary

About the author

Godfrey Pearlson, award-winning physician and researcher, completed his medical degree at Newcastle University in the United Kingdom and a graduate philosophy degree at Columbia University in NYC. He trained in psychiatry and was subsequently a psychiatry professor at Johns Hopkins University. He is currently a professor of psychiatry and neuroscience at Yale University and directs a brain research center at the Institute of Living in Connecticut. He has published over 700 peer-reviewed scientific articles and his marijuana research is funded by the National Institute on Drug Abuse and the National Highway and Traffic Safety Administration. Dr. Pearlson is an expert in the fields of marijuana and of psychosis. He believes that any political platforms regarding cannabis should be based on scientific facts rather than emotion.

You can find out more about him at the website GodfreyPearlson.com

Foreword

Cannabis is the noisiest plant on earth. Between the divisive politics of legalization, the cyclical giddiness and despair over fortunes made and lost in the "dot. bong bubble," and the steady fear-mongering over the plant's supposed dangers, the din grows louder, which makes discerning the fictions from the facts challenging, if not at times, impossible.

That's why this book is so valuable. Godfrey Pearlson has managed to turn down the volume while increasing the scrutiny of the science. In doing so, he has produced something that every legislator, policy maker, medical professional, and truth-seeking individual should read. It couldn't be coming at a better time.

The truth is, cannabis *is* a confounding plant. Does it induce a state of euphoria or unhappiness? Does it quell anxiety or ramp it up? Does it lessen concentration or intensify focus? The answer to all of these questions? It does both depending on the dose, the variety of cannabis, the quantity consumed, how it is administered, and the still mysterious "tone" of each person's endocannabinoid system. We may blithely dismiss it as a mere weed but this plant is a complicated chemical factory of over 500 compounds that combine to produce different effects in different bodies.

In addition to the plant's botanical and chemical complexities, the politics of prohibition over the last decades have made things even more complicated. The largest obstacle to uncovering the science of cannabis is the Schedule 1 restrictions that US (and international) laws have placed on studying the plant since the 1970s. Long before there were Russian hackers and disinformation specialists, the US government knew that the best way to sow confusion was to limit scientific inquiry and simultaneously foment its own disinformation campaign. Schedule 1 is the perfect instrument to accomplish this. It states that cannabis has no medical use (despite the multitudes of people who rely on it to treat conditions ranging from epilepsy to pain to PTSD) and creates a tangle of red tape that keeps the even most intrepid American researchers away. This is tragic when you consider how much more we would know about the health benefits of cannabis if investigation had been encouraged rather than impeded for the last half century.

This is not to say we know nothing about the science of cannabis. In fact, as Dr. Raphael Mechoulam, the original OG of cannabis chemistry and the man who isolated THC and CBD in 1964 told me, we know quite a bit. A quick scan of Pub Med will reveal some 21,000 studies, many of which have been conducted by an impressive cadre of dedicated researchers from around the globe. Americans, who at times view the world through a very solipsistic lens, tend to discount these studies because they weren't performed within our hallowed borders. But discoveries about the endocannabinoid system, the endogenous cannabinoids our brains produce, or the effective medical uses of cannabis, come from studies conducted abroad. When viewed as a whole, this body of research provides a multitude of clues about how this plant works. But you have to be willing to look for and piece together those clues.

This is exactly what Pearlson, a professor of Psychiatry and Neuroscience at Yale University School of Medicine, a doctor, and a principal investigator at NIDA, NIMH, NHTSA and NIAAA, has done. He has combed through years of research, old and new, and evaluated the design and conclusions of hundreds of investigations to determine what science actually tells us (and what it hasn't yet told us) about this plant. This is a long overdue and impressively coherent sweep of history, botany, epidemiology, and biochemistry, blended with a bit of culture and lore, collected into the one volume you are holding in your hands.

You might expect as much from a dedicated scientific sleuth. But the great surprise about *Weed Science* is the writing. Most readers, myself among them, struggle to slog through the numbing jargon of journal articles—they can often be as dry as that old bag of marijuana that was stashed in the garage 3 years ago and forgotten. Dr. Pearlson is a fine writer with a well-honed sense of humor that adds some sheen to material could come off as dutifully academic. Even though he takes science seriously he also makes it enjoyable—no easy feat.

Pearlson's authority is further enhanced by the fact that he is an unapologetic child of the 1960s and doesn't hesitate to share some of his own personal experiences, positive and negative, with the plant.

He describes the Holi festival in India where entire regions of the country drink bhang lassi (cannabis dissolved in yogurt) in celebration of the god Shiva. With the benefit of cannabis he has happily communed with nature in the Atlas mountains of Morocco and come to a more intuitive understanding of the music of Karlheinz Stockhausen and Olivier Messiaen. He also ate far too many pot brownies and suffered the uncomfortable consequences of over medicating, thus casting doubt on those who claim that there are no negative consequences.

In other words, he has been both pleasantly high and highly spooked by cannabis. This strengthens his observations and makes him a far more credible reporter than other so-called experts who raise all sorts of alarms that have little bearing on reality. This is crucial to cutting down the noise. It wasn't that long ago that institutions, medical journals, and an unquestioning media were frothing about cannabis causing men to grow breasts, or blaming it for leading to heroin addiction, or reducing sperm counts. As recently as 2019, a fear-baiting book by Alex Berenson called *Tell Your Children* made headlines by distorting research, which falsely claimed that cannabis inevitably leads to psychosis and increased violence. Even when the authors of the original study denounced Berenson's mangled misinterpretation of their findings, the mainstream media continued to repeat them, further obscuring the important questions of how we, as a society, should regulate and police intelligent adult use.

It also enables him to better deconstruct the experience of being high and explain why so many millions around the world turn to cannabis as a relatively safe way of altering their consciousness. For example, it's definitely true that cannabis temporarily limits thinking to the here and now, but how this is interpreted means everything. Certain cannabis-naïve researchers would consider this a "cognitive decrement." But Pearlson understands that these effects are temporary and that intense attention to the here and now can also lead to deep absorption in a task—ask any coder in Silicon Valley about the focus they say it provides in their work. Ditto the way it "distorts" time. In one experiment, subjects who were high were asked to estimate how much time had passed between the sounding of two musical tones. One person responded, "A billion, trillion, quadrillion milliseconds," which is a far more accurate (and poetic) depiction of the gooey way cannabistime passes than "3.5 seconds." (It also sheds light on why so many musicians say it allows them to feel the space between notes).

But the motivation or creativity bump that comes from being high is not the point of this book. The points are gaining a deeper understanding of how cannabinoids affect the brain's Default Mode Neural Network, or how being under the influence affects motor skills like driving (and the challenges of studying this), or knowing why the relationship between cannabis and psychosis is not causal, or how one plant can treat illnesses as diverse as arthritis, epilepsy, Parkinson's, anorexia in AIDS, IBS, or chronic pain. With this scientific knowledge behind us, we can proceed more sensibly, more calmly, to using this plant to its maximum potential.

That's why I invite you to turn the page and dig in. You may never call it "weed" again.

Joe Dolce
Author, *Brave New Weed: Adventures into the Uncharted World of Cannabis*

Acknowledgments

Thank you to all the beta readers who slogged through early drafts of this book and provided useful feedback. This dedicated crew included Mirjana Domakonda, John Gartland, Shari Cantor, Jean Vitale, Chris Grosso, Phil Salfield, Will Shadboldt, and Michal Assaf. Where possible inaccuracies and rough edges persist, those are my responsibility.

Portions of the book were helped greatly by conversations with Glenn Treisman on the nature of and vulnerability to drug addiction, and with Deepak Cyril D'Souza on dozens of aspects of cannabis science and policy.

Alan Alda's workshop on communicating science to the public started me on a journey that I am still continuing.

Anna Bertoldo patiently wrangled all the references, helped with the proofing process, and sent chapters back and forth to the publisher.

Listening to presentations by Ethan Russo on the history of cannabis, the importance of terpenes and on cannabis-related archaeology was eye-opening. Several books were particularly helpful in pushing me to think about different aspects of cannabis research. These included David Nutt's *Drugs Without the Hot Air*, Michael Pollan's *The Botany of Desire*, Charles T. Tart's *On Being Stoned: a Psychological Study of Marijuana Intoxication*, and John Charles Chasteen's *Getting High: Marijuana through the Ages*. Sue Sisley provided encouragement and introduced me to a number of interesting sources in the cannabis field. My academic chair, John Krystal, was very supportive of my writing a book on cannabis. I would like to acknowledge that Hal Leonard LLC has granted permission to the song mentioned in Chapter 8, page 159. The details are as follows: **Stop Breakin' Down Blues**, Words and Music by Robert Johnson, Copyright © (1978), 1990, 1991 Standing Ovation and Encore Music (SESAC), Under license from The Bicycle Music Company, All Rights Reserved, *Reprinted by Permission of Hal Leonard LLC*.

And last, but certainly not least, special thanks to my book buddy Doreen Stern for being happy to meet at the crack of dawn for our regular discussions over almost 3 years, during which she has provided consistent valuable advice, patient feedback and encouragement to nudge me away from dry science writing and toward a more personalized narrative, (a process which continues to evolve).

Chapter 1

Introduction

In 1970, perhaps as a prelude to their subsequent one giant leap for mankind, NASA fired a bunch of frogs into space. The amphibian astronauts were on a one-way mission designed to investigate the effects of weightlessness on the inner ear. Hearing of this event one evening, a regular at the local pub speculated what it might be like for a helpless creature with little sense of self to be blasted at enormous speed into the silent darkness of space. I was the only one at our table who could address that question from personal experience.

Earlier that year, on my 20th birthday I had baked a mini chocolate birthday cake employing, alongside the usual ingredients, a fat, 2.5 g cube of crumbly Lebanese hashish carefully dissolved in Danish butter. I was unfamiliar with the potency of baked cannabis, but well-acquainted with my sweet tooth, so I ate a generous serving and proceeded to wait. Half an hour and no identifiable effects later, I concluded I must not have consumed enough and finished the entire cake. It tasted surprisingly good.

Forty minutes later, I was seated at the dining room table chewing a piece of bread when my mouth became extremely dry. "This bread is as dry as a stone" I said, and spat it out. The barely moist bread flew to my plate and rattled like a handful of beach pebbles. I was astonished. Unbeknownst to me at the time, I had just experienced what neuroscientists refer to as synesthesia, a phenomenon where sensations from one sensory system produce signals in another (i.e., the ability to taste colors or see music).

Feeling tired, I lay down on my girlfriend's bed. Within minutes, I had the vivid, unpleasant, and increasingly frightening feeling of being shot from a cannon at supersonic speed and traveling faster than any roller coaster into the blackness of starless space. My ego and sense of self dissolved into the pure sensation of a barely sentient being, devoid of an identifiable physical location. I was accelerating ever-faster into the void, terrified by the uncontrollable fragmentation of the familiar "me," body and mind, my otherwise inseparable companions in life. Now, there was only the elemental, explosive experience of my inner being hijacked by a malevolent, uncontrollable bat-out-of-hell.

Occasionally, I would become aware of my unpleasant reality, and during those brief moments of quasi-lucidity, I was certain I had gone permanently mad. My girlfriend sat by my side and held my hand throughout as she became increasingly anxious. Amid one of my fleeting returns to earth, I asked her how I could break the news of my emergent psychosis to my family and medical

Weed Science. http://dx.doi.org/10.1016/B978-0-12-818174-4.00001-X

school supervisor. Her brutally honest, albeit practical, response was "if you've really lost your mind, I'm not sure I can stay in a relationship with you." Familiar with Kafka's *Metamorphosis*, I easily identified her instinct to move away from unpleasant transformations, but was nevertheless saddened by the implied distancing. And then just like that, I was back in interstellar space.

Luckily for me, my sense of self re-emerged within a few hours. I regained control of my thoughts and bodily sensations. My previously disparate mind and body rejoined to manifest my physical presence within my current reality. Time ticked at a familiar pace, and I was no longer convinced I'd gone crazy. Instead, I now knew precisely what it felt like to be a helpless creature blasted into outer space, unprepared for the mental and physical toll of the journey.

This experience did not permanently alter my psyche, but helped cement my career choices and academic focus. I developed a keen interest in cannabis' effects on the brain and body and chose to pursue a career as a psychiatrist and clinical researcher to further explore these powerful forces. The psychological sequelae of my experience confirmed my prior interest in psychotic disorders, such as schizophrenia, which echoed my transient experience with cannabis. Most importantly, my brief stint as a psychonaut helped me connect and empathize with people who are chronically psychotic, lost in a frightening alternate reality, and unable to navigate home.

This book is about the science of cannabis. In the first part, I examine the origins of the cannabis plant, its chemical structure and composition, and its interactions with humans throughout the centuries and the internal cannabinoid system of the body. In later sections, I will explore how plant-derived cannabinoids produce the subjective "high" and medicinal effects experienced individually, to the plant's collective impact on economics and healthcare, with a focus on how scientific scrutiny has contributed to the contentious debate about marijuana.[a]

My apologies if you are a lonely farmer in the Castro Valley who purchased a book titled *"Weed Science"* in hope of learning the latest research advice on eliminating thistles and other pesky weeds from his or her property. This book is not for you. Nevertheless, the thematic thread of "What is science?" and "How can we understand the medicinal properties of marijuana?" are woven throughout the book and will likely interest clinicians and cannabis cultivators alike.

Beginnings

I will rewind events here to relate how I became interested in cannabis, psychosis, and neuroimaging of recreational drugs. The first time I ever encountered cannabis was as a high schooler in 1966. The local British police, in some early precursor of "Just Say No," or D.A.R.E. visited our grammar school to

---

a. I use the terms "cannabis" and "marijuana" interchangeably throughout the book to signify THC-containing chemovars of cannabis, and the word "hemp" when I'm referring to non-THC bearing plant varieties used primarily for fiber or seed production, (although the latter contain CBD)

lecture on the evils of drugs, that is, every known pharmaceutical compound excluding alcohol and tobacco. Needless to say, questions were not encouraged. Police gospel was that all recreational substances were indistinguishably bad and equally damaging, if not lethal. To provide concrete examples of this deadly potpourri, law enforcement personnel, as embodied in two large, wearied policemen, passed around illustrative examples of "Satan's manna" contained within a dozen or so thick glass pharmacy jars. One contained a variety of brightly colored pills and capsules, while another displayed tarry dark chunks of hashish that appeared to have been stored in the dusty basement of a Victorian druggist's emporium for the last 100 years. From our perspective, the police might as well have circulated sampler jars of delectable Willy Wonka ambrosia. To the fascinated audience, this was less of a parade of bad examples and more of a challenge for ingenious young minds. Mischievous schoolboy hands were soon tugging and twisting determinedly, yet inconspicuously, to pry open one or more of the goodie jars. I had my eye on a particular target and, while feigning innocence, gave the cannabis Indica vessel my most determined grip and wrench. Alas, it was glued shut tighter than a pharaoh's tomb. Enviously, one bright young lad managed to extract a small handful of pills from another bottle that were later, disappointingly, revealed to be seasickness tablets.

I was introduced to smoking marijuana a few years later at a high school party and can still vividly recall the scene and sensations. The excitement of the unchaperoned party was palpable, food tasted wonderful, conversation bubbled along delightfully, and I heard Robert Johnson's "Hell Hound on My Trail" in his unforgettably spooky voice for the first time. Whereas alcohol simplified my world and enveloped me in sleepy splendor, cannabis jolted me awake, and enhanced my imagination's subtleties, dialing up the inherent humor and friendliness of my familiar universe. I continued using marijuana occasionally in medical school and viewed it as a happy and harmless substance. That is, up until my birthday cake experience. That episode of delirium sparked both a new respect for cannabis (thereafter equivalent to a large, generally friendly dog that occasionally bites and needs to be leashed) and a persistent lifelong interest in mind-altering substances. How do they work their magic? What do they tell us about brain function and dysfunction? How can we safely test these effects? During my psychiatry residency with the inspirational Paul McHugh and Philip Slavney at Johns Hopkins in the 1970s, I tried to answer these questions by exploring the psychoactive effects of catnip on felines. Perhaps catnip was to house cats what cannabis was to humans, a perfectly reasonable, but as I soon discovered, woefully inaccurate hypothesis.

For my postdoctoral fellowship in 1981, I worked in a research laboratory serendipitously situated next door to Joseph Brady, a pioneer investigator into the behavioral effects of varied abused substances on animals, ranging from rats to rhesus monkeys. A powerful-looking, ruddy-faced man, Joe's research encompassed everything from writing guidelines for ethical research to preparing monkeys for space flight. He loved his work and told anyone who would listen about his most recent project, why he was engaged with it, and what he was discovering. As part of his research portfolio, Joe also collaborated with NASA and conducted month-long experiments on groups of human volunteers who lived on a remote research unit, isolated from outside contact. The purpose of these studies was to predict how groups of astronauts might behave on long space voyages in the event of an equipment failure or aeronautical emergency. His animal research, on the other hand, predicted successfully which newly discovered drugs were liable to be abused and ultimately how substance abusers might be treated. I found all of his work, regardless of species, fascinating.

A couple years later, Joe recruited the behavioral scientist Marian Fischman to join the Hopkins faculty. Marian, who resembled a small, forceful version of Elizabeth Taylor, was a pioneer in safely administering recreational drugs to experienced users in controlled laboratory settings, a nice complement to Joe's animal work. Through persistence and careful preparation, Marian was able to convince human subjects safety committees that it was important and feasible to conduct qualitative studies on the effects of these drugs in individuals who were already using them recreationally. She reasoned that by systematically studying these experienced users, scientists could assess the effects of these substances on the human body in a naturalistic manner, rather than relying on guesswork or extrapolating effects from laboratory rats. Her research subjects were paid, voluntary study participants who were admitted to the hospital or a special research unit for the duration of the study (typically 1–2 weeks). They could leave the investigation at any time, and if they requested a referral to a substance use treatment program, they received it. Marian Fischman taught the world much of what we know about the clinical effects of cocaine in humans. Her agile mind probed for details and sought rigor, an asset that made her a valuable, if strict and tough-minded advisor. Despite an untimely death in her early 60s, Marian nevertheless taught a generation of clinicians how to perform careful, quantitative human drug research, while respecting patient autonomy and minimizing harms. I was lucky to have her as a mentor.

In addition to my burgeoning interest in the neurological effects of substance use, I was equally fascinated by a complementary scientific thread. Several years earlier, Eve Johnstone and Tim Crow had used the newly developed technique of X-ray computed axial tomography (CAT scanning) to take vivid, precise images of the brains of chronically hospitalized patients with schizophrenia. To the surprise of physicians everywhere, the patient's brains appeared significantly shrunken, with large fluid-filled spaces showing prominently on the images. This finding revolutionized thinking about schizophrenia, reclassifying it as a genuine brain disease with obvious neurological hallmarks, rather than a mysterious behavioral disorder unrelated to the nervous system's anatomy. Inspired by this discovery and its implications for patients with severe mental illness, I began talking to radiologists at Johns Hopkins about how I could use neuroimaging to understand what might be affected in the brains of my own patients.

My twin interests in psychosis and abused substances have been intertwined with the use of different neuroimaging techniques for almost 40 years now. I've been especially interested in how recreational drugs, such as marijuana and alcohol, can alter performance on complex tasks (such as driving a motor vehicle), and their simultaneous effects on the brain. Functional magnetic resonance imaging (fMRI), in particular, has recently emerged as a tool that allows scientists to visualize which brain regions are more metabolically active, or receive more or less blood flow, during a behavioral task. In layman's terms, we are now able to observe which areas of the brain "light up" or "turn down" in response to environmental cues, such as visual or emotional triggers. The problem with assessing complicated behaviors is that there may be so many things going on simultaneously that the entire brain can illuminate like an over-amped Christmas tree. At that point, sorting out which circuit is linked to a specific behavior seems unsolvable. For example, driving either a real car on the highway or a virtual car inside of an MRI scanner activates similar brain circuits connected with spatial memory and judgment (e.g., "I need to take the next exit on the right, and keep the right distance from the car in front.") and holding facts online (e.g., "the speed limit here is 65 mph," "It's raining so my braking distance is longer."). Yet other circuits need to simultaneously coordinate and integrate eye, hand, and foot movements, respond to unexpected events, and focus attention on what's happening on the road and not the squabbling kids in the backseat. Faced with the dauntingly difficult task of disentangling which circuit was doing what, when, in response to which behavior, Vince Calhoun in my laboratory at Hopkins discovered how to apply a novel statistical approach to separate out distinct brain circuits that simultaneously switch on or off. The technique, known as independent component analysis, began in the narrow fMRI arena of imaging drunk driving, but is now one of the most widely used analytic methods in fMRI experiments in the world. Beginning at Hopkins with measuring the effects of different doses of alcohol versus placebo on driving, and some parallel early experiments with cannabis, my laboratory moved to Connecticut in 2002 and began to study in earnest marijuana's effects on complicated behaviors, including driving.

That brings us to the present. Now, as Connecticut and many other states ponder legalization of recreational or medical cannabis, it's rare to chat with friends or browse the media without cannabis-related issues emerging. In the process of writing this book, I've talked to marijuana law reform advocates and opponents, epidemiologists, geneticists, politicians, dispensary entrepreneurs, psychologists, cannabis consumers, cannabis growery owners, and marijuana botanists. I've interviewed clinicians who treat cannabis-using adolescents who became entangled in the juvenile justice system, care for patients with psychosis related to cannabis use, or run medical marijuana drug trials in the Veterans Affairs (VA) system for patients with PTSD.

The most striking observation I've made is how polarized peoples' opinions are on marijuana-related issues. This goes hand-in-hand with the findings of a recent Pew Research Center survey of over 5000 US adults that revealed members of the public struggle to distinguish between factual (able to be confirmed or disconfirmed by objective evidence) and opinion-based (expressing their views and attitudes) news and views [1].

I argue that scientific inquiry, by its inherent reliance on objective evidence, can help resolve many of these seemingly stark yes/no, right/wrong arguments. In addition, science can help predict the likely outcome of specific actions, such as legalizing recreational cannabis (e.g., increasing drug dependence or motor vehicle accidents), allowing for more rational decision-making around these issues. If you've ever used marijuana, thought about using it, wondered why others would use it, or are concerned about how the legalization of cannabis might affect society as a whole, and you think that addressing these questions scientifically might be informative, then I invite you to go on a road trip with me to explore these issues. (But please don't drive stoned.)

Reference

[1] Mitchell A., et al. Distinguishing between factual and opinion statements in the news, 2018. Available from: https://www.journalism.org/2018/06/18/distinguishing-between-factual-and-opinion-statements-in-the-news/.

Chapter 2

Good science and bad science

"To make matters worse... the opposing factions in the cannabis debate often interpret the same scientific evidence differently to suit their own purposes."

Leslie L. Iversen The Science of Marijuana,
second edition. Oxford University Press; 2008, page 5.

"Reefers and rhythm seem to be directly connected with the minute electric waves continually generated by the brain surface. When the rhythm of the music synchronizes with the rhythm of the brain waves, the jazz fans experience an almost compulsive urge to move their bodies in sympathy. Dope may help the brain to tune into the rhythm more sharply".

Chapman Pincher, Dope: Is there a link between dope
and hot jazz dancing? Daily Express, November 28, 1951 [1].

It is almost impossible now to plug into the media without being confronted by news about marijuana, or to give it its less contentious name cannabis. Strong contradictory claims about the drug are made touting opposite conclusions. Cannabis stimulates creativity and has no effect on IQ or conversely causes significant memory loss and lowers IQs. Legalizing cannabis leads to a measurable drop in opioid prescriptions, and fewer opioid-related deaths, or on the other hand an increase in both of these measures. Data show purportedly that states legalizing recreational cannabis experience significant increases in arrests for driving while intoxicated and motor vehicle accidents, yet other analyses of the same data sources yield precisely the opposite conclusion. With society poised to make such important decisions about cannabis legalization on a state and national level what information are we to believe? And from which sources? What are the criteria to help us decide who and what to believe? Answering these questions involves bringing science into the debate, so that we can evaluate these competing claims skeptically and help reach our own, hopefully rational, scientifically informed conclusions. But without multiple degrees in toxicology, epidemiology, psychology, ethnobotany, public policy, neuroscience, sociology, and more, what might be a starting point to think intelligently about these issues in an informed way? These types of questions provided the initial push for me to begin writing a popular science book about cannabis. The idea was

Weed Science. http://dx.doi.org/10.1016/B978-0-12-818174-4.00002-1

to air these debated issues within a scientific framework and in a manner comprehensible to the average person. Popular science is a harsh taskmaster. Lean too far in one direction and what you write is impenetrable or textbook-dry. But move too far to the other side, and you are at risk of dumbing down the material, over-sensationalizing it, or purveying opinions rather than facts.

Which leads immediately to the question, what is science? I'm not just referring to "hard scientists" in white coats, wielding test tubes and beakers within white-tiled laboratories, or reaching with arcane instruments into the bowels of charged particle accelerators, although we will certainly encounter such people in this book. Those individuals will include plant geneticists, botanists, neuroscientists, toxicologists, and chemists. But many more of the scientists we will hear if about belong to the so-called "soft" sciences. These include sociologists, physicians conducting clinical trials, legal policy researchers, epidemiologists, psychologists, and economists. What do all of these diversely specialized scientists have in common?

Science encompasses both a worldview or manner of thinking about problems and a process of searching and evaluation. Its purpose is to try and explain and ultimately to understand reality—that is, what's out there in the universe (including the universe inside our heads) and how it works. In practical terms; science involves using a set of attitudes and practices. In the attitude realm, the key stance is one of skepticism. In other words, using critical thinking skills. All of us are inclined to accept conclusions that fit in with our prior prejudices and beliefs. It's a wonderful thing to have our preconceptions confirmed, but we tend to reject contrary evidence without properly examining it. Skepticism involves not reflexively accepting purported facts on faith, even if we happen to like the conclusions drawn, but trying to review the supporting evidence critically whenever possible. In other words, science is based on evidence. It is also helpful to extend the process of weighing facts backward into re-examining whatever evidence led you to your original belief. Another aspect of skepticism is the practice of neither automatically accepting as truthful "what everybody knows," even if it's from a "leading scientist" nor the last thing you read on Facebook, posted by Karen. As the saying goes, "stay open-minded, but not so much that your brain falls out."

So, how can we become thoughtful skeptics regarding the information on cannabis that bombards us every day? When you the reader are confronted with some apparently scientific claim, such as "Scientists show that CBD oil will make your skin smooth & beautiful and cause your wrinkles to disappear," or "Researchers prove that smoking marijuana causes schizophrenia," or "Gateway drug marijuana sparks addiction and violent crime wave in USA." How can you begin to verify such claims? Let me suggest the following as a starting point. However apparently strong a statement, the first question to ask oneself is, "is this purported information even scientific to begin with?" What is the quality of the evidence? Many spurious pseudo-scientific claims can masquerade as science. Some are easy to identify as such, and can be unmasked by simple logic. So, for example, when my friends from the 60s tell me earnestly that because cannabis is a "natural herb," it must therefore be harmless at worst and is likely good for you, it's easy to refute that argument with examples of the equally natural herbs deadly nightshade, tobacco and poison ivy. And if smoking tobacco is bad for you, then smoking tobacco with cannabis in a blunt or spliff probably isn't good for your body either. Non-scientific logical chains quickly fall apart when probed. A further aspect of skepticism is to try and unmask any potential ulterior motive of the person reporting the factoid in question. When such motivation is present, it can often be a clue to underlying poor science or "spin." The skeptical reader should ask him or herself who exactly published these data, who gathered them, how were they analyzed and whether the parties involved have an overt or perhaps covert agenda. For example, when a report is released purporting to reveal the supposed good or bad consequences of cannabis legalization, ("motor vehicle accidents soar," or "opioid deaths fall dramatically") it's important to know who paid for the study and any possible motivation to impose a particular spin on the conclusions. Was it a lobbyist for a large liquor distributing firm anxious that cannabis legalization will cut into their profits, a drug cartel worrying about their own bottom line, a lobbying group that believes that all cannabis is bad under all circumstances, or a state legislature eagerly eyeing a potential source of budget-balancing cannabis tax revenue? Study size is often a key to demonstrating the likely reality of a particular result. Scientists refer to "belief in the law of small numbers." This is the statistical fallacy that something that occurs (often by chance alone) in, for example, a small group of individuals or a small number of coin tosses, scales up directly to the same result in large groups. Chapter 7 will hopefully reinforce the point that "the plural of anecdote is not data." A final aspect of skepticism is one emphasized by Yale professor Dan Kahan, that he calls "science curiosity." People who are science-curious like to challenge themselves by poking around in new evidence, whether or not, it contradicts their belief system. If you're a Fox news reader who doesn't mind checking out CNN or the Washington Post, (or vice versa) then you probably meet some of the criteria for what we might call skeptical curiosity.

So, now that we're clear on cultivating skepticism to allow us to better understand both ourselves and the world around us, let's move on with our minds open and questioning, from scientific attitudes to scientific practices.

The latter consist of an informal rulebook, whose contents advise on first defining the questions you want to ask about some little corner of reality and making one or more predictions about what you expect to find there. Does Halley's Comet whizz by Earth every 75 years? Then we should see it again in 2061. What you're hunting for are predictably re-occurring patterns, and rules and principles that can then be tested experimentally. These practices are an essential part of the activity of science. Questions usually follow logically from an explicitly stated hypothesis about what might be going on. Procedures include carefully defining whatever process you're interested in, determining the appropriate techniques or technological tools to be used in measuring it, and specifying exactly what those methods and concepts are. For example, if you want to study 3-inch fish, then don't use a 4-inch mesh net to capture your sample. Scientists strive to be precise and to make fine distinctions. Scientific conversations often invoke metaphorical sharp-edged tools used to dissect facts, from using Occam's razor (simpler explanations are generally better than convoluted ones) to not having axes to grind, (see earlier on motivation) so that concise, dispassionate descriptions are favored. For example, not "the mice were angry and thirsty," rather "on average, the mice fought more with intruders introduced into their cages and consumed 50% more from their water bottles." If your convoluted explanation for an observation involves more than minimal hand-waving and is closer to Occam's chainsaw than his razor, you may have veered way off course, scientifically speaking. If fuzziness is undesirable, so conversely is the false certainty of easy and definite black-and-white answers; scientists generally speak in probabilities.

To the possible extent, scientists try to set up experiments that rely on unbiased observation to gather what they believe are the most relevant facts to be collected, and then analyze them statistically. Once the researcher has made the observations, she or he wants them to be empirically confirmed by repeating everything at least one more time and obtaining the same result. In other words, are your measurements reliable? As well as reliability, validity is another issue that concerns people performing experiments. If my bathroom scale consistently registers my weight as 120 pounds, I'm flattered but know that the measurement, however reliable must be invalid. Study number three later in this chapter gives an example of a failure of validity, where a test isn't actually measuring what it purports to. Related examples are concluding that a particular cannabinoid is impairing memory when it is in fact making people extremely drowsy. Humans seem predisposed to explain coherence in nature, but unfortunately what seems like common sense can easily lead to false interpretations without double checking everything relevant. Unexpected confounding factors can bollix up the most carefully planned experiments. For example, results of an experiment might lead one to conclude that "the mice drank significantly more water after we increased their daily dose of THC." But first, check that at the same time you did not accidentally change the thermostat setting so that the room was 10° hotter.

Science also asks for us to come up with logically consistent predictions based on whatever new hypothesis you have to explain your observations, and demands that you test your hypothesis over and over again. And the best way to test a hypothesis is to design some experiment that can potentially prove it false, no matter how dear it may be to your little scientist's heart. For example, as we will see in Chapter 7, a research group that reported that long-term cannabis use was associated with falling IQ over time in a long-term community population sample rechecked to see whether they could replicate these data in a completely different study group. To do that, they examined pairs of twins who had been intelligence-tested over several years, where one twin used cannabis and the other did not. The effect did not replicate.

Open communication plays a big part in this process, so you want every step that you went through to reach your particular conclusion to be laid out transparently, not unlike a well-documented recipe, and you will share your data for others to examine. In the process of peer review, other scientists in your specialty area are able to pick through your work, provide unvarnished feedback and request missing details before allowing publication in reputable scientific journals. They act as gatekeepers to exclude junk science. A positive outcome of this peer review process is that your experiments can be re-created, and your data can be straightforwardly and independently rechecked by scientists as skeptical as yourself. If your work is provocative, there are usually doubters who are itching to see whether, if they repeat exactly what you did, they will obtain the same results. And if they do, you may have convinced them.

Operational definitions of variables should be as clear-cut as possible. Thus, potentially ambiguous terms such as "drug user" should be defined rigorously and transparently. For example, a "cannabis user" could refer to either someone who takes a single bong hit once a year, or spends most of their waking existence inhaling budder fumes from a rig (we learn more about these items in Chapter 9). Scientists who study associations of alcohol with various health outcomes use standard alcohol beverage units that can equate a bottle of beer and a glass of wine. But given the different formulations, strengths, and methods of consumption for cannabis, we run into the vexing question of what constitutes a "standard unit." For example, how do we equate a bong hit of 35% THC Purple Kush with inhaling a sesame-seed sized morsel of 85% THC and 15% CBD shatter (we learn more about these items in Chapter 9) from a dab rig? And how do those stack up against a hash brownie? If you want to show that larger doses of a drug have proportionately more effect on a measure such as memory or pain relief, then you need a reliable and valid measure of dose. Another problematic example is that of "marijuana-impaired vehicle drivers." Does this designate individuals pulled over by law enforcement for driving dangerously while simultaneously observed to be smoking cannabis in their car? Or on the other hand do we include all individuals stopped at a routine police check point who test positively for THC metabolites, even if they last used cannabis 2 weeks ago? Even the term "cannabis" is used to refer to both pure CBD oil containing no THC, or 1:1 CBD-to-THC concentrate, and to both very high and very low THC-containing marijuana buds. So try to be specific.

Another sensible scientific practice is striving to remain consistent in the way that we count things. For example, when a state begins to measure cannabis constituents such as THC in the blood of intoxicated drivers, and switches to defining intoxication based on the results of blood tests rather than on roadside behavior as previously, then it's not possible to directly compare rates of driving under the influence of cannabis before and after the change in measurement practice. The size of the sample and its representativeness of the population as a whole is another important aspect of any study. When conclusions are drawn about drug effects we want to know that sufficiently large numbers of representative individuals were examined so that the findings are likely generalizable to other populations. Asking 10 of my neighbors whether recreational cannabis should be legalized is less useful than conducting a nationally representative survey of 35,000 adults, although the former it is likely to be much more entertaining.

In addition to the way we count things, *when* we count them may be critical. Oscar Morgenstern illustrates this point by noticing that over a 10-year span, the Bulgarian pig population seemed to more than double. But this apparent increase was illusory, based on the nation's change from the traditional Russian Orthodox to the modern Gregorian calendar, causing New Year's Day to shift by 2 weeks. This dislocation in dates resulted in the first set of pigs being counted *after* Christmas day, (when the number alive and kicking was small, as most had been slaughtered for the holiday) while the second pig census now tallied pigs *prior* to Christmas, when none had yet reached the chopping block (so that their apparent number was much larger). In fact the overall number of pigs hadn't budged [2]. So, why am I mentioning a faulty 100-year-old Bulgarian swine census? Well, this problematic counting issue applies to Alex Berenson's recent [3] claim that marijuana is responsible for rising crime trends in states where cannabis has been legalized: "Violent crime has also soared in the legalized states since 2013." This claim has received strong rebuttals. In January 2019, a group of criminologists claimed in the Seattle Times that they had found "no increase in violent crime that can be directly attributed to marijuana legalization" [4]. Beatriz Carlini, a senior research scientist at the University of Washington's Alcohol & Drug Abuse Institute (ADAI), commented in the article on the supposed connection drawn between marijuana legalization and crime increases in Washington state (a 17% rise in aggravated assaults between 2013 and 2017). The problem, she says, is that there was a 1-year decrease immediately prior to 2013, so after that year "the numbers are just creeping back up to where they were before." So that picking a starting date of 2013 is stacking the deck statistically speaking, because your measurements are likely to trend upward anyway because they are beginning in a trough. Back to the pigs then. Where you pick your starting point to demonstrate a change can be crucial. Otherwise your findings may be hogwash.

To summarize what we've learned so far, since any particular hypothesis is theoretically falsifiable at any point, the job of science is never done, explanations are never complete, and you can never stop metaphorically kicking the tires. Particularly in cases when a scientific finding seems unexpected or provocative, then it needs re-testing and confirming in a new population. For example, reports that common genetic differences between individuals explained a large proportion of risk for experiencing psychosis after adolescent cannabis use seemed initially convincing and scientifically plausible, but these results have been extremely hard to replicate in fresh samples (something that we will discuss in Chapter 8).

Thus, science is an endless process, which I mean in a positive way, rather than how you felt personally about science at 5 p.m. in 10th grade chemistry class. Ideally, the theoretical framework you're working with was arrived at rationally, based on solid evidence, is maximally simple, useful, and has an ability to predict events, observations, or circumstances that no one has examined previously.

Scientists are keen on words such as impartial, logical, rational, unbiased, neutral, and objective. Some of the procedures I describe in this book hew to the sort of scientific model I just described. These include such things as measuring cannabinoids and terpenes in chemical laboratories, performing genetic analyses of human or plant DNA, figuring out what wavelengths of light cause cannabis plants to flower, and conducting clinical drug trials of cannabinoid compounds. Other parts of the book venture much more into social sciences, where we are examining things like subjective mental states, people's intentions, feelings, and thoughts. Chapters that examine what it feels like to be stoned, or why people tend to use recreational drugs in the ways that they do, fall more into this second category.

Human behavior does not obey laws of nature in the same way as subatomic particles in an accelerator do. Causation for our behaviors and consequences is often hard to establish clearly, and the tight, rigorously controlled experimentation possible in the world of chemicals, and even laboratory rats, is never feasible when it comes to people. For example, in an animal laboratory we can administer precise doses of THC or placebo to adolescent mice under carefully controlled conditions and track their behavior as they make their way slowly and dreamily around tiny mazes, and then kill them and measure precisely their brain chemistry and structure. However, nobody in their right ethical mind would ever propose identifying all the 15-year-olds in a large city and randomly assigning half of them to smoke cannabis containing a precise amount of THC every day for 10 years, and giving the other 50% a similar placebo, in order to document how many people in each group developed psychosis. And even if improbably we could try to perform such an ethically blighted study, many of the teens would inevitably engage in behaviors that will thwart our would-be experiment, bless their little hearts. These might include drinking various quantities of alcohol, surreptitiously sharing the study drug with one another, using their own cannabis supply in addition to or instead of what we assigned them, dropping out of the study, not answering our survey questions honestly, or using magic mushrooms immediately before answering them. Human beings in other words are refreshingly ornery. So, we might ask, is the type of social science that confines itself to description really science at all? Philosophers of science regularly engage in amusing spats in debating these types of questions. Maybe the best we can do under such circumstances is to apply whatever feasible elements of scientific logic and strategy that we can to the world of people and to gather as much worthwhile information as possible. For example, it is often practicable for researchers to make careful observations of naturally occurring events. They can obtain parental consent (and the child's assent) to follow a representative community sample of a thousand 10-year-olds every year for 10 years, before the children have ever used alcohol or recreational drugs, and ask them to complete IQ tests and fill out confidential surveys about their drug use. Then they can measure relevant differences among groups with different patterns of subsequent substance use. These latter types of observational studies are not planned to intervene in natural circumstances, only to track and report carefully what's already occurring. We will discuss the results of several studies that used this type of design.

Humans differ from laboratory rats in other ways. To make experimental results standardized, laboratory rats are essentially clones, or at least one rat differs minimally from another. Humans differ from laboratory rats in the massively scaled-up complexity of their brains and the substantial differences between most individuals. Furthermore, humans and rodents differ in many aspects of basic physiology. Experimental drugs that produce dramatic therapeutic responses in rats and mice are mostly ineffective in humans. The Twitter account @*justsayinmice* appropriately re-tweets extravagant science claims that inappropriately extrapolate to humans from rodents with the headline "IN MICE."

Finally, the science hymnbook frequently invokes the refrain "correlation does not imply causation." Thus, if A happens, followed by B occurring, this does not automatically mean that A caused B. For example, in the summer months, ice cream sales and cases of drowning both climb in tandem day-by-day. This doesn't lead to frantic calls to limit ice cream sales, because the average person intuits that both occurrences are due to the daily weather. Similarly, if adolescents who smoke cannabis have lower IQ scores than their non-cannabis smoking peers, then this cognitive difference may be due to environmental factors (e.g., lack of parental supervision, socioeconomic factors, attending different schools) that might explain both relevant facts both about IQ and a tendency to smoke cannabis. Thus, smoking the cannabis didn't lead directly to the IQ difference, but the same environmental factor explains why both of them occurred. To disentangle this situation, we might want to know whether IQ scores were lower in the cannabis users before they even began using the drug. We could assess whether individuals who smoked more cannabis had significantly lower IQs than those smoking small amounts, so that higher drug doses were associated with larger effects on IQ. Many of the points discussed earlier are relevant to the various case examples that we will discuss in the remainder of this chapter.

Good science, bad science, and how scientific data can be reframed

In writing this book, I would like to help readers re-examine not only what they think about cannabis use and legalization, but more importantly to continue weighing and examining related future questions for themselves, based on their cultivation of logical and skeptical aspects of scientific thinking. With those considerations in mind, let's consider four separate cannabis-related science reports that illustrate different points on a conceptual continuum of scientific quality, plus the case of a recently published book.

Case study number one, from *notallowedto.com*, is headlined "*THC found on meteorite from outer space.*" "NASA-affiliated scientists at the University of Hawaii discovered trace amounts of tetrahydrocannabinol (THC) on a meteorite found in the Nevada desert in 2010........ Tetrahydrocannabivarin (THCV) were (sic) also found in a meteorite fragment in 2009 by a research team from the University of Mexico but the findings were dismissed at the time because of the "controversial nature of the discovery"... astrophysicist James Hun of the University of Hawaii is quoted: "...If psychoactive elements are found outside of this planet's atmosphere, what does it say about the rest of the universe? ..." what role then has (sic) cometary impacts played on the human species? .... This discovery ultimately leaves us with more questions than answers. It also gives a whole new meaning to the term getting high" [5]. This story has legs—I have been asked about these "findings" by students several times, and it's been extensively republished elsewhere. So how do we judge its veracity? First of all, the source, *notallowedto* [6] is not a peer-reviewed scientific journal, but the internet equivalent of a supermarket entertainment tabloid; a representative headline is "Midget stuck in mailbox after falling in while mailing letter." The THC meteorite story has been shared hundreds of thousands of times, so it has proved to be effective clickbait, garnering traffic for the website, and therefore handy for their advertisers. Following the dictum "extraordinary claims require extraordinary evidence" this scientific claim, unfortunately was advanced with no valid supporting evidence at all. For example, the astrophysicist whose name is cited as the source doesn't check out on the University of Hawaii's astrophysics faculty list. The analytic method by which the THC was identified is never mentioned. Conclusion: not science, but obvious "fake news." Scores an A for chutzpah, F for scientific content [7].

Case study number two derives from an article published in the *South African Journal of Science* in 2001 by Francis Thackeray and colleagues. The gist is that clay tobacco pipe fragments from Shakespeare's garden tested positive for THC metabolites, and similar fragments from the same English town tested positive for traces of cocaine and several other recreational drugs, in addition to the expected nicotine. The author's conclusion was that Shakespeare smoked cannabis for inspiration [8]. The back story is that the Shakespeare birthplace trust in Stratford-upon-Avon allowed Thackeray and his fellow-researchers to analyze 24 17th-century pipe fragments from its museum collection. To accomplish this, the scientists employed the analysis technology of gas chromatography, mass spectrometry (GC-MS, explained in study number 4 later), whose necessary equipment resided in a South African narcotics crime laboratory. The tobacco pipe specimens derived variously from the site of Shakespeare's garden from a home that he owned in later life, from his birthplace, and locations elsewhere in the environs of his hometown of Stratford-upon-Avon. There was no evidence that any of the pipe fragments belonged to Shakespeare himself, but only that they were excavated from places in and around where he lived. Evidence for the presence of THC or related chemicals was less than convincing. The 2001 study stated that "unequivocal evidence for cannabis has not been obtained." The researchers did detect GC-MS mass-to-charge ratios of compounds that were indicative of those derived from marijuana, but not in quantities sufficient for proof. The authors argued that the lack of unambiguous evidence for THC may be "associated with the effects of heating, and problems in identifying traces of cannabinoids in old samples," but ultimately concluded that "the results are suggestive but do not prove the presence of cannabis" [9].

Essentially the same findings were reanimated a decade and a half later, discussed in an editorial [10], and in newspaper articles written by and about Thackeray and his findings [11] and [12] so that the story is still alive and appears regularly on the Internet. The idea that the immortal Bard, who invented the word "addiction" (Othello), was a druggie, is apparently too entertaining to let go.

When I discussed the "was Shakespeare a stoner" thread with my laboratory group at one of our weekly marijuana science meetings, it elicited an immediate shower of cannabis-influenced Shakespearean alternative play titles, including *As You Light It, Much Ado about Puffin', Titus Hydroponicus, The Hempest, the Taming of the Shroom* and *Henry and the "Eighth."* But the claim seemed to have sufficient plausibility to avoid dismissing it out of hand, so let's review the science with a skeptical eye.

The lead author of the 2010 paper was John Francis Thackeray, a distinguished PhD paleoanthropologist at the Evolutionary Studies Institute at the University of the Witwatersrand in Johannesburg, South Africa. Outside of his primary specialty in studying early fossil humans, Thackeray has many divergent scientific and artistic interests, including English literature, mathematical pattern analysis, ancient climate change, and African art. The paper's chemical analysis method of GC-MS yielded data that the authors themselves acknowledge were suggestive but insufficient for proof for the presence of cannabis. So Thackeray sought additional support for his hypothesis not from the world of science, but that of literature. He bolstered his claim with Shakespearean quotes. For example, he notes correctly that Shakespeare's sonnet 76 refers to "a noted weed" (...."Why write I still all one, ever the same,/And keep invention in a noted weed...") and later in the same sonnet writes of an aversion to "compounds strange." The consensus of Shakespeare scholars is that "weed" in this context is synonymous with clothing, not drugs, as in a "widow's weeds." Use of the slang term "weed" for drug cannabis first occurred around 1910 in the United States. King James I's contemporary *Counterblaste to Tobacco*, for example, refers to the latter as "herbe," not weed, which would have been a more likely contemporary term for a drug plant. There is also a lack of plausibility in multiple aspects of Thackeray's account. Nobody knows who owned the pipes from which the fragments derived. There's no evidence that Shakespeare possessed or smoked them. Where might the cannabis purportedly smoked in them have originated? While non-drug hemp has been grown in the United Kingdom since the Viking period for rope and sail making, drug cannabis as an intoxicant on the other hand didn't reach England until the 1830s from India (via Dr. O'Shaughnessy with the East India Company). As has been pointed out, if Shakespeare was indeed smoking cannabis, he failed to record this other than in one ambiguous line in a single sonnet. And cannabis was not an illegal drug in Shakespeare's time, albeit an unknown one so that he would have had no disincentive to describing his use of it. Similar skepticism applies to claims that traces of cocaine were found in some of the pipe fragments analyzed. Cocaine hadn't yet reached England from South America in Shakespeare's time. Smoking or vaporizing coca leaves (erythroxylum coca) yields no psychoactivity, because the concentrations of active alkaloids in the leaf are extremely low and smoking would destroy by combustion any that were present rather than releasing any intoxicating chemicals. While coca leaves can be chewed or brewed in tea to produce mild stimulation, smoking the plant to get high never developed in traditional culture, (unlike marijuana or tobacco, for example). Using modern chemical techniques unavailable in Elizabethan England, one can now extract highly concentrated forms of the cocaine alkaloid and process them into smokable product such as "crack," but this was unknown until 400 years after Shakespeare's time. So the supposed cocaine traces remain very hard to explain.

What additional evidence might make Thackeray's initial claim more believable? One strategy might be to test other pottery from the same garden such as kitchen cookware as a control, to see if this also yielded positive result for traces of cannabis, cocaine, and the other drugs claimed to have been detected by GC-MS. If pretty much any type of pottery revealed the same chemical signatures, then that would make the specific findings in the pipe fragments seem less likely. For example, a source of THC-like compounds in the soil could conceivably come from a relative of one of the recently discovered moss-like plants that manufactures cannabinoids. From there the cannabinoids could leach into anything porous in the environment. Along the same lines, one could test contemporary miner's clay pipes from the other locations in the United Kingdom. Working class men from the same time period would be most unlikely to have used any of these intoxicating substances, so that showing their tobacco pipes bore no drug traces would be important as a negative control. Unlike the more recent use of GC-MS to reveal chemical relatives of THC in Chinese braziers recovered from gravesites by archaeologists, the South African instrument seems to have pushed technology to its limits 20 years ago and perhaps beyond, using equipment that was designed for larger specimens confiscated by then current day drug enforcement police.

The overall conclusion one draws from this story is that despite a startling initial claim, no substantive new supportive evidence has emerged [13]. This publication receives strong ratings as a great after-dinner speaker's story, but is not very strong science, and without further evidence is low in plausibility. But unlike the first story, it actually is science.

Case study number three. The newborn nursery at UNC Hospitals began using a revised drug testing protocol in February 2011 to screen infants who may have been exposed to abused substances during their mothers' pregnancies. Urine samples for testing were collected in various ways, but usually by swabbing or squeezing the inside of the infants' diapers. Alert nurses reported that the number of positive screens for THC in the infants had shot up to nearly 20% since the new protocol was instituted. Were they facing an epidemic of cannabis abusing mothers? If so, then this had potentially serious consequences in terms of mothers being charged with child abuse or social services being called in, or even infants being removed from their parents. (As one article has pointed out [14], drug testing of the children of "at risk" mothers is performed much more often on single, poor, non-white women). A team of laboratory scientists, physicians, social workers, and nurses led by Catherine Hammett-Stabler, a UNC professor of pathology and laboratory medicine, finally figured out that a baby wash product was causing false positive readings on the cannabinoid/THC immunoassay. More precise testing with mass spectrometry and other laboratory techniques on the positive infant urines came back with uniformly negative results on the same specimens that had previously tested positive on the nursery's new commercial drug testing screen. The team next added small amounts of various baby wash products to clean urines and found that many specimens then came up positive on the commercial baby pee screening test [15].

Bottom line; in this study, thoughtful hypothesis-guided detective work, aided by properly used laboratory techniques, prevents a hospital from leaping to false conclusions.

Case study number four will be discussed in more detail in Chapter 4. It concerns the recent report from Meng Ren and colleagues at the Institute of Vertebrate Paleontology and Paleoanthropology at Beijing's Chinese Academy of Sciences [16]. This group examined the funeral braziers and cannabis specimens from a 2500-year old cemetery in Northwestern China. As in the Thackeray study, the analytical method used to identify cannabinoids was GC-MS.

A brief methodologic diversion: so, some readers are asking, what is GC-MS, and how is it able to detect and identify specific molecules? GC-MS is an analytical test procedure that when used properly identifies chemical substances with 100% specificity. Its many uses include airport security screens to detect explosives, forensic drug detection, and coincidentally in planetary probes, for example, screening samples from the Martian surface. So theoretically it could have been used to examine the Nevada meteorite, had it existed, in case study number one. The technology has been around since the late 1950s and continues to improve, so that the 2019 Chinese study would have benefited from incremental technological advances over the 2010 South African investigation, an issue separate from the significantly better design of the later study.

GC-MS is a two-step procedure that uses two completely separate technologies hooked together in series. The first step, the gas chromatograph, consists of a capillary column whose overall dimensions and the material packed inside of it are designed to differentially slow down and separate different molecules in a sample that is injected into one end of the column and wafted through it by a stream of inert gas such as nitrogen. Particular chemicals make it out of the far end of the column first, in a kind of molecular horse race based on their composition, and thus can be analyzed one at a time as they enter the mass spectrometer in the next stage of the analysis. The MS device smashes each molecule that enters it into electrically charged (ionized) fragments by bombarding it with electrons emitted from an electric filament, similar to the one you would find if you broke open an old-fashioned light bulb. The resulting fragments bump into an electron multiplier that converts them into a quantifiable electrical signal. These electrically charged molecular bits have characteristic mass-to-charge ratios that depend on the energy beaming out of the filament, and under standard conditions (i.e., knowing how powerful the filament's electron energy was), the ratios can be used as a fingerprint (spectrum) to identify the original molecule that gave birth to them, very specifically.

Whereas the experimental details are relatively skimpy in the Thackeray report, the Ren study provides much more comprehensive information on how their analysis was carried out, as well as a lengthy supplement showing the specific chromatograms of each GC-MS analysis from the charred wood and burnt stone in each brazier, plus a photograph of an easily identifiable cannabis plant taken from one of the tombs. This degree of detail makes the results not only easy to understand, but also more believable because of the high degree of specificity. Whereas the first study detected mass-to-charge ratios of compounds that suggested cannabinoids, but "not in quantities sufficient for proof," the second conclusively found measurable quantities of cannabinol, (the cannabinoid that THC decays into), along with CBD. Thus, the Ren study meets criteria for high-quality, scientifically rigorous research.

The final case example is not a study, but a recently published polemical book, Alex Berenson's "*Tell Your Children*," [3] that in my opinion and that of many others (see further) "spins," the scientific literature that he cites to exaggerate the dangers associated with marijuana.

My first experience with how science can be "spun" was as a 9 year old. A local political candidate, part of whose platform was "making our streets safe for children," spoke repeatedly about the fact that several local candy stores were selling liquor-filled Italian chocolates. He was instigating a brouhaha based on the supposition that local children would purchase these for purposes of intoxication, and that unnamed local "experts" had assured him that this would result in damage to childrens' brains. Furthermore, as a valiant protector of innocent young lives, he would assuredly root out this scientifically proven menace to our young folk and ban the perfidious Italian chocolates. This piqued our curiosity. Several of us pooled together sufficient funds from our pocket money to buy an over-priced box of said chocolates. Within each we discovered, in the center of the confection inside of a candy coat was a tiny amount of sickly sweet liquor. Breaking-open the entire boxful and decanting off the contents barely filled a teaspoon. We figured out that it would have required eating dozens of boxes of the chocolates to reach a barely perceptible level of intoxication, and that therefore the politician was full of it. Collectively, we wrote an indignant letter to the local newspaper pointing this out, that was dutifully ignored. This episode made me forever skeptical of substance-related scare tactics, ranging from "LSD will make you stare at the sun and go blind" to "marijuana is a gateway drug that shrinks your brain, damages your chromosomes and makes you sterile."

Thus, when I encountered Berenson's *"Tell Your Children,"* that relates how marijuana use is causing an epidemic of psychosis and violence, my dubiousness index went up several notches. Berenson is a former New York Times reporter who has written clearly and thoughtfully on a variety of topics, toured Iraq twice to cover the war, and subsequently retired from journalism to write 10 or more successful spy novels. His new book makes many exaggeratedly strong claims regarding ill-effects of cannabis legalization. The book's central argument is that "marijuana causes psychosis. Psychosis causes violence. The obvious implication is that marijuana causes violence." "The black tide of psychosis and the red tide of violence are rising together on a green wave, slow and steady and certain" [3].

I believe that it's worth devoting some time to discussing his published arguments critically, because Berenson interviewed "some of the world's foremost experts on marijuana and mental illness" to reach his conclusions and gathered large amounts of evidence. I believe that his assessment of marijuana's propensity to increase risk for psychosis generally and schizophrenia in particular is correct but significantly exaggerated. Evidence does not suggest to me that marijuana either inherently, or through its admitted relationship to psychosis is a major risk for violent behavior to anything like the extent that he portrays it, and that this aspect of Berenson's argument in particular is greatly overemphasized. So, it's important to understand how scientific findings can be spun in support of a particular agenda, as I believe they are in his book. More explicitly, I would further argue that many of the arguments employed by Berenson are skewed, statistics in support of his major thesis cherry-picked, contrary facts ignored, and correlation confused with causation (remember the ice cream and drowning). But, I don't want to devote an entire chapter to an extended book review, so that I'll tackle a single issue here, the question of whether, as Berenson suggests, cannabis use is leading to a schizophrenia epidemic.

Thus, my problem with his book is twofold: my first concern is that readers will be unnecessarily panicked by Berenson's magnification of cannabis' harms. My second worry is in some sense the opposite: that people will discount his alarmism and invalid arguments and therefore dismiss the entirety of Berenson's book, including his genuine assertions. These latter include, for example, the data that there is an increased risk of psychosis associated with the drug, (albeit a much smaller one than he posits [17]; see Chapter 8), or that part of the reason for reduced perception of cannabis' harm includes pressure from dubious commercial interests. But when, as is the case here, a case is significantly overstated and claims run ahead of the evidence, it's easy for readers to categorize the author as the boy who cried wolf, so that the valid points are tossed out with the hype. The book has been editorialized by Berenson himself [18,19], and reviewed multiple times both favorably [20–24], critically [14,25–34], and debated in a roundtable [35]. I acknowledge a debt to these authors and debate participants and others [36,37] for first raising many of the issues I summarize further. So let's examine Berenson's major points regarding schizophrenia.

Does Marijuana cause psychosis?

Berenson makes much of the 2017 National Academies [38] report statement that "cannabis use is likely to increase the risk of developing schizophrenia and other psychoses; the higher the use, the greater risk." But he pushes this too far in concluding that cannabis smoking is directly responsible for very large numbers of new cases of schizophrenia that would not otherwise arise. During the Marshall Project's debate [35] Berenson says "Marijuana causes psychosis. This is an established medical fact, not open to debate.... The mainstream literature and the physician-scientists who have done the most work on the issue also believe it is responsible for some cases of schizophrenia that otherwise would not have occurred-that is to say, that it can cause schizophrenia, especially when used regularly to heavily by adolescents." Later, in Chapter 7 I will discuss the substantial evidence that supports an association between cannabis use and serious psychosis that extends beyond the phase of acute drug exposure, (as explained in Chapter 8). But this relationship is far more nuanced than Berenson's stark statement that cannabis causes schizophrenia. In discussing schizophrenia risk, the NAS report was much more balanced than Berenson's summary. It stated that the relationship between marijuana and psychosis was "multi directional and complex" and that many other, non-cannabis related factors, are involved in its genesis, including genetic background and family history.

This issue was clarified by pharmacologist Ziva Cooper, an expert on effects of cannabis and cannabinoids. She is the research director of UCLA's Cannabis Research Initiative and one of the authors of the National Academies 2017 report that summarized what was then known about the relationship between cannabis and psychosis [38]. Dr. Cooper tweeted the following: "In response to the recent New York Times editorial on cannabis and as a committee member on the NASEM cannabis and cannabinoid report, we did NOT conclude that cannabis causes schizophrenia. We found 1) an association between cannabis use and schizophrenia and 2) an association between cannabis use and improved cognitive outcomes in individuals with psychotic disorders (not mentioned in the editorial). Since the report, we now know that genetic risk for schizophrenia predicts cannabis use, shedding some light on the potential direction of the association between cannabis use and schizophrenia [39]. We also now know that under placebo-controlled conditions, cannabidiol (CBD) improves outcomes in patients with schizophrenia when given as an adjunct med, showing that cannabinoids (not necessarily cannabis) improve symptoms."

Ziva Cooper's comment regarding genetic risk refers to findings from a group of Dutch investigators, who in 2018 examined the heritability of cannabis use, (which is known to run in families) [40]. They found 21 genetic variants in the form of commonly occurring alterations in a single DNA coding "letter" (as detailed in Chapter 7) which explained 11% of the variance in this heritability, actually a very high proportion in these sorts of analyses. They then performed a complex analysis that showed evidence for a positive influence of schizophrenia risk on cannabis use. In other words, this finding provides some evidence that genetic risk for schizophrenia may influence marijuana use, so that the causal path is not necessarily one leading directly from cannabis to schizophrenia. In other words the two are confounded. Although the Dunedin study discussed in Chapter 7 showed evidence of schizophrenia following youthful cannabis use [39], the more recent Di Forti paper [41] did not. Maria Di Forti found that while cannabis potency and frequency of use were strong risk factors for the development of psychotic illness, age of first use was not. So, this portion of the story remains somewhat unresolved for now. To summarize, as HL Mencken said: "For every complex problem, there is an answer that is clear, simple and wrong". Berenson's assertions I believe are not so much as wrong, but overly simplified at the expense of ignoring rather subtle and complex disease risks.

Another major point in the "marijuana causes psychosis" argument is that acute anxiety and paranoia after using the drug (as illustrated in the description of Arjun in Chapter 3), plus delirium and transient psychotic events are very different from cases of schizophrenia. The latter illness characteristically involves persisting positive (e.g., hallucinations) and disabling negative (e.g., apathy) symptoms plus a deterioration in life course (see case of Janet in Chapter 8). In Berenson's book however "psychosis" and "schizophrenia" are not consistently distinguished, acute versus chronic psychosis-like symptoms are not always taken into account, and one set of terms tends to elide with the other. During the Marshall debate Mark AR Kleiman, Professor of Public Policy at NYU [35] alluded to this. "In addition to wanting to know what sort of "psychosis" cannabis might cause, it's sensible (to) want to know how often these bad things happen: both what fraction of psychosis is attributable to cannabis and what the probability is that any given pattern of cannabis use will lead to psychosis... (or schizophrenia) ...however defined.... I think the problem is more the difference in professional practice between journalism and science."

What is the evidence that rates of schizophrenia are rising because of marijuana use? Let's try and parse that statement. Outside of isolated testimony in his book from various emergency room physicians, there are no nationwide data to suggest that there is an upswell in psychosis incidence in the

United States. Nationwide incidence rates of serious mental illness are hard to track accurately without systematic, multi-city household surveys such as the Epidemiological Catchment Area study [42]. There is not mandated reporting for psychotic illnesses as is required, for example, in newly diagnosed cases of tuberculosis or HIV. But this statement regarding 18–25-year-olds seems inconsistent with Berenson's other claim elsewhere in his book that many cases of supposedly cannabis-caused psychotic illness are actually appearing in marijuana users who are *outside* of the standard age of risk, in older, previously stable individuals.

Finally, as an investigator who is personally administering typical doses of cannabis to volunteers in four separate federally funded research studies, among hundreds of doses provided to our subjects that they have rated as "typical of what I would use by myself or with friends." I have seen zero instances of psychotic symptoms. This is not at all to claim that such events never occur, but that they are nowhere as common as implied in Berenson's book.

Berenson says of those who deny the connection between marijuana, psychosis, and rising violent crime rates, that "The tricks can be hard to find- and journalists who are almost never trained in science or statistical analysis often parrot the results unskeptically, especially when the findings confirm their own biases toward ideology or sensationalism. Once car accidents and violent crime are involved, the results can be deadly." I think that this criticism could equally well be reflected back on his own arguments. This point was emphasized in a February 2019 "*Letter from scholars and clinicians who oppose junk science about marijuana*" [37]. "The vast majority of people who use marijuana do not develop psychosis and schizophrenia, nor do they engage in violence, thus making Berenson's claim far-reaching and exaggerated" [37]. A further point is Berenson's dismissal of the claims of David Nutt, a British professor of neuropsychopharmacology, who has researched extensively and written cogently on diverse drug-related topics as well as publishing with other distinguished researchers (Chapter 8) on the topic of how to rank drug-related harms. Berenson dismisses Nutt's work as the product of some slightly loopy fringe scientist. But this brushing aside of Nutt's work stems perhaps more from the latter's ranking of cannabis significantly below that of alcohol in terms of its score on an empirically derived relative harm index, an example of Berenson's cherry picking.

As well as the *practices* of science that help us to decide whether a particular study has incorporated the procedures and safeguards that meet the criteria for believable science, cultivating a skeptical scientific *attitude* is essential to help us think clearly through cannabis-related issues, such as questions regarding legalization and its consequences. Such modes of approaching problems help us avoid the extremes of cannabis boosterism and alarmism. This then enables, for example, more rational planning of harm reduction policies and dispassionate comparisons of marijuana-associated risks to those of other widely used recreational drugs.

References

[1] Pincher C., Dope: is there a link between dope and hot jazz dancing?, Daily Express; 1951.

[2] Morgenstern O. On the Accuracy of Economic Observations. 2nd ed. Princeton, New Jersey: Princeton University Press; 1963.

[3] Berenson A. Tell Your Children: The Truth About Marijuana, Mental Illness, and Violence. New York: Free Press; 2019.

[4] Blethen R. New Yorker article about marijuana strikes nerve with pot researchers; 2019. Available from: https://www.seattletimes.com/seattle-news/marijuana/new-yorker-article-about-marijuana-strikes-nerve-with-pot-researchers/.

[5] Marijuana in Space –NASA discovers THC on meteorite fragment. Available from: https://notallowedto.com/marijuana-in-space-nasa-discovers-thc-on-meteorite-fragment/.

[6] https://notallowedto.com/.

[7] Weinberg B. THC from outer space? Um, No. High Times. 2016. Available from: https://hightimes.com/culture/thc-from-outer-space-um-no/.

[8] Thackeray FvdM, Nikolaas, Van der Merwe TA. Chemical analysis of residues from seventeenth-century clay pipes from Stratford-upon-Avon and environs. S Afr J Sci 2001;97:19–21.

[9] Thompson H. Did Shakespeare smoke pot? Smithsonian Magazine. 2015. Available from: https://www.smithsonianmag.com/smart-news/did-shakespeare-smoke-pot-180956223/#Kv08xESKr7b10sOR.99.

[10] Thackeray F. Shakespeare, plants and chemical analysis of the early 17th century clay 'tobacco' pipes from Europe. S Afr J Sci 2015;3:7–8.

[11] Thackeray F. Was William Shakespeare high when he penned his plays? The Independent. 2015. Available from: https://www.independent.co.uk/arts-entertainment/theatre-dance/features/william-shakespeare-high-cannabis-marijuana-stoned-plays-hamlet-macbeth-romeo-juliet-stratford-10446510.html.

[12] Mabillard A. Did marijuana fuel Shakespeare's genius? 2000. Available from: http://www.shakespeare-online.com/biography/notedweed.html.

[13] Delman E. Hide your fires: on Shakespeare and the 'Noted Weed', The Atlantic. 2015. Available from: https://www.theatlantic.com/entertainment/archive/2015/08/shakespeare-marijuana-nope/401087/.

[14] Devitt-Lee A. Data distortion: use & abuse of cannabis science, Project CBD. 2016. Available from: https://www.projectcbd.org/science/data-distortion-use-abuse-cannabis-science.

[15] Cotten SW, et al. Unexpected interference of baby wash products with a cannabinoid (THC) immunoassay. Clin Biochem 2012;45(9):605–9.

[16] Ren M, et al. The origins of cannabis smoking: chemical residue evidence from the first millennium BCE in the Pamirs. Sci Adv 2019;5(6). peaaw1391.

[17] Di Forti M, et al. High-potency cannabis and incident psychosis: correcting the causal assumption—authors' reply. Lancet Psychiatry 2019;6(6):466–7.

[18] Berenson A. Weeding out dubious marijuana science, Wall Street Journal. 2019. Available from: https://www.wsj.com/articles/weeding-out-dubious-marijuana-science-11557090088.

[19] Berenson A. What advocates of legalizing pot don't want you to know, NY Times. 2019. Available from: https://www.nytimes.com/2019/01/04/opinion/marijuana-pot-health-risks-legalization.html.

[20] Gladwell M. Is marijuana as safe as we think? The New Yorker 2019. Available from: https://www.newyorker.com/magazine/2019/01/14/is-marijuana-as-safe-as-we-think/amp.

[21] Mencimer S. This reporter took a deep look into the science of smoking pot. What he found is scary, NY Times. 2019. Available from: https://www.motherjones.com/politics/2019/01/new-york-times-journalist-alex-berenson-tell-your-children-marijuana-crime-mental-illness-1/.

[22] Berenson A. Is smoking marijuana safe? Why we can't let Big Weed bury the risks the way Big Tobacco did, NBC News. 2019. Available from: https://www.nbcnews.com/think/opinion/smoking-marijuana-isn-t-inherently-safe-we-can-t-let-ncna970681.

[23] Neese D. PROVOCATIONS: Ignoring marijuana's bad news, Trentonian. 2019. Available from: https://www.trentonian.com/opinion/provocations-ignoring-marijuana-s-bad-news-david-neese-column/article_e2433e58-9b6c-11e9-866e-d338d76bc886.html.

[24] Jaquiss N. In an Op-Ed, Former New York Times Pharmaceutical Reporter Raises Concerns About Legal Recreational Weed, 2019. Available from: https://www.wweek.com/news/2019/01/06/in-an-op-ed-former-new-york-times-pharmaceutical-reporter-raises-concerns-about-legal-recreational-weed/.

[25] Nathan DL, Elders J, Adinoff B. 21st century reefer madness, Psychiatric Times. 2019. Available from: https://www.psychiatrictimes.com/cultural-psychiatry/21st-century-reefer-madness.

[26] Szalavitz M. All the things the new anti-weed crusade gets horribly wrong, Vice. 2019. Available from: https://www.vice.com/en_us/article/pa5j48/all-the-things-the-new-anti-weed-crusade-gets-horribly-wrong.

[27] Hart CL, Ksir C. Does marijuana use really cause psychotic disorders? 2019. The Guardian. Available from: https://www.theguardian.com/commentisfree/2019/jan/20/marijuana-cannabis-health-effects-issues-mental-health-disorders-science.

[28] Lartey J. Popular book on marijuana's apparent dangers is pure alarmism, experts say, The Guardian. 2019. Available from: https://www.theguardian.com/society/2019/feb/17/marijuana-book-tell-your-children-alex-berenson.

[29] Way K. What fearmongering about pot tells you about mainstream marijuana coverage, The Nation. 2019. Available from: https://www.thenation.com/article/alex-berenson-marijuana-legalization-tell-your-children-review/.

[30] Levitan D. Reefer madness 2.0: what marijuana science says, and doesn't say, 2019. Available from: https://undark.org/article/reefer-madness-marijuana-science/.

[31] Singal J. Did marijuana legalization really increase homicide rates? New York Magazine. 2019. Available from: http://nymag.com/intelligencer/2019/01/no-pot-legalization-probably-didnt-increase-homicide-rates.html.

[32] Lopez G. What Alex Berenson's new book gets wrong about marijuana, psychosis, and violence, Vox. 2019. Available from: https://www.vox.com/future-perfect/2019/1/14/18175446/alex-berenson-tell-your-children-marijuana-psychosis-violence.

[33] Hopper T. Is pot really ruining all the men? 2019. Available from: https://www.thegrowthop.com/cannabis-news/cannabis-legalization/is-pot-really-ruining-all-the-men.

[34] Sullum J. Does legalizing marijuana cause 'sharp increases in murders and aggravated assaults'? 2019. Available from: https://reason.com/2019/01/09/does-legalizing-marijuana-cause-sharp-in/.

[35] Project T.M. How dangerous is marijuana, really? 2019. Available from: https://www.themarshallproject.org/2019/01/14/how-dangerous-is-marijuana-really.

[36] Lopez G. The benefits and harms of marijuana, explained by the most thorough research review yet, Vox. 2017. Available from: https://www.vox.com/science-and-health/2017/1/14/14263058/marijuana-benefits-harms-medical.

[37] Ashford R. Letter from scholars and clinicians who oppose junk science about marijuana, 2019. Available from: http://www.drugpolicy.org/resource/letter-scholars-and-clinicians-who-oppose-junk-science-about-marijuana.

[38] National Academies of Sciences, Engineering, and Medicine. The health effects of cannabis and cannabinoids: the current state of evidence and recommendations for research, 2017: Washington (DC).

[39] Arseneault L, et al. Cannabis use in adolescence and risk for adult psychosis: longitudinal prospective study. BMJ 2002;325(7374):1212–3.

[40] Pasman JA, et al. GWAS of lifetime cannabis use reveals new risk loci, genetic overlap with psychiatric traits, and a causal influence of schizophrenia. Nat Neurosci 2018;21(9):1161–70.

[41] Di Forti M, et al. The contribution of cannabis use to variation in the incidence of psychotic disorder across Europe (EU-GEI): a multicentre case-control study. Lancet Psychiatry 2019;6(5):427–36.

[42] Sharifi V, et al. Psychotic experiences and risk of death in the general population: 24-27 year follow-up of the Epidemiologic Catchment Area study. Br J Psychiatry 2015;207(1):30–6.

Chapter 3

Holi

Alles, was ich erzähle, ist erfunden.
Einiges davon habe ich erlebt.
Manches von dem, was ich erlebt habe, hat stattgefunden.
(Everything I narrrate is invented. Some of it I actually experienced. Some of what I experienced actually happened....)

'Raumpatrouille' Matthias Brandt [1]

How can science (as i just defined it) help us think rationally about cannabis in a way that can guide public policy? To set up many of the key issues that we will revisit in different parts of this book, I'd like to start with a narrative that raises questions that many people have about cannabis. These topics include "What are typical motivations for people to use cannabis?," "What methods do people use to get high?," "What are typical drug effects?," "Why do different individuals have such different experiences after cannabis use?," "Why do some people have long- or short-term problems after using the drug and not others?," and finally "What are some of the issues society will need to cope with, and what key decisions are to be made, if cannabis is legalized at a federal level in the US?" I think that the best way to begin considering these issues is to start with a story that helps lead us through them, in part by understanding the motives and experiences of ordinary people using the drug. The remainder of the book then addresses those major topics.

Okay Jeopardy! fans—here's a question for you. "This city is associated with Day-Glo colors, music, dancing and large-scale public marijuana consumption." Any guesses? If you answered "What is Portland Oregon?" think again: the correct answer is "What is Varanasi?" Cannabis-wise this North-Indian city is a couple of thousand years ahead of us in the United States, so presumably they have much to teach us. Let's dive in and explore. Oven-like in the summer, but surprisingly chilly for half of the year, Varanasi, previously known as Benares, is one of Hinduism's seven holy cities, famous for its 88 sets of slabbed stone steps (sufficient for more than seven 12-step programs). Some of these flights of stairs (known as ghats) are spectacularly beautiful. All lead down to the Ganges River, on whose North bank the city was constructed. Visitors cruising by the steps in brightly colored boats observe bathers

Weed Science. http://dx.doi.org/10.1016/B978-0-12-818174-4.00003-3

immersing themselves in the river that is sacred to Hindus, whose cremation ceremonies, and various meditative and religious rituals, including evening fire worship, all unfold on the ghats. Although many Hindus believe that dying in this holy city leads to Nirvana, the state of bliss we will focus on here is an earthly one, associated with the large amounts of cannabis that have been consumed in Varanasi during the annual festival of Holi since its beginning at least 1700 years ago. The plant is called *Cannabis indica* due to its ceremonial use in India, which here in particular is ancient, and thus provides a useful window into the drug.

Given the many religious and historical legends linked with Varanasi and Holi, trust me for a minute to take you figuratively by the hand and lead you on a mini-pilgrimage down the steps of myth. One of the 88 ghats is intimately associated with the Hindu deity Lord Shiva, regarded by some as the supreme God in the Hindu pantheon. Shiva is not only linked in legend with Varanasi, but the city was dedicated to him sometime in the 8th century. Most relevant for us, among the pantheon of deities Shiva is traditionally depicted as enjoying the many forms of cannabis, including bhang, or ground cannabis that is blended into drinks or sweetmeats. Legend has it that he dwells with his family and followers in a distant Himalayan peak, where he can only be reached by crossing a burning ghat and jettisoning one's mortal body. Using cannabis is a traditional means of unity with the spiritual world and communing with Shiva, aided by his favored drug, through enhanced meditation.

Several varieties of cannabis are consumed most often during the Hindu celebration of Holi, the festival of bright colors, play and laughter, thanksgiving and springtime, celebrated through much of India and Nepal. It's useful to distinguish among the terms for cannabis used here, principally ganja (the Sanskrit word for dried cannabis buds and leaves), charas (cannabis resin, handmade hashish), and bhang (marijuana paste made from cannabis leaves and stems added to edibles, principally bhang lassi and bhang thandai). Ganja and charas are smoked, traditionally in a chillum or conical pipe. Charas/hashish is created by repetitive rubbing of cannabis flower between the palms or over cloth to strip off the resin-bearing hairs (known technically as trichomes), yielding yellow-green-brown, crumbly or putty-like hashish. Typically charas contains around 6% THC compared to perhaps 1%–2% in bhang. The charas preparation process always evokes for me the hand rolling of fine Cuban cigars.

The custom of Bhang consumption during Holi is strongest in North India, appropriately enough in Varanasi, Shiva's devotional city. Bhang preparation on the ghats is common during Holi festival time celebrating the end of winter and beginning of spring in late February or March. Appropriate, wise consumption of cannabis is believed by devotees to enable unity with Shiva, the shedding of sins and avoidance of future hell by entry into his sacred circle. The last two points are particularly ironic in contrast with the view of marijuana use in some Western cultures. Casual Holi revelers use the substance as a way to get buzzed and to enhance their overall enjoyment of the sights, sounds, celebration and bonding with friends and family that are an essential part of the festival. Cannabis use in India as a sedative and mind-altering substance dates back to at least the 1st century BC.

****

Three expectant Holi revelers sit on a bus carrying them across the Ganges from their small village to the big city of Varanasi. They are generally feeling lighthearted and looking forward to the festivities, except for Arjun, who is a tad glum and preoccupied. The others poke gentle fun at him because he is a little superstitious and a can be a worrywart, but well-liked because he is Mr. Responsible, sensible, and pretty much guaranteed to keep an eye on details. If someone forgets to fill their water bottle or bring a warm jacket, Arjun, tall and curly haired is the first to notice and remind them. Vikram, the oldest and short-est, is really looking forward to Holi this year. As a youngster he was fascinated by itinerant holy men who wandered through his town and the many legends of Hindu deities his grandmother would share with him. He was entranced by tales of the powerful Shiva. Vikram sometimes imagines what it would be like to ded-icate your whole existence to seeking transcendence as a wandering Sadhu, and how different that life would be from the professional career he has embarked on. The third musketeer, Devendra, with his faint, wispy moustache is the joker in the pack. Lighthearted, always enjoying the moment and outright silly at times, D is always ready with a quip. Family members and teachers accuse him of not taking life seriously enough, but he feels at 16, that there is plenty of time for that in his future, just not now. Right now life is about having a good time and hanging out with his friends.

Just before dawn, ushered in by the embers of the previous night's bonfires and torches, Holi begins with a faint remnant of the smell of burnt wood. The little band stops at a stall to sip some sweet chai, so painfully saccharine that Devendra jokes that the dentist back home must be in cahoots with the skinny stall owner. The group rapidly runs into a mad crush of hustling and bustling people. Chatty tourists with long-lensed cameras, stall owners touting their wares, brightly beturbaned locals, orange-and saffron-robed religious figures with colorful patterns drawn on their foreheads, white-bearded religious figures, beggars looking imploringly from mats on the sidewalk, Vedic astrologers and razor-wielding barbers looking for customers have already made it to the river-side and are congregating around the ghats. It's not just the mass of humanity, the individual bodies with a common purpose, but in the half-light by the pea-green, syrup-slow Ganges flowing from the South there is abundant evidence of animal life. Roosters are crowing, chickens clucking, pigeons flapping, dogs barking and snuffling on the shore, placid, long-eared sad-eyed cows wandering, and the occasional monkey soliciting scraps. Adding to the symphony, someone is chanting a prayer in a raspy, slightly off-key voice that Devendra claims was stolen from a frog. Vikram and Arjun chide Mr. D for being disrespectful.

Along the streets, wood fires flare irregularly and help guide the travelers. Some flames are heating cauldrons of delicious smelling food. This is a religious city and the fires remind everybody that a few streets over, in the two special riverside burning ghats, corpses are being cremated. Here in the holy city their former owners are thus freed from the eternal cycle of death and rebirth. The elderly and ill are drawn to Varanasi to live out their remaining days before being released from earthly suffering in clouds of smoke. The river itself cleans and purifies the living through immersion, and a little later, pilgrims will be dunking themselves in the healing waters. Legend has it that at the beginning of time, Shiva let the Ganges burst forth from a lock of his hair in a cataract of clear, holy water.

Arjun, Vikram, and D reach the broad river shoreline exactly as the sun rises on the opposite bank. Vikram claims he can see their hometown way over there, amid general skepticism from the others. Hundreds of tourist camera shutters click and chatter at the same time as the huge red orb is reflected in the river. The sun begins to pick out clumps of people boarding bright-colored boats on the shoreline, and women beating clothes on planks and rocks to clean them in the river as discarded floral garlands float by. Having served their purpose, the lamps that exist to help guide spirits home, craning their necks over the river on long wooden poles, are extinguished, and all over the city endless rituals continue in a thousand temples.

The three friends wander away from the ghats and climb the slight hill back to the throngs of people who are already enthusiastically eating and drinking, sampling pastries, sweetmeats and occasionally alcohol, chattering away amid the colors of joy and excitement. In the streets containers of flowers and garlands are displayed on plates. Street merchants stand by enormous sacks of colored pulverized cornmeal, whose rolled back-tops reveal interiors already intensely dyed by the contents. The brightly hued powders have already been partially spooned into plastic bowls ready for weighing and bagging. As bicyclists leisurely cycle by and pedestrians jostle, the stall holder arranges the bowls, scales, weights, cups and ladles to make his wares more attractive and competitive. Nearby, on the stone blocks in front of a temple, flanked by chairs, red cloth flags on the tips of long poles wave over the jostling throngs of men, women and children. The wind flaps the banners above the crowd stirring the complex smell of sweat, tobacco and ganja smoke, cooking food, charcoal, goat blood, and animal dung.

As the boys make their way down the street, they are met by chromatic explosions of brilliantly colored powders and the most vivid hues, resembling a cottage garden on acid. Vivid purples, yellows, blues, greens, scarlets and oranges are sometimes singular expressions of colored joy. At other times they are mysteriously coordinated, erupting vertically from 1000 places in the huge crowd simultaneously, swirling together in the air. There is a Day-Glo ocean of particles and rainbows of psychedelic dust, as devotees fling handfuls of colored corn starch heavenward. Others toss tinted water-filled balloons or spray dyed water at each other. All the colors mix and spiral in the resulting puddles beneath your feet that reflect the temple, the sky and the crowd. Dull brown sparrows splashing in the bright-hued liquid look as striking as parrots for a few gorgeous moments, before they shake themselves off and fly away.

The bedazzled crew wanders over to a large stall where bhang infused drinks are being concocted from scratch. Like amateur foodies watching cooking shows on TV, the process is inherently entertaining. To prepare the bhang, boiled female cannabis plant leaves, buds and stems are being ground by a young member of the stall owners family into a dark olive-green paste. This traditional grinding process uses simple slabs of worn stone. The higher-end stalls with fancy prices two streets up use expensively crafted mortars and pestles but the end result is identical. The resulting smooth cannabis paste is mixed with slurry of nuts and spices. Slapped together by hand, the mixture is mounded into baby volcano shapes or rolled into tennis ball sized spheres and rubbed and filtered methodically through a fine cloth net into steel buckets. Because THC is so fat-soluble, cannabis-laced drinks are prepared with ghee (clarified butter) or milk in some form to aid its absorption. The resulting gallons of milky, slightly murky green and mildly psychoactive fluid await.

The stall owner asks if they're actually going to buy anything or just gawk, so each of them purchases a generous cup full of cool, fragrant, and refreshing bhang thandai. This drink is made from boiled milk, plus a ground-up pot-pourri of crushed rose petals, sugar, cardamom, dry nuts, peppercorns, fennel, poppy and melon seeds and aniseed. And of course the main attraction, bhang. The concoction is ladled from a large jug into bright copper cups. Saffron strands float invitingly on top. Vikram and Devendra order "medium strong" and Arjun pays a tad extra for "strong." Traditionally thandai and lassi are drunk with the head back, mouth wide open and poured down one's throat without lips touching the vessel. All three of them somehow manage this procedure without spilling a drop.

A little knot of mildly confused but friendly tourists in their early 20s, attired in brand-new safari clothes, wanders over to ask for an explanation of all things bhang. Arjun enters professorial mode and explains that there are two main bhang-based beverages consumed during Holi. Thandai is made from boiled milk and is exotically flavored. "Like everything in your mom's spice cabinet mixed together" jokes Devendra. Bhang Lassi is similar, but the base is yogurt not milk, so the taste is pleasantly sour. Then there are the green-colored foods. Halwa is a nutty sugary sesame confection, blended with bhang for the holiday, while Bhola are chewy, grape leaf-colored ping-pong ball-sized spheres of bhang, and other mysterious ingredients. Finally if you're itching to get high but are not hungry or thirsty, you can always smoke ganja leaves or charas, which is hash, Arjun explains. Unlike bhang, ganja and charas are illegal, but during the festival this ban doesn't seem to be very strictly enforced. The tourists introduce themselves, ask many questions, take many selfies, purchase bhang lassi at the stall, and hang out with the Three Musketeers.

Everyone mellows out as the morning unfolds and Holi swirls around them. The frantic rush seems to settle down. Everyone's feeling gently chill, soothed, easy-going, and slightly silly. Devendra is the catalyst for the latter. He launches into a recent online news item about wild parrots 500 miles west in Madhya Pradesh who are addicted to opium. Flocks of these brightly colored birds dive into the poppy fields to nibble on the ripe pods, or even snip them off to fly away with their illicit booty. Devendra has the group in hysterics with a passable imitation of a stoned and wobbly parrot, with his wings/arms outstretched, colliding with an imaginary tree. One of the Canadians launches into a riff about wanting to be reincarnated as a bird, which somehow elides into birdseed back home being made of hemp and the phrase "stone the crows." Devendra claims that Arjun has already been reincarnated as a vulture, because he's always hungry. Arjun agrees, and to illustrate the point, wanders off in search of a snack, bringing back a small plate full of bhang-stuffed bholas to share with the group. One of the tourists takes a bite, and after chewing the confection describes it as "weird stoner marzipan." Devendra makes the outlandish claim that the main bhola ingredient is toad. For one reason or another no one else seems tempted, so Arjun, channeling his inner vulture polishes off the remainder. One of the tourists has a kazoo, which sparks group hilarity about loud, cheap, plastic and terminally annoying Vuvuzelas, and their kinship with the clay chillums in which the holy men smoke ganja. This is followed by much speculation about smoking marijuana through varied musical instruments from saxophones to gigantic William Tell-style Alphorns.

Somewhere in the proceedings, the plateful of unwisely eaten Bholas kicks in for Arjun, who notices that his heart has speeded up and is pounding unpleasantly like a tabla drum in his chest. An uncle died of a heart attack a few months back and he dwells on this memory, feeling simultaneously sad and anxious. He hopes his own heart is okay. Maybe it's not. Things are becoming steadily less comfortable, and his chest much comfortable, so that he excuses himself and quickly wanders away from his old and new friends to chill out and get a little alone time.

As he makes his way down the street, the whole festival begins to feel oppressive with the colors being a little too bright. People seem to be looking at him and giggling. His thoughts swirl. Everything is off-kilter, as if the world has changed in some unpleasant and scary way. He clearly recalls that he had this same feeling the last time that he used cannabis. This time around though, the state is much stronger and harder to rein in. There has been a lot of talk in the press recently about a new satellite launch in India—suddenly A begins to worry that maybe the satellite is somehow tuning into his thoughts and broadcasting them to people in other places around the world. Part of him knows that that makes no sense. A suddenly becomes aware of fearful looks on the faces of young women in the throng who seem wary of the glances they're getting from some of the men. He's worried that he might be responsible for this in some way. His thoughts are bouncing around a little fast and randomly so it's hard to keep track of them. The pervasive tone is edgy and unpleasant, with little rivulets of upset turning into a powerful waterfall of negativity. Arjun has the strange feeling that he can see his own soul and the souls of those all around him. Each soul is a small, fragile white worm, like the wriggly grubs he had seen as a kid in a sack of old lentils. This idea feels very strong and important, but makes him feel small, insignificant, and vulnerable. He could easily be squashed. His swirling thoughts return to his dead uncle. When he looks into a nearby puddle Arjun does not see his own reflection and it occurs to him that he might have died. This leads his mind to Vantara, the river full of human bones, sharp fingernails and filth where sinners are tormented. Everything's spinning a little out of control. The rational part of his brain kicks in and he makes his way to the emergency medical tent at end of the street. Everyone there is helpful and reassuring. They talk to him calmly, asking him what's going on, measure his pulse and blood pressure and tell him that he's feeling the effects of a few to many sweetmeats. They give him a cup of hot tea and make him feel relaxed. Slowly, and it's hard for him to say exactly how long it takes, Arjun notices that everything is coming back to normal, his thinking is clearer and after a brief check with the staff, he wanders back a little shakily to find his friends.

Meanwhile, shortly after Arjun wandered off so abruptly, Vikram waves goodbye to the group too. He is in search of one of the many groups of Shiva devotees. The Hindu community provides free marijuana for these holy men or sadhus who neither work nor have family ties and are supported by donations from the community. The sadhus cover themselves with a simple loincloth and paint their bodies with sandalwood ash or dye. They tame their bodies through yoga practice or assume painful postures, sometimes for extended periods, even years, tolerating suffering. Thus by their prayer, austere bare-bones lifestyles, fasting and bodily deprivation, they atone for their own sins and those of the entire community. Part of Vikram feels respect and admiration for the Sadhus, in part because he knows that he could never pursue such a life himself. He respects the fact that as a type of human scapegoat, they achieve self-liberation through worship, yoga, meditation, reflection on the divine, and of Shiva in particular. As part of this mode of life, smoking ganja frees the mind to aid calm contemplation of the infinite. And Shiva himself legendarily consumed marijuana as a blessing.

According to the British Indian Hemp Commission report of 1894, [2] (which we will discuss later), "These religious ascetics, who are regarded with great veneration by the people at large, believe that the hemp plant is a special attribute of the god Siva, and this belief is largely shared by the people. Hence the origin of the many fond epithets ascribing to ganja the significance of a divine property, and the common practice of invoking the deity in terms of adoration before placing the chillum or pipe of ganja to the lips. There is evidence to show that on almost all occasions of the worship of this god, the hemp drugs in some form or other are used by certain classes of the people" (p. 160).

Vikram perambuates along a narrow brick-lined street toward an area with many temples. He passes a pink-robed, benecklaced man with a pink turban, shoulder-length black hair, and a long windswept-looking white beard. Vikram continues his thought that after smoking ganja, everyday feelings and concerns disappear for the holy men, who become one with the universe. The space left by the loss of their individual self-identity and earthly cravings, fills with religious insights about Shiva, and concentrating on God is more easily attainable. Lost in thought, bareheaded women in red and purple silk scarves give him a wide berth, but he presses on and guided by the sound of religious music, finds his way to a large room in a shabby temple complex.

He wanders tentatively onto the threshold to be greeted by the sight of an elderly bearded man seated at a large wooden table chopping light brown ganja leaves on a copy of the local newspaper. The old fellow is using a double-edged metal knife very proficiently. The sunrays from a window way up on the wall shine on him directly as he slices methodically at the dry material. One of his companions helps mix the leaves with shredded cigarette tobacco and layers the contents into an 8-inch long clay chillum, or vertical clay smoking pipe. Vikram thinks back to the vuvuzela jokes and smiles. A man who could be anywhere between 30 and 70 with dreadlocked hair braided into a loose birds-nest topknot beckons him in. The man's forehead is painted with wide stripes and dots of yellow and white paint contrasting with his brown skin. Someone has lit the chillum, wrapped its narrow lower part in rags and is taking giant leafy puffs. Eventually an arm with a well-worn pink string wound around a hairy wrist, hands it to Vikram. After inhaling, he too passes it on to the person next to him. This man has dilated pupils in deep set eyes under a set of bushy black eyebrows, and a long, seemingly never-combed beard. He acknowledges the gift of the pipe and soon he too is huffing and puffing.

On the other side of the room, a group of traveling musicians is performing devotional songs. Some strike rows of coin-sized silver cymbals on metal bars others perform on saucer-sized brass cymbals, tabla drums, or sitars. Seated on the floor, several singers accompany the instruments as the light streaming through windows highlights the details of their robes, colored turbans, and painted faces. Some of the musicians smoke ganja to concentrate on Shiva, who is thus honored, as they sway and clap to the rhythm. With the ganja, their timescape slows and they can focus keenly on both the individual parts and overall patterns of the changing music.

Vikram is absorbed inexorably into everything going on in that room. A man with thousand-year-old eyes and a golden-yellow spot between his eyebrows smiles beneficently at him. He is attired in a reddish cloth wrapped around his waist. Vikram notices the man's long fingernails and dirt-covered bare stomach. One holy man lights the chillum for his neighbor. Friendly brown faces around the room bear variously wrinkles, paint, noses ranging from long and aquiline to snub, and expressions ranging from rapt to blissful. One Saint looks to be aged about 70 with bags under his rheumy eyes. He has a thermometer-shaped red streak running the entire length of his forehead, then plunging beneath his headscarf. Vikram becomes aware that the chillum has a white turban of ash sitting on top of it. A younger sadhu dressed in a white silk shirt puffs furiously before inhaling. The rag is now nicotine-stained, but doing its job of preventing burned fingers and lips. And along with a smooth pebble at the bottom of the chillum, it is stopping the user from injudiciously sucking burning coals of weed into his gullet. Vikram's mind drifts slowly to some other place where he feels the benevolent gaze of Shiva penetrating and illuminating everything. He is both detached yet completely part of everything going on around him. He knows that the earth is made of music and that its rhythms paint and disclose our world and in turn, reflect the rhythm of the universe. High above in the heavens, the great iron wheel of cause and effect turns slowly and portentiously. Its hub is in the center of the room with them all, thrumming and vibrating. Vikram sees himself seeing and feels the inevitability and liberation of death and rebirth under the eternal Eye of Shiva.

Back in the street near the Bhang Lassi stall, Devendra is still relaxing with his new friends. He has slipped comfortably into his usual role as court jester and can intuit what will get the now-inebriated little group giggling and snorting with laughter. They sign onto his suggestion to open a stall selling T-shirts emblazoned with the slogan "I drank bhang lassi in Varanasi." His body feels comfortable and he's utterly at ease with his companions. He's happy to be out here as part of the milling crowds celebrating Holi. Devendra observes that the festive colors have a depth and brightness that seems unusually magnified. He can almost reach out and touch them with his eyes. He is able to distinguish subtle differences among the various shades, and seems to perceive at least 10 hues of green. Sounds and colors and tastes blend together and change places. When he takes a final sip of the Bhang Lassi, the flavors echo the color-bursts and surges. Time is slowing down and the thrown colors appear blissfully suspended in midair for what seems like an entire minute. Inspired by these dancing hues Devendra stands up and begins to pirouette comically, but the combination of being slightly uncoordinated and not noticing a slippery patch on the cobbles result in an unexpected pratfall onto his butt straight into a roadside pool full of brightly colored Holi-tinted water. Everybody in the group, including Devendra, finds this hilarious. Sitting down comfortably, if a little damp, a streetside dandelion catches his attention. The golden-yellow wildflower, relatives of which he has seen a thousand times previously, is unfamiliarly absorbing and utterly beautiful. How could he have never noticed the subtle toothyness of the petals, or the delicate almost pulsating aura that perfuses the bloom? He's engrossed by this enchanting speck of creation for what seems hours, only interrupted when a woozy Arjun wanders back from the medical tent.

****

By now, you're either googling "do-it yourself bhang lassi" or shaking your head in horror, wondering what kind of parents would allow their teenage sons to go to a pot and powder party. No matter which, what can we learn from this description of Holi? I like to think that one aspect of the festival is to provide a gateway of sorts to marijuana-related issues. After all, this is a venue where cannabis has been used for 3000 years, for the most part safely. Over this long time period Indian society has worked out how to minimize problems associated with the drug. And in the past 120 years historical commissions and reports have attempted to understand and document these issues. Although there persists a wide variety of opinions regarding the drug and people who use it, nearly everyone tolerates the festival and alongside that celebrations of which cannabis consumption is an inherent part. In turn, public policy makers in the West can learn from, and usefully take into consideration, the experiences and policy decisions of a culture that has done a creditable long-term job of accommodating recreational and religious cannabis use. In other words Holi provides anecdotes that are important subject matter for scientific inquiry.

Here are some of the major issues raised by our snapshot of Holi. The celebration is a cannabis-themed festival that has been celebrated for thousands of years. But how long have humans used cannabis, and for what reasons? This book will a conduct a very brief historical tour through different civilizations that have used the hemp plant, with its THC-containing sibling cannabis, to provide food, fiber, medicine, religious insight and recreation at different times and places. We will explore where the plant first grew, when humans encountered it, and helped spread its seeds around the globe.

One key question is how the drug produces its psychoactive and medicinal effects. We investigate the complexities of the brain and body's endocannabinoid system in Chapter 5. Why did all three boys experience something different? An important observation is that among individuals using intoxicant-containing chemical varieties of the plant for recreational or religious purposes, different people can have markedly varied subjective encounters with cannabis. The rather disparate individual bhang-provoked experiences of our three albeit unavoidably somewhat stereotypical protagonists mirror typical reactions of many cannabis users. Devendra got stoned and enjoyed a good time with his new friends—an experience that probably outweighs the other outcomes in frequency by a considerable margin. Arjun's anxious and mildly paranoid reaction to the drug is the rarest. Besides, individuals who have unpleasant experiences when using marijuana will not willingly choose to use the substance again. Vikram's religious awakening is also relatively less common, although may be the closest to the original use of religious ceremonial drugs such as cannabis. Chapters 6 and 8 will explore common short and long-term experiences of those using the drug, both pleasant and unpleasant.

Our story also illustrates that there are many different means by which one can consume cannabis. Edibles, in India for millenia, were subsequently adopted in the Islamic world and later France in the 1890s but are relatively new to the United States. I interviewed two Indian psychiatrists regarding their experiences after drinking bhang lassi and using cannabis edibles as revelers at Holi. Both agreed that compared to smoking marijuana in the United States, the effect of the drinks or sweetmeats was distinguished by more of a feeling of mellow happiness and calm relaxation, with more of a body high and less "buzz." One attributed this to the festival setting, while the other speculated that the higher CBD-to-THC ratio of cannabis used typically to prepare these edibles was responsible. Another relevant factor is the total dose consumed; bhang's THC content is scanty.

Every recreational drug has its Yin and Yang, risks and benefits, good and dark sides. I try to examine both aspects of the drug as open-mindedly as possible. Chapter 10 probes how recent research illuminates what we know (and mostly don't know) of cannabis' emerging medicinal properties. The 1893 Indian Hemp commission [2] was the first to enumerate many of these risk/benefit issues, and a controversial recent book has focused more specifically on particular cannabis-associated risks [3]. But what is the magnitude of overall risks associated with cannabis, and how do these stack up when compared directly to those of say alcohol or nicotine? Chapter 8 will address these issues.

Earlier we learned that Arjun partook of significantly more drug than the others, which likely contributed (along with his pre-existing mental set) to his brief but unpleasant, out-of-control cannabis experience. Chapter 6 discusses which effects of cannabis are most likely to occur in particular dose ranges and why. Deepak Cyril D'Souza, a psychiatry professor at Yale University medical school, is both a clinician and researcher with considerable depth of knowledge in the areas of both psychotic illnesses and cannabis. His research in India examined psychiatric hospital records, and revealed a significant upswing in first-episode psychosis admissions to the huge regional mental hospital at Ranchi that occurred in the month following Holi, compared to the month prior to the festival. This observation illustrates what we know from other sources; that part of the dark side of cannabis is its property of raising the risk for psychotic experiences of various kinds. These range from the frightening but relatively harmless, to the enduring and destructive. We will parse these various distinct states in Chapter 8, and explore their underlying mechanisms. That portion of the book, alongside Chapter 7, probes how common such experiences are. The book also examines the chicken and egg question of whether cannabis use provokes psychosis or vice versa, as well as issues related to cannabis addiction, and the drug's short and possible long-term effects on cognition and IQ.

The dangers of driving under the influence of alcohol are well known. But how about those of cannabis? Driving on Indian roads can be hazardous under normal circumstances, but the Delhi-based NGO Community against Drunken-Driving, CADD, says that many people die in road accidents on Holi day, even more than on New Year. Many people believe that the accident rate rises because revelers consume both bhang and alcohol [4]. But how dangerous are cannabis-intoxicated drivers compared to drunk ones? And what happens to driving when cannabis and alcohol are taken together? Chapter 8 tries to answer these questions. Finally, Berenson's recent book [3] makes explicit (if debatable) links between cannabis and crime. Again, Holi encompasses this issue. Female students complain of harassment by intoxicated men during Holi. When they protest, the standard response from those pestering them is "don't feel bad it's Holi" [5]. Whether the links between cannabis and crime are an inherent consequence of using the drug, or more attributable social/criminal policies related to cannabis (and their differential enforcement when it comes to ethnic minority groups and people of color) is a further issue that I will deal with.

Finally, we should note that ganja, charas, and bhang used to be available without constraints in India, until they were banned under the country's Narcotic Drugs and Psychotropic Substances Act in 1985, "under diplomatic pressure from Western countries including the US." As a result, ganja and charas were declared illegal for both medical and recreational purposes. But cunningly, bhang was exempted from the definition of "cannabis" and as a result can still be purchased in some Indian states from government shops, perfectly legally. Recently, there have been increasing calls for full-on cannabis legalization as India undergoes an agrarian crisis [6]. After all, this is the country for which *Cannabis indica* is named, where the drug has been consumed for millennia and literally grows like a weed across the subcontinent. Thus, Holi lets us revisit the question of national cannabis legalization and encourages the careful review of the scientific evidence for and against this major decision that needs to be addressed in both India and in the United States. Lawmakers need to contend with these issues if they hope to stay ahead of the cannabis curve.

References

[1] Brandt M, Raumpatrouille; 2018: Kiepenheuer & Witsch GmbH.

[2] Kalant OJ. Report of the Indian Hemp Drugs Commission, 1893-94: a critical review. Int J Addict 1972;7(1):77–96.

[3] Berenson A. Tell Your Children: The Truth About Marijuana, Mental Illness, and Violence. New York: Free Press; 2019.

[4] Pandey G. Move to make Holi festival safer; BBC News. 2008; Available from: http://news.bbc.co.uk/2/hi/south_asia/7306383.stm.

[5] *India female students take on 'Holi harassment';* BBC News. 2018. Available from: https://www.bbc.com/news/world-asia-india-43252444.

[6] Datta D. *Legalising cannabis could be one solution to India's agrarian distress;* 2019. Available from: https://qz.com/india/1632037/modi-must-legalise-cannabis-to-tackle-indias-agrarian-distress/.

Chapter 4

Ethnobotany, botany, and archaeology

"I am very glad to hear that the Gardener has saved so much of the St. foin seed, and that of the India Hemp.... Let the ground be well prepared, and the Seed be sown in April. The hemp may be sown any where".

George Washington to William Pearce, 24th of February 1794.

The cannabis plant is a busy chemical factory. What are the different compounds it manufactures? Are all of them psychoactive? Why do these substances exist in the plant to begin with? Do they occur in other plants? When did humans first encounter cannabis and how did they use it—food, versus fiber source, versus intoxicant. Is cannabis a "camp follower" that has travelled with humans across the globe? How has digging up ancient tombs and caves, analyzing pollen, and using modern genetics helped answer these questions?

My visit to the cannabis growery: It's a hot, sticky afternoon in Denver as I pull into the car park. Nearing my destination's unassuming but tightly secured front door, fugitive tendrils of a characteristic musky green odor betray what lies beyond. I press the door buzzer. A tall, middle-aged linebacker in acid-washed jeans beckons me in, checks my ID, and has me sign for my visitor's badge. This is Andy Williams, director of Medicine Man, often described as the "OG of the legal Colorado marijuana business" and the brains behind this 1300 square-foot MedPharm facility operation. Chatting with him before we tour the place, he seems slightly awed by the functional beauty of his own sleek, well-staffed new building. As we enter the complex, immediately confronting the visitor is something I mistook for an instant as an animal exhibit at a high-end zoo. This is a wall-sized, tinted inset glass frame behind which looms a room full of lush semi-tropical greenery. I half expect to glimpse a bug-eyed lemur or sinuous python gazing back at me through the glass, but this room is devoid of any creatures. Here live the cannabis mother plants from which clones are harvested on a regular basis and whose ancestors hailed from Northern California and Amsterdam. The genetically identical clone cuttings are the most reliable source of healthy, all-female, delta-9-tetrahydrocannabinol (THC)-rich flowering shrubs.

A well-known cartoon from the British caricaturist James Gillray portrays the Duke of Bedford looming tall in a stockyard, with the digits of his dominant

Weed Science. http://dx.doi.org/10.1016/B978-0-12-818174-4.00004-5

hand proudly grasping the flank of a gigantic ox that he had helped breed. Alongside the physical resemblance, there is something about Andy's gratified stance in front of his mother plants, juxtaposed with a parallel example of the triumph of selective breeding, which immediately echoes the confident Duke in this early 1800s print. In this case, the stock is selected for beefed up THC content rather than bulky meat on the hoof. Contrary to my initial assumption, the mother plants are not an immortal clone source. Instead, they are ultimately removed to some horticultural slaughterhouse, yielding their place to a brand-new, healthy maternal generation.

Cannabis groweries seem to evoke a palpable tension between the new and the old, perhaps because they exemplify the balancing act between nature and our ever-advancing technology. I appreciate the gleaming, silver-and-glass, high-tech, HEPA-filtered, UV-treated, root-hormone-fueled, carbon dioxide-charged modern ambience. This vibe is somehow familiar. I first encountered this business-meets-science dynamic at a high school visit to a British chemical factory, yet this environment is different. Where I'm standing, a modern academic science lab has been grafted directly onto Hogwarts. On one hand, every activity is on a precise calendar driven schedule. The whole facility is meticulously planned out, ranging from the exterior felon-deterrent security bollards to the no-sweat wall paint that discourages potentially plant-threatening molds. The character of the staff is of a piece with the building design. There are plenty of white coats and labwear. Attitudes are polite—businesslike yet enthusiastic with nary a stoner reference or pot joke. On the other hand, I also note the semi-medieval alchemical recipes comprising rosemary and coriander scented bug-repellents, sprinkles of insect poop fertilizer, and other traditional hand-me-down hippie strategies for driving out thrips and powdery mildew.

Over 60% of MedPharm's floor space is dedicated to growing cannabis plants. The healthy little 4-inch babies are snipped carefully from their mother plants under the watchful eye of the security camera domes that peer constantly at every interior and exterior surface. Their infant stems are dipped in powders to help them grow strong roots and then they are whisked away to a room to be potted (by definition), sorted into black trays, and eventually arranged by strain and by medical versus recreational fate. This is necessary because different state rules and regulations apply, depending on the product's ultimate purpose. Soon after these plants begin to show growth, their peaty roots are embedded in cream-colored grow cubes and the size and heft of a building brick. By the end of 9 or 10 weeks, many of these grow-bricks are decorated superficially with green marble swirls of algae that can be handily sliced off at the next stage where the cubes are embedded in what looked like bucket-sized plastic wraps full of artificial soil. I spot hundreds of these beefy, healthy looking 2-foot high plants in individual white baggies within their new home at one end of greenhouse number one. All horticultural activities throughout the cannabis growery are tightly time-driven to fit precisely five and a half complete growth cycles into a calendar year. Irrespective of the outside temperature, which today is sweltering, in this segment of the operation, it's always 78°F. And in the brave new world of the cannabis greenhouse, every single plant is a pampered alpha.

By the time the plants are ready to move to greenhouse number two, they look much more mature. They have been "schwazzed" Andy tells me. In this process, many of the fan-shaped leaves are stripped off, in a method also known more formally as trimming or pruning. This procedure forces the plant to concentrate on building a strong "core" of beefy roots and robust stems. Later, the process is repeated, this time to push the plants to concentrate on growing luscious buds and flowers, rather than large, THC-poor fan leaves. In the controlled environment of this mega-grow greenhouse, there are top-to-bottom and cross-ceiling pipes of all shapes, colors, and sizes: white-insulated, bare metal, or mixtures of polished metal and black cladding. One way or another, they maintain the desired atmosphere and temperature. I'm shocked by this coolness—I was expecting a tropical orchid house-like muggy heat, but of course that would kill the cannabis plants. Everything here is controlled by those pipes, including the light, temperature, humidity, and carbon dioxide concentration. The lights (some blue, some red at this stage) are turned on for 18 hours a day to bamboozle the plants into thinking that it's springtime and that they need to grow quickly. Later, the light wavelengths are tweaked to be progressively redder (it's Fall, guys!) to encourage flowering. The cannabis thrives in the high humidity, but so do mildews, so a classic garden battle is waged against the potentially devastating powdery mildew under the relentless glare of the specialized lights. As well, various rots, bugs, and other critters that can attack the plants and must be guarded against in this never-ending war.

Andrew, the head gardener, is in an adjoining room, supervising huge polythene vats containing hundreds of gallons of water and blends of murky brown nutrients. These are the extracts that are mixed prior to watering the burgeoning plants one by one with special wands. Andrew is a polite, intense individual in his late 20s, full of arcane and fascinating information about cannabis cultivation. While keeping a careful eye on the processes behind him, backlit by the somewhat glaring and surreal grow lights, he peppers his conversation with specialized gardening terms such as canopy depth, pythium root rot, and spider mites. As we all gaze appreciatively over the serried ranks of clones, I realize that Andrew is not simply tuned into the overall esthetic, but is a proud father doting on his clonal babies (there's a reason they call it a nursery). In addition I sense in him the slightly wary attitude of a watchful epidemiologist, his ever-observant CDC antennae twitching for a potential Patient Zero to appear in his greenhouse. Such and intruder will require a quick squirt of this or that treatment to halt the epidemic.

Next door, a young woman wearing a black baseball cap, black medical gloves, and navy medical scrubs, is supervising the drying room. This is where Christmas tree-sized plants hang upside down on racks like aging beef, curing to 3% humidity under gentle heat. When they are ready, she will remove the buds and leaves in the adjacent strip room, and from there, transport them to a nearby space for bagging the former and grinding the latter. The buds are sorted on heavy, institutional battleship-gray plastic carts topped with trays. From these, the dry buds are shuttled efficiently into translucent greenish polysacks, roughly the size of the leaf bags seen on northern lawns in the fall. I want to ask whether the bagger is tempted to take the occasional sample home, but the ever-present domed security cameras answer my question. The leaf grinder is a sleek, steel, funneled device on shiny metal legs that would not look out of place in a very high-end coffee roastery. From its belly protrudes a flexible tube that conveys the precisely chopped lower-THC plant parts to the neighboring extraction room, the habitat of Devon. A slightly -built wispy-bearded man in his 30s, Devon runs the arcane, 6-hour carbon dioxide-driven extraction process. This necessitates pressurized gas being forced through gleaming, steel blue-topped canisters, each of which holds around 15 pounds of plant material. The carbon dioxide assumes some super-critical meta-state between gas and liquid, driven by a huge lime-green painted pump, that is quietly chugging and puffing in one corner. This is the single device in the entire operation that is distinctly non-21st-century. It would not be out of place in a 19th-century fun fair, powering a merry-go-round or calliope. The cannabinoid-rich extract removed by the CO_2 is filtered and part of it is "winterized" by refrigeration. The rest of the resulting brew is then swished with alcohol and transferred behind the large safety glass panel of an expensive-looking 5-foot-tall laboratory-white machine. On the other side of the glass, a mysterious, slowly rotating foot-wide orb is ensconced in a steel vat of precisely temperature-controlled water. Its purpose is to gently boil away the alcohol under a vacuum, separating it from the remaining molasses-colored plant distillate. The orb's surface mirrors a tiny spinning reflection of the room's inhabitants and the overhead fluorescent lights.

Once all the alcohol wafts away, the remaining valuable chemicals are separated into their key constituents. The finished process yields a variety of compounds that are decanted off. One such product resembles a large honey-filled mason jar. When this glass container is tipped upside down however, the contents with their luxurious, glassy, almost amber-like consistency are reluctant to trickle down the side of the vessel, barely moving at all. This 700 g of almost pure THC can't decide if it's solid or liquid; clearly it already has its own built-in couch lock. Another pure distillate is a small vial of fresh green, almost fluorescent liquid terpenes. Sniffing it, I expected to be overwhelmed by a big harsh green smell akin to an entire pine tree in a vial. Instead, I'm surprised by the pleasant, subtle, low-key vaguely lemony verbena scent, with the barest hint of evergreens. The terpenes can be blended back into the purified cannabis compounds such as THC for both flavor and for "entourage effects," above and beyond those of the familiar cannabinoids. Yet another jar in the chemical goodie cupboard looks to me like it's full of solid cream-colored coconut oil; these plant waxes are yet another by-product of the extraction process.

Next, we enter the test and formulation labs, where young, brown-eyed, white-coated Stephanie shows us around. In this sterile room, mandatory quality control of the various products is carried out on a high performance liquid chromatograph and mass spectrometer (LC-MS). Up to 10 different cannabinoids can be assayed here in less than 15 minutes. The different compounds travel at unique speeds through the device and the height of the resulting peaks generated by the machine correspond to their individual concentrations. Stephanie peers carefully at this mini mountain range through her glasses, and we move on to the gas chromatograph/mass spectrometer (GC-MS) whose purpose is to perform additional assays. Around the corner is yet more equipment, dedicated to detecting possible microbial or fungal contamination.

We continue into the formulation room that boasts the contents of a manufacturing pharmacy. There's an ointment mill, a tincture blender, a machine for making pills and most interestingly a process for making water-soluble THC powder. This off-white fluffy material resembles the result of somebody whacking the pumice stone in their bathroom into coarse chunks. As Stephanie holds up the jar, her pale pink nail polish provides contrast against the 23% THC material. She explains that a patented encapsulation process birthed this fluffy stuff. This involves mixing pure THC with a secret powder in a volatile solvent, and then vaporizing off the solvent to leave behind the chunks of residue. These can be ingested without the need to dissolve them in anything oily or fatty, merely by stirring them into water or juice. I sniff the selection of cannabinoid lotions and creams compounded here, and we learn about various research projects and experiments to develop injectable compounds, to improve the vape formula and to formulate new tinctures. I inquire about edibles and I'm surprised to hear from Andy that MedPharm does not venture into this area. Legislators are wary of them, particularly of the possibility of accidental ingestion by children, and consequently the regulations and labeling requirements for edibles are apparently stringent and ever changing. Finally, we end up in the packaging room. While I'm appraising the container-loaded pallets, my phone buzzes to remind me that I have a plane to catch, so I thank everybody and head out away from the cannabis and back on the road.

****

What is Cannabis? What's the difference between cannabis and hemp, how do we classify them, and why does it matter?

Cannabis is an annual plant belonging to the Mulberry family and related to hops, that dies at the end of each growing season. It propagates itself through seeds grown from the female after fertilization by pollen from the separate male specimen. Some cannabis leaves (like those on deciduous trees) turn golden brown in the fall. At this time of year, the flower clusters too can be shot through with purples, pinks, reds, and oranges. Female flower clusters can fuse together into long, lumpy aggregates that are sometimes called "colas" (from the Spanish word for tail, not the soft drink). In their song "Hotel California," the Eagles sing of these little cannabis fox tails or "colitas." Hidden within these female flower clusters are the wannabe seeds, one to a flower. Each potential seed is swaddled by a concentric spiral of small, modified leaves that are packed full of THC. There is far more THC here than in any other part of the plant. Here's how it has played out. Historically, the original indigenous cannabis plant from a particular location has been cultivated by humans for a particular purpose. For cannabis there are four separate such reasons: long tough hemp fibers that can be spun into yarn, ropes or textiles, nutritious seeds that yield oil, and finally chemicals with both intoxicating and healing properties. Humans generally choose the best specimens of useful plants to propagate; in this case, those with the strongest longest fibers or most fragrant and intoxicating properties, and improve them over time by breeding. Being the messy species that we are, human-bred strains travel alongside us as we migrate, and they escape back to nature as we drop seeds or neglect to pick them up when harvesting. These mislaid seeds adapt to the local conditions in terms of soil, water, weather, etc. and some are selected by nature to thrive best in the local environment. Otherwise, the natural ability of the cannabis plant to disperse itself over large distances is pretty limited. Michael Pollan has suggested that cannabis has relied on humans to spread it successfully far and wide. After several generations these inadvertently sown plants establish themselves to thrive in the spot where chance deposited them, in what botanists term "landraces." To further complicate things for botanical classifiers, these newcomers may interbreed with their original wild ancestors. That tendency has not deterred modern-day cannapreneurs from trying to hunt down the ancestral native cannabis flower. Back in the 1960s and 1970s hippy pioneer marijuana breeders travelled to central Asia, India and Afghanistan to look for traces of the original cannabis parent plants. They collected the landraces growing there to breed back in the United States, much as gardeners collect heirloom tomatoes as living representatives of older classic varieties. But whether the hippy-gathered landraces were true ancestor plants or merely feral offspring of long-ago human agriculture is anybody's guess.

The fact that hops and cannabis are "cousins" botanically speaking is of interest. Hop plants produce the terpene compound Humulin that is found in some cannabis chemovars that we will encounter in Chapter 9. This terpene may be the reason that hops is said to possess mild psycho-active properties and that some natural products aficionados sleep with hop-filled pillows, claiming that it cures their insomnia and encourages vivid dreaming. Hops is used to flavor beer, (it's what gives India Pale Ale (IPA) that characteristic bitter taste). When I lived in an old farm in Maryland, one of my friends came by, to present me with an interesting proposition. By his calculation, because of the close family relationship between cannabis and hops, one could grow cannabis and hops seedlings, and when the time was auspicious horticulturally speaking, cut off the cannabis leaves and graft the hop stem and leaves onto the cannabis rootstock. According to him, this would produce a hybrid that ostensibly resembled a regular hop vine, but that would covertly synthesize the same quantities of cannabinoids that would have been manufactured by the original cannabis plant. Because it resembled a hop plant above the ground, it would fail to draw attention from law enforcement. Surveying my 2-acre field appreciatively, he said "And once those hops are ripe, we will have enough weed for ourselves and a couple of hundred buddies." I told him politely that, while I sincerely valued his friendship, that didn't extend to going to jail on his behalf.

Where and when in the Earth's history did cannabis first emerge as a distinct plant?

Analysis of ancient pollen sheds light on this question [1,2]. Every known plant produces a one-of-a-kind pollen grain. Under a microscope these motes can look as different as spiky chestnuts, corals, bath sponges, bats, bowling balls, death stars, donuts, and seashells. Worldwide, there are over one third of a million distinct pollen types. Cannabis pollen is dull-colored and small, resembling a tiny, olive-green navel orange. Individual grains are far too small to be seen with the naked eye. Half a dozen of them end-to-end would just cross the width of an average human hair. Each one is the right size to be wafted by a barely perceptible breeze into a cloud of its million-fold companions [3]. Pollen experts are known as palynologists. Pollen grains can be used by forensic palynologists to construct the "travel history" of an item, based on its exposure to the unique regional plant mixture characteristic of various global locales. The US State Department uses pollen to check the origin of imported illicit drug samples. Unique blends of pollen grains have even been proposed to be added to firearm ammunition cartridges for forensic identification purposes. Because pollen grains can endure for thousands of years, archaeologists use them to reconstruct the flora of long-vanished environments and of ancient crops [4]. Prior evidence for the origins of cannabis based on the distribution of wild-type plants suggested that it is indigenous to somewhere in central Asia. Tracking the unique pollen of cannabis back to its origin by combing through ancient soil sediments would seem to be the way to figure out the answer to this question. But hops pollen is confusingly similar to that of its botanical cousin cannabis and easily throws researchers off the trail. Now, using an ecological proxy analysis, that is, deciphering those archaeological and geological sites where cannabis pollen co-occurred with open grassy treeless terrain (different than the locations where hops prefer to flourish) scientists can distinguish the two. Using this approach, the palynologist McPartland [2] cleverly separated cannabis' divergence from the "joint" family with hops to discover that the two plants split into distinct species nearly 28 million years ago. The evidence takes us all the way back to a region in Northeast Tibet near the Steppe community around Quinghai Lake, a region located 10,500 feet above sea level on the Tibetan plateau. From there, cannabis dispersed east to East China by 1.2 million years ago, west to Europe by 6000 years ago, to India 32.6 thousand years ago and to Japan approximately

12 thousand years ago. This dispersal happened by natural spread and via people adopting the plant and traveling, taking the seeds with them on their journey.

The present-day climate in the Quinghai Lake region where ancient cannabis originated is harsh and unforgiving. It has an average altitude of over 4000 m, low levels of oxygen, low temperatures, strong bone-chilling winds, and intense ultraviolet light. Or, if you prefer to believe the puffery of local tourist brochures, this idyllic region located not far from the Silk Road is centered around an enchanting, elliptical blue-green lake teeming with easy-to-catch fish and surrounded by mountains, with innumerable varieties of lovely birds flying overhead. According to an article in China daily Xinhua [5] archaeologists have confirmed human activity around this lake dating back over many thousands of years. Dozens of stone tools were found at the same site, showing that prehistoric humans lived, hunted and made tools in the harsh and chilly environment over 10,000 years ago. So much for the stereotype of stoners relaxing on the couch in the comfort of their own centrally heated basement. Curiously, Qinghai is located only a couple of hundred miles away from the Denisovan cave where 160,000 years ago humans distinct from both modern humans and Neanderthals dwelled during the Ice Age. Cannabis pollen has been found in the layers of soil excavated from that cave [6]. These various observations inevitably give rise to speculation about when humans first began to pay attention to the plant and begin cultivating it deliberately.

When did the Cannabis plant translate into human agriculture?

Cannabis is one of the world's oldest cultivated plants [7]. There's an ongoing debate about which humans and where first deliberately cultivated it, but one group of scholars defends the viewpoint of earliest use in central Asia or China. This cabal fights it out with others stating claims for East Asia and the Steppes. The earliest archaeological evidence for cannabis pollen occurring next to crop pollen, or where cannabis seeds (from a botanical point of view, fruits or achenes) have been found in archaeological sites, is in Japan from 12,000 years ago attached to shards of pottery. Next, it appears in China. Recent analysis makes a strong claim for even earlier use [8]. According to a recent survey from the German Archaeological Institute in the Free University of Berlin, by about 11,000 years ago evidence of its use appears in Europe and East Asia. Nomadic traders known as the Yamnaya seem to have entered Europe and the Middle East from their homeland in the eastern Steppes in current Ukraine and Russia about 5,000 years ago. They arrived bearing knowledge of metal crafting and herding as well as participating in a lively trade in cannabis seeds, millet, wheat, barley, and horses across the Steppe zone. The Yamnaya people also left traces of their genes in most of the populations that live currently in Europe and South Asia [9–11]. This dating is inferred from a systematic trans-regional review of evidence from archaeology, and analysis of distribution data from cannabis pollen, seeds, and fibers. For example, the German archaeologists found plant remains and residues in numerous Yamnaya dwelling sites across Ukraine and Russia.

Hemp has been cultivated for its strong, durable fiber and nutty nutritious oily "seeds," (achenes). The desirable long, tough fibrous threads inside the stems are harvested traditionally by soaking them in water until the outer stem decomposes, in "retting ponds," leaving the fiber behind. These ponds are remarkably malodorous and the smell of rotting cannabis stalks is definitely an acquired one. Fiber can also be gathered by crushing the stems. Hemp plants produce cannabidiol (CBD), mainly in their leaves, but manufacture no THC and are therefore not intoxicating. In essence, you can smoke hemp till the cows come home and never get stoned. The hemp threads are traditionally woven into mighty ropes, often used aboard ships, spun into strong cloth used in ships sails (the word "canvas" derives from cannabis), or into somewhat coarse material that can be worn as clothes. Hemp also makes strong, durable paper and banknotes can be printed on it. Hemp has been grown in the United States for many years including by George Washington and Thomas Jefferson. During World War II there was a short-lived "Hemp for Victory" campaign to produce material for naval ropes.

Nowadays, processed hemp fiber and hemp seed oil are imported from Canada and Europe, as is sterilized hemp seed for bird food. Spaced a few feet apart, hemp grows like little bushes; planted in close proximity the plants crowd each other and compete for light. They attain this by reaching skyward, concentrating on developing long, thin bamboo-like stems packed with dense, lengthy, tough fibers. Hemp seeds are full of nutritious oil and protein. They were grown as food thousands of years ago and oil was extracted from them by crushing. Hemp seeds can still be found in health food stores today; I use them in my morning smoothies. These seeds, sterilized so that they will not produce baby plants when they fall to the ground, are used in inexpensive bird food in the United States. Hempseed oil can also be used to manufacture paints and varnishes. Hemp oil rapidly becomes rancid, so when used for food it needs to be consumed quickly after production.

Growing cannabis crops for drug harvesting is a completely different proposition. It relies on different plant varieties that have been bred to contain intoxicating THC. Recall that there are separate male and female plants. Both hemp and drug cannabis are pollinated by wind-blown pollen, dispensing with the necessity for bees, moths, birds, or bats to do the job. The male flowers resemble miniscule pale-green bells. Because the female flowers don't have to signal gaudily via bright colors or seductive fragrances to attract any pollinating critters, they can afford to be inconspicuous, and are green and tiny, consisting of what botanists refer to as stigmas. Delicate, fuzzy, and sticky-tipped, these pollen traps hang together in clusters, poking a mere fraction of an inch out of what look like shambolic clusters of mini leaf tips and untidy flower clusters at the top of the female plant. These latter are the colas we referred to above. If you want to see what these structures look like under a microscope, you can admire their sticky green beauty in Ted Kinsman's wonderfully photographed book 'Cannabis: Marijuana under the Microscope [3]. Close-up and personal, the stigmas are the palest of pale green, resembling spindly, twisted hummingbird tongues reaching out in all directions to seek pollen. When it comes to THC and its intoxicating band of chemical relatives, the lion's share of these cannabinoids forms in the female plants, specifically as sticky resin in their flower buds. In the wild, the minimalist male flowers shed pollen that's puffed by the wind to the eagerly awaiting female flowers.

Since wind pollination is extremely inefficient, the male cannabis plants must make far more pollen than ever reaches the females. Having evolved to float efficiently and to stay aloft in wind currents, these billions of pollen grains waft on the breezes until gravity or rain ultimately gets the best of them. Whatever fails to be captured by sticky cannabis stigmas, ultimately falls to the ground, from where as we have seen, it can be unearthed thousands of years later by eager botanist/archaeologists. As soon as the female flowers are pollinated, they cease secreting cannabinoids and switch all of their energy into producing hundreds of thousands of fruits, strictly named "achenes," but informally called seeds. So, if you're growing hemp for seeds, having male flowers around is fine. But when you're cultivating cannabis for medicinal or intoxicating cannabinoids, then male plants are public enemy number one and must be culled. Depriving the female plants of pollen encourages them to up the production ante by increasing their numbers of stigmas and sticky resin, like desperate Tinder users swiping indiscriminately to find a mate. Ultimately this process runs out of steam and the plant is left at each branch tip with a fat unfertilized flower bud/cola packed full of red-brown deceased stigmas and sprinkled in a frosting of sticky, shiny cannabinoid-rich crystals. This familiar nugget is known as sinsemilla, the Spanish for "without seeds." Unlike cannabis growers, hemp farmers don't give a fig about buds, to mix botanical metaphors. They grow male and female plants together, but harvest everything early before the pollen disperses, so that the desirable fibers are not degraded post-fertilization. Or they avoid mowing a portion of the field to let fertilization do its thing and hemp seeds to develop.

Nobody really knows why the female flowers produce resin. One hypothesis is that it protects the female flowers from drying out in hot dry climates, where cannabis resin production is maximized, and traditionally where drug-bearing strains have been cultivated, (a theme we will revisit in Chapter 9.) Resin is not secreted after the fruits have ripened. Drug and hemp varieties are presumed to share a common origin, but have been driven to diverge by farmers employing different cultivation approaches over millennia [12]. These differences in cannabis plants draw us inevitably into a botanical discussion. For years cannabis horticulture has been the province of hands-on growers more concerned with practicality than accurate terminology, to the detriment of the latter. For example, the term "bud" to cannabis consumers and growers refers specifically to a cluster of female marijuana flowers, but to botanists this is a far less specific label that designates any newly appearing bump that signals an emerging plant part of any type from leaf to twig.

Humans have been successful in consciously tinkering with nature and trying to improve selected wild living things for human benefit. We've tamed the friendliest wolf cub in the pack, bred the cat that's the best mouser, or propagated the corn plant with the biggest kernels. Thus our species has kept the undesirable wolves from the door (and the new, domesticated ones inside) and the greedy rats out of the better, plumper crops. Selective plant breeding in particular goes back thousands of years. Wine grapes, corn, peppers, tomatoes, tobacco and potatoes, and more recently ornamental plants have all thrived under our attentions. Michael Pollan has described much of this process masterfully in books such as *The Botany of Desire* [13]. As he points out, marijuana crops have been historic beneficiaries of such human attention. We have selected features such as plumper seeds, stronger fiber or more potent psychoactive properties. More recently we have imbued fruits and vegetables with characteristics more desirable to supermarket consumers. And in our past lives as hunter-gatherers we've capitalized on the novel characteristics in random genetic mutations or "sports" that Nature occasionally throws our way. This makes sense: no primitive agriculturalist wants to be the tribal farmer responsible for providing a pitiful crop of tiny hemp seeds that will not tide the group through a freezing winter. Equally undesirable is the sub-par shaman doling out low-grade schwag in place of powerful, insight-inducing weed at an important religious ceremony. (Exactly how shamans would test for cannabis potency ahead of time leads to interesting speculation, but I'm sure there was never a shortage of eager volunteers).

When and where was cannabis first used for its mind-altering properties?

Part of the answer to this question relies on ancient texts and later, retrospective reports of ancient practices. Until recently, questions regarding the earliest use of marijuana as a psychedelic led to much speculation, but few definitive answers. If archaeologists discover a handful of ancient hemp seeds, without more evidence it's hard to tell whether the associated plants were used as a seed crop, grown for their fiber, or for medicinal and psychoactive properties. Now surprisingly even during the period of time that I have been writing this book, archaeology has been filling in the gaps with actual scientifically convincing physical evidence. Discovery of such evidence beats the odds, given the challenge of preserving archaeological traces of cannabis use activities, or even of plant samples whose THC can be analyzed.

According to John Chasteen, the Indian Atharva Veda manuscript mentions ceremonial use of the bhang plant 3000 years ago, in a context suggesting intoxication, that is, alongside mention of the sacred (but currently unknown plant species) *soma*. Thereafter, written texts from India are spotty and relatively uninformative for literally thousands of years until the early 1300s. Chasteen suggests that this absence is in part because much early literature was written on palm leaves, which decay in the hot, damp climate [12].

Just over 10 years ago, among the Yanghai tombs in the Turpan district of Northwest China, in the Xinjiang desert region to the north of Pamir, a large amount of cannabis plant material radiocarbon-dated to 2,700 years ago was found in the tomb of a Caucasoid male in his mid-40s. The man is assumed to have been a priest or shaman of the Gushi culture, based on artifacts found in the tomb such as a rare and distinctive harp. Almost a kilo of chopped and pounded cannabis was buried next to the shaman's head and feet in a large leather basket and wooden bowl. Because of the deep burial and "old and cold" climate conditions, the cannabis was superbly preserved [14,15]. According to Professor Ethan Russo, these cannabis remains consisted almost exclusively of the more psychoactive parts of the plant with the large stalks and branches removed. A multi-disciplinary international team confirmed the identity of the cannabis through botany, plant chemistry, and DNA analysis. They showed that it contained significant amounts of THC. The still-green plant material, unusually well preserved in the dry, cold environment of the Gobi desert, was immediately identifiable as cannabis as even the characteristic glandular trichomes were obvious, along with the characteristic seeds. The latter were plump, characteristic of cultivated rather than wild cannabis. Nothing in the tomb betrayed how the cannabis might have been consumed.

Additionally, in a burial site excavated in 2015, the 2,400–2,800 year old Jiayi tombs (also located in the Turpan region), an even more startling discovery was made. This ancient cemetery had been discovered by chance when Chinese workers cleared an area close to an ancient riverbed, ironically to construct a new graveyard. In all, 240 tombs were excavated, with the area yielding ceramic pots, bows and arrows, and bones of domesticated animals including horses, goats and sheep. In only one of these tombs the archaeologists found something truly unexpected. Surrounded by red colored earthenware pots, and lying on a bed made of wooden slats, with his head resting on a reed pillow, a 6 foot tall man in his mid-30s had been buried with a bouquet of 13 cannabis plants spread diagonally across his upper body like a shroud. Due to the dry climate, the 2 to 3 foot tall plants were easily identifiable, although they had turned yellowish-brown. All were locally grown female plants bearing newly ripe seeds. The cannabis still had roots attached and a microscopic view confirmed the presence of densely packed golden-colored glandular trichomes [16].

The story continues. In 2019, archaeological and biochemical collaboration revealed the earliest chemical evidence to date of human psychoactive use of high-THC cannabis via smoking and told us definitively for the first time how the cannabis was consumed [17]. The plants in question were cultivated and used in funerary rites 2,500 years ago in central Asia, as shown by a cemetery excavation that took place at the foot of the Pamir mountains in Northwestern China, ten thousand feet above sea level in the area known as "the roof of the world," near the border of Tajikistan and China. The cannabis chemical remains found at the Jirzankal burial site bore high levels of THC, as shown by gas chromatography-mass spectrometry (GC-MS), the technique that we encountered in Chapter 2. The plants and their associated artifacts were excavated from ancient circular graves buried under rounded mounds, surrounded by rock borders. Among the skeletons were interred harps, pieces of silk and fascinatingly, 8 of the tombs housed a total of 10 ancient charred, bowl-like wooden burners carved from logs, containing ping-pong ball-sized stones that had obviously been exposed to fire. Beijing scientists analyzed material from these burned wood braziers and the stones and detected not only CBD, but more importantly cannabinol, a distinct chemical by-product of heated and air-exposed THC. The cannabinol was more highly concentrated than in most ancient wild hemp samples and larger than any previously discovered at an ancient site, implying that the plants had been bred specifically for this psychoactively linked property. The analysis isn't able to tell us precisely how much THC was in the original, long since vanished cannabis plant material, only that it had to have been much higher than that found in wild hemp. The authors speculated that the increased THC levels could have been produced in response to environmental stressors such as higher levels of UV light, sunshine, or high altitude, conditions which boost THC levels in modern-day cannabis. Inferentially, fire-heated stones had been dropped onto herbal cannabis material (or vice versa) to produce intoxicating smoke that was inhaled by the mourners, perhaps inside of small tents to confine the smoke. The artifacts found at the site are fairly typical for practitioners of the ancient Zoroastrian faith that originated in Persia, versions of which still persist today. Mark Merlin, Professor of botany at the University of Hawaii and co-author with Robert Clarke of the massive *Cannabis, Evolution and Ethnobotany*, [18] stated that this discovery "…adds to growing indications of an association among many cultures of cannabis with the afterworld and death (and) ….used to facilitate the body communicating with the afterlife, the spirit world" [19]. The harps suggest that music was used as part of the ritual. This obvious connection between weed, music, rock, and the Dead was made in a New York Times article [20].

Even though this is the earliest archaeological evidence discovered to date, it certainly is not the last word on whether this geographic area was where the practice began. The flat plain below the Pamir Mountains where the cemetery is located is right on the burgeoning Silk Road that formerly connected trade and travel from central and southwest Asia with China. Ritualistic use of high THC cannabis is speculated to have started even earlier in regions from Syria to China, with the practice spreading alongside exchange of religious and cultural ideas and practices, along with trade goods including crop seeds, up and down the Silk Road. Convergent evidence of cannabis use in this time period derives from ancient graves elsewhere in China.

While the scientist/nerd part of me is fascinated by the archaeological details, my inner psychiatrist wants to know what the mourners at the time of this ancient funeral thought and felt as they took a hit of cannabis smoke from the smoking, terpene-rich Juniper wood brazier. Presumably the ritual helped them contact a parallel spirit world where they could commune with the gods of nature and the dead person's soul. The ritual chanting and music combined with the cannabis may have helped shape their thoughts and feelings of grief at losing an important member of the community as well as aiding some form of communication with the deceased. The general outline of my speculations on funeral rites squares roughly with the one ancient historical account of ceremonially inhaled cannabis recorded by the Greek historian Herodotus (known as the "Father of History"). Written around 450 BC, this description occurs in his account of the Scythians, (a group of Central Asian tribes that lived beyond the Araxes river). This roughly 700-mile river of the Caucasus begins in Turkey, running through current-day Armenia, Azerbaijan, and Iran, coincidentally close to the Azerbaijani city of Ganja. Herodotus detailed how the Scythians sought an altered state of consciousness as a communal experience using cannabis, a plant whose appearance he describes, remarking that it both grows wild and is cultivated. He relates vividly how the tribe kindled a bonfire, tossed on cannabis, inhaled the smoke to the point of intoxication, and danced and burst into song. He also relates that in a purification ritual after a burial, the Scythians cleansed themselves, then leaned three poles against one another around a dish and draped the frame with felted woolen blankets. Next, red-hot stones from a fire were placed in the dish, cannabis seeds tossed onto the glowing stones, and the Scythians crawled under the blankets to inhale "the dense smoke and fumes... causing them to shriek with delight." This latter description of enjoyment doesn't imply that the cannabis was being used recreationally and separately from the funeral rites; anyone who has been to a traditional wake will understand that the phrase "celebration of the deceased's life" can have a quite literal meaning.

The descriptions of Herodotus jibe remarkably well with our current understanding of the Scythians, and recent archaeological evidence of their use of cannabis. These nomadic peoples were groups of Eurasian steppe-dwelling herders, who controlled huge areas of land all the way from Siberia to Eastern Europe. They were skilled horsemen, archers, and crafters of strikingly beautiful jewelry who because of their nomadic lifestyle, left behind no cities or permanent settlements. The tribe travelled and traded widely between China, the Middle East and Eastern Europe from as early as the ninth century BC over an era that lasted 1,200 years. The Scythians held elaborate funerals and buried their deceased nobility in large elaborate mound-like tombs.

From one of these structures, a frozen burial mound in Pazyryk located in the Altai Mountains of Siberia close to the Mongolian border dated somewhere between the fifth and third centuries BC, early 20th century archaeologists unearthed a frame made of wooden poles that once belonged to a small tent, and a copper/bronze cauldron or censer containing burned hemp seeds and stones dated to around 2,500 years ago [21]. Close by was a fur-lined leather bag filled with cannabis seeds and clothing made of hemp. The tent frame presumably belonged to a long-vanished cannabis inhalation tent exactly as described by Herodotus. Later, in 1970, another archaeological group excavated a leather Scythian flask containing hemp seeds and asserted that smoking cannabis may have been part of everyday life, not merely a funeral rite. Those seeds were minute and marbled, resembling tiny turtles, consistent with observations of Vavilov for a wild cannabis strain, and recent agronomic analysis.

In 2013, Andrei Belinsky excavated an elite 2400-year-old Scythian tomb that contained golden artifacts including exquisitely detailed gold vessels portraying violent images, inside of which adhered sticky residues that local criminologists confirmed had contained cannabis and opium [22].

Hemp was first used in China 12,000 years ago for fiber, and cannabis seed was a staple grain in the region from ancient times. The pictographic representation of cannabis (Ma) is easily recognizable as a depiction of stalks of hemp hanging upside down in a shed to dry. The Chinese recognized that plant grew in male and female forms, and that male plants were superior in providing fiber, while the females were preferred for intoxication. For the latter, there are various descriptions of how to prepare the plant, dating back to the third century BC.

How do we classify different kinds of cannabis plants?

The question of how to classify cannabis is inextricably linked to techniques for growing the plant: a case of horticulture meets nomenclature. In our modern age of dispensaries and selective marijuana propagation by cloning, we've retained and updated many of these ancient plant-breeding principles. Unfortunately, alongside the ancient business of producing ever-better weed, we have added a lot of unnecessary perplexity to the process, including confusing and misapplied terminology. Everyone from breeders to budtenders chats knowledgeably but often imprecisely about the nugs in question. Back in the 1970s, the Firesign Theater comedy troupe released an album entitled "Everything you Know is Wrong." This is an apt summary of the world of cannabis classification. Let me show you why. Many people, even seasoned marijuana consumers, speak about "varieties" and "strains" of cannabis. Walk into the average dispensary and you will see listed for sale a selection of "strains" with exotic names such as Blackberry Kush, Blue Widow, Sour Diesel, Super Lemon Haze, Girl Scout Cookies, Purple Kush, and Lamb's Bread. Dispensaries sell "varieties" with an assigned name such as "Rainbow Gummeez" and some basic information, for example, "Indica strain; 15% THC 5% CBD, has energetic effects." What's wrong with this picture? First of all, "strain" is not a valid botanical term. Second, the boundary between cannabis indica and sativa has been blurred by plant breeding to the point of obscurity. Third, we often don't know exactly who is using what kind of assay to come up with that "15% THC" figure. Fourth, what subjectively makes one person feel "energetic" may well make another sleepy, or paranoid or anxious. And finally, are the "Rainbow Gummeez" buds I'm purchasing from dispensary A, the same as those I bought 3 months ago from dispensary B? The bottom line is what useful information do we need to distinguish a nugget of Three Bears from one of Acid Rain? And what's the correct terminology we should be using to pose these questions? To address these important issues, we will need a botanical detour to explain the differences between strains, species, cultivars, landraces, chemovars, and varieties. My apologies if your head is soon whirling and buzzing from fact overload, before you even take a toke of the weed we are talking about.

In the biological world, the term "Species" refers to individuals that can reproduce sexually with members of the same species and their progeny in turn can also reproduce sexually and produce offspring, or what my friends south of the Mason-Dixon line refer to as "chirren." For example, horses and donkeys are related but separate species that can mate to produce mules, but mules are sterile. Lions and tigers are another example of separate species that can mate and produce "tigons" or "ligers." But when these mate they are unable to produce offspring. All three types of cannabis, indica, sativa, and ruderalis that we encounter below, easily cross-pollinate, producing viable plants that are a mixture of the different types and that easily produce viable seeds themselves. So if I had to choose, I'd agree with J.C. Chasteen that different groups of farmers have created separate varieties over the years from a single species, by selecting for separate features of ancestral cannabis, whether long, tough fiber, buxom seeds, or plump resin-rich flower buds. Sixties-era hippies and their plant-breeding successors have blurred the lines even more by crossing varieties to boost THC. If you concentrate on developing a breed of chickens that are productive egg layers, they are unlikely to be terrific meat producers. Thus, cannabis varieties bred to be high in THC are generally low in CBD. This changing ratio has health consequences relevant to psychosis, as we shall see a little later. Although species were initially described by exact appearances, classification now relies on chemistry, genetics, microscopy, and other techniques and branches of science, as we will see.

The species *Cannabis sativa* (fancy Latin for cultivated cannabis), was initially described by Leonhart Fuchs in 1542, the German physician and founder of modern botany, in his classic book *The New Herbal*. Fuchs is commemorated in the naming of the plant and color fuchsia. The dual moniker of *Cannabis. sativa* was adopted by the Swedish botanist and physician Carl Linnaeus in his exhaustive list of plant species descriptions in 1753, to describe the European hemp plant. Thirty years later, Jean-Baptiste Lamarck, French soldier, biologist, and plant nerd described *Cannabis indica* as a possible distinct species originating in the Indian subcontinent, differing in being shorter and bushier than *Cannabis sativa* with broad leaf parts. Finally, we encounter the Rodney Dangerfield of the cannabis world, a species that many botanists claim doesn't really exist, *Cannabis ruderalis*, often insultingly referred to as "ditch weed." This was named by the Russian botanist Janischewsky in 1924, who discovered a "new," non-cultivated wild cannabis plant. The term "ruderal" derives from the Latin word for rubble. Ruderal plant species colonize land after it's been cleared. These quick-fire opportunists are generally tough and scrubby plants, matching their role as hardy botanical pioneers. There has been an ongoing battle of the plant classifiers, with some advocating for a single cannabis species and others for distinct and separate species.

How then can we settle this botanical bickering and definitively classify the plant? Classically cannabis varieties are named by morphology. Indica plants are short and squat with bigger leaves, wider leaflets, and are favored by indoor growers, due to their short 6–8 week lifecycle. By contrast, sativa are tall and narrow-leaved, reputed to have stimulating and cerebral properties. Cannabis consumers speak knowledgeably of differences between indica and sativa parsed by their subjective psychoactive effects (e.g., body high, sedative, couch lock). These designations are used widely in dispensaries, by users and online, for example, by WoahStork [23]. Logically we could perform this distinction using a plant's outward appearance, but as a reminder on the potential limits of classification by morphology, Danny Devito and LeBron James do not belong to different species. A second method to distinguish species might be by place of origin, but this remains pretty much lost in the mist (or smoke) of history. Additionally the plant has been so extensively crossbred that the botanical cat's cradle of different types is almost impossible to disentangle so many years down the line. Ancestral forms have likely long since disappeared and may no longer even exist in the wild, as the landrace-seeking hippies of the 70s discovered. Furthermore, the distinction between indica and sativa, if there was one to begin with, has become increasingly meaningless due to extensive crossbreeding. Examination of the genes of multiple sativa and indica "strains" clearly shows this blurring. Yet another way to classify the plant might be through its effects, either subjective ones (e.g., sedative, activating) or its medicinal properties—for which we would ideally need clinical trials. But here we would have to account for the fact that different people may have different subjective responses to the same cannabis sample. Another approach might be to use genetics and to describe the key genes determining various types of vital plant characteristics, from the shape of its leaves to the amount of THC it can manufacture. (More on these genetic topics a little later). Chemistry too plays a role in classification, for example, in distinguishing THC to CBD ratios.

"Strain" is the term used correctly by microbiologists to describe bacteria, and incorrectly by some cannabis breeders and dispensaries to describe varieties of the plant. In the movies when you see a white-coated scientist take a wire loupe, poke it into a bacterial colony and then swish it across the face of a virginal petri dish, to grow a new colony, strains refers to what they're endeavoring to multiply in the growth medium. In this context, a strain is defined as the descendent of a single bacterium. This term is not employed in botany but used casually to refer to variations within cultivars and their offspring. So we can cross "strain" off our list of terms to ID cannabis varieties.

"Variety" on the other hand is a bona fide term. It refers to the unique characteristics of a plant that occurs in nature. Plant varieties are cultivated or "bred" by humans, in a process that involves selecting deliberately for particular traits, such as larger flowers buds, more attractive flowers, more robust plants etc. The potatoes that reach your local McDonalds have been bred precisely for large size and optimal starch content. The process of creating new varieties also relies on sharp-eyed plant breeders deliberately picking a "sport" (the name for unusual plant that appears randomly in nature by genetic mutation); or by hybridizing two plants with distinguishing features and breeding them together into a new plant that marries their two desirable characteristics. Examples might be taller plants with bigger flower buds. Although, true varieties will produce comparable copies of themselves when their resulting seeds germinate, hybrids may not.

In plant and animal biology, a "hybrid" is the offspring that results from genetic mixing, when two organisms of different breeds, varieties, or species that have different qualities are combined through sexual reproduction. Hybrids may display the different qualities of each parent, a blend of both, or even exhibit new qualities such as "hybrid vigor" where the offspring is bigger or tougher than either parent. Hybrid vigor is a guiding principle in animal breeding, where hybrids such as cockapoos (cocker spaniels mated with poodles) are sometimes referred to as "designer crossbreeds." But there are large elements of trial and error and even pure blind luck in creating a successful hybrid. Hybrids of indica and sativa such as White Widow, are generally described based on approximate proportions of their parents (e.g., 60% Indica, 40% sativa) and by definition exhibit traits from both. The hope is that hybrids co-mingle the desirable parental traits, but as any plant breeder will tell you that doesn't necessarily occur. A classic illustrative example of an unfavorable hybrid from outside of the plant world, is of the proposed pairing of George Bernard Shaw, the homely Dublin-born Nobel prize-winning playwright, polemicist, gadfly, political activist, and cultural critic, with Ellen Terry, the classically beautiful, accomplished and notably histrionic actress with whom he had a lengthy correspondence. When apocryphally, Terry suggested that they have a child together, "You have the greatest brain in the world, and I have the most beautiful body. Just think of it, we will produce the most perfect child". Shaw in his curmudgeonly way, responded with "Yes my dear, but what if it inherits your brain and my looks?"

Back to cannabis breeding. Commercial cultivars often use auto flowering plants that automatically switch from vegetative to flowering stage with the plant's age as opposed to ratio of dark to light hours, as occurs with photo-dependent strains. In nature indica and sativa cannabis plants switch to flowering as summer changes to autumn and the ratio of light to dark hours approaches one. Crossing or hybridizing ruderalis with other varieties is common because the ditchweed flowers uniquely develop independent of photoperiod, propelled only by the plant's maturity. To create the original hybrid you only need pollen from a male plant to fertilize the female. Ruderalis is thus used to breed so-called "auto-flowering" varieties. Commercial crossbred hybrids are usually a mixture of all three types of cannabis and are both auto-flowering and fast growing. Male plants are excluded from groweries to obtain resin-rich sinsemilla.

In contrast to simple varieties, "cultivars", short for "cultivated varieties," (sometimes termed varietals) are plants purposely selected for their desirable characteristics. Unlike varieties, seeds of a cultivar will produce something that differs from the parent plant. Leaving things to chance is usually a bad bet when you're starting off with something good whose desirable traits you want to preserve. Thus, profitable cultivars are bred true to the parent by a series of methods that include propagating cuttings, cloning, grafting, or tissue culture, all of which are employed in marijuana cultivation. These procedures produce exact copies of the parent plant through a method other than seeds. Essentially, cultivars originate from nurturing various sized pieces of the parent plant to produce precise copies or clones of themselves. In the animal world, Dolly the sheep was a famous clone. Cannabis breeders are not a particularly risk-averse group. But given the vagaries of propagating, a successful variety once you have put so much time and effort into creating it, making generations of numerous exact copies is achieved most predictably by taking cuttings from the mother plant and propagating the resulting clones. Carefully tended, mother plants can be wheedled into surviving way beyond their designated annual lifespan. When clones mature, an important question for growers is how to name your resulting babies in order to market or "brand" their products. Most often they are given catchy titles, whose purpose is to telegraph key descriptive information such as size (big bud), color (rainbow, purple train wreck), geographic origin (Kush, Hawaiian), smell (sour, diesel, lemon, skunk), or taste (pineapple, strawberry). But in order to benefit fully from the fruits, or in this case the buds, of your plant-breeding labors, then cultivars must be legally registered. Because cannabis is illegal in most jurisdictions, currently this almost never happens. Using these definitions, it's clear that a cannabis plant can have both a variety and a cultivar.

This chapter has provided some key information on where and when cannabis evolved, archaeological hints as to when it began to be used for intoxicating purposes, and schemas for classifying the cannabis plant botanically. But we will learn the essentials of how cannabis synthesizes intoxicating and other cannabinoids and terpenes and how to classify the plant chemically in Chapter 9.

References

[1] Weisberger M. We may finally know where the cannabis plant originated, 2019. Available from: https://www.livescience.com/65517-cannabis-asia-originated.html.

[2] McPartland JM. Cannabis in Asia: its center of origin and early cultivation, based on a synthesis of subfossil pollen and archaeobotanical studies. Veg Hist Archaeobot 2019;28(6):691–2.

[3] Kinsman T. Cannabis: Marijuana Under the Microscope. Atglen, PA: Schiffer; 2018.

[4] Miroff N. Pollen 'nerds': U.S. government enlists scientists to track drug loads, crack cold cases, Washington Post. 2019. Available from: https://www.washingtonpost.com/national/pollen-nerds-us-government-enlists-scientists-to-track-drug-loads-crack-cold-cases/2019/08/30/96f50024-cacb-11e9-8067-196d9f17af68_story.html.

[5] XinhuaNet. Available from: http://www.xinhuanet.com/english/.

[6] Jacobs Z, et al. Timing of archaic hominin occupation of Denisova Cave in southern Siberia. Nature 2019;565(7741):594–9.

[7] Russo EB. History of cannabis and its preparations in saga, science, and sobriquet. Chem Biodivers 2007;4(8):1614–48.

[8] Long T, et al. Cannabis in Eurasia: origin of human use and Bronze Age trans-continental connections. Springer; 2016;**26**(2): pp. 245–258.

[9] Lawler A. Cannabis, opium use part of ancient Near Eastern cultures. Science 2018;360(6386): 249–50.

[10] Lorenzi R. Pot dealers traced back 5,000 Years, 2016. Available from: https://www.seeker.com/pot-dealers-traced-back-5000-years-1932290630.html.

[11] Hale T. Archaeologists think they've discovered the world's first marijuana "dealers", 2016. Available from: https://www.iflscience.com/editors-blog/archaeologists-think-theyve-discovered-the-worlds-first-marijuana-dealers/.

[12] Chasteen JC. Getting High: Marijuana Through the Ages. Lanham, MD: Rowman and Littlefield; 2016.

[13] Pollan M. The Botany of Desire. 1st ed. New York: Random House; 2001.

[14] Russo EB, et al. Phytochemical and genetic analyses of ancient cannabis from Central Asia. J Exp Bot 2008;59(15):4171–82.

[15] Karasavvass T. High times in ancient China: 2,700-year-old marijuana stash found in Shaman Grave, 2017. Available from: https://www.ancient-origins.net/artifacts-other-artifacts/high-times-ancient-china-2700-year-old-marijuana-stash-found-shaman-grave-021722.

[16] Jiang H, et al. Ancient cannabis burial shroud in a central Eurasian cemetery. Econ Bot 2016;70:213–221.

[17] Lawler A. Oldest evidence of marijuana use discovered in 2500-year-old cemetery in peaks of western China, Science Magazine. 2019. Available from: https://www.sciencemag.org/news/2019/06/oldest-evidence-marijuana-use-discovered-2500-year-old-cemetery-peaks-western-china.

[18] Clarke RC, Merlin MD. Cannabis Berkley and Los Angeles, California: University of California Press; 2013.

[19] Willingham E. Using marijuana to get high dates back millennia, Scientific American. 2019. Available from: https://www.scientificamerican.com/article/using-marijuana-to-get-high-dates-back-millennia/.

[20] Hoffman J. Scientists find ancient humans used weed 2,500 years ago, too, 2019. Available from: https://www.nytimes.com/2019/06/12/health/cannabis-anicent-dead-marijuana.html.

[21] Rudenko SI. Frozen tombs of Siberia; the Pazyryk burials of Iron Age horsemen. Berkeley: University of California Press; 1970.

[22] Curry A. Gold artifacts tell tale of drug-fueled rituals and "bastard wars", National Geographic. 2015. Available from: https://news.nationalgeographic.com/2015/05/150522-scythians-marijuana-bastard-wars-kurgan-archaeology/.

[23] Available from : https://www.woahstork.com/.

Chapter 5

Neuroscience

"... Providing compelling evidence that these compounds serve as a new and additional class of endogenous signaling molecules involved in a plethora of physiological functions"

D. Chanda, D. Neumann and J.F.C. Glatz, The endocannabinoid system: overview of an emerging multi-faceted therapeutic target [1].

"Your body is teeming with weed receptors"

Scudellari M. [2].

This chapter covers some key aspects of how the psychoactive constituents of cannabis interact with our brains to produce the characteristic experiences associated with the drug's "high." Beyond that, I will explore the body's complex and naturally occurring neurochemical internal cannabinoid signaling system that interacts with plant-produced chemicals found inside the cannabis flower. This so-called endocannabinoid system (ECS) inside of our brains and immune cells has important but complicated interactions with other brain signaling systems. We will challenge the idea that cannabidiol (CBD) is not psychoactive, and explore what pharmaceutical companies are doing in terms of designing drugs to make the brain's cannabinoid system behave in finely tuned ways for therapeutic purposes, and how rogue chemists in China design super-chemicals to coax the same system into producing super-highs.

Revelation number one. Imagine if 50 years ago explorers had suddenly stumbled upon a vast, lost continent, populated by people who spoke languages similar to our own and whose rules of living paralleled to ours, but whose entire existence had been previously completely unsuspected. This discovery sparks a wave of investigation by geneticists, linguists, geographers, behavioral scientists, and research investigators of every stripe, seeking to understand every detail of this mysterious land and its populace. After a while they feel pretty confident that they have all the details straight, the key measurements complete, and that they comprehend the major facts and interrelationships; basically they have everything pretty much figured out. Revelation number two. Now imagine that the investigators gradually realize that the rules that govern the new land are more complex, bizarre, and contradictory than anything in their prior experience. Principles that apply in one part of the territory have corresponding direct opposites nearby. Many of their early predictions about the continent turn out to be incorrect, initial

Weed Science. http://dx.doi.org/10.1016/B978-0-12-818174-4.00005-7

expectations are confounded and they come to the realization that what they once believed they knew was simplistic and undermined by counter-examples.

This little fable encapsulates the unexpected discovery of the complex and ubiquitous ECS 50 years ago, by a scientific team seeking to understand only how marijuana makes us high, and our subsequent attempts to delve into the system's underlying neuroscience. The more research that is undertaken on this network, the more byzantine and convoluted its details appear to be. Telling this story necessarily contains a fair number of technical details. But part of the narrative also includes tales of scientists traveling on buses with shopping bags stuffed with large chunks of Turkish hashish, and details of black truffles, spiny puffer fish, mutant rats, peanuts, sea squirts, and volunteers getting buzzed in research labs. So stay with me for the ride.

How the brain works

In order to understand questions such as how marijuana gets us high and how CBD and THC exert their medicinal effects, we need some basic understanding both of the body's ECS and how the human brain operates. That's a tall order and involves some rather complicated biology, but let's give it our best shot, starting with the brain.

The human brain is probably the most complicated and elusive thing in the universe. Explaining how it works is something that still remains mostly unsolved and that has preoccupied some of the finest scientific minds for centuries. Nobel prizes have been awarded and lengthy research careers devoted to this topic without yet really grasping many of the fundamentals. Given all of that, let's boil this entire, massively convoluted topic down to a 2-minute elevator pitch that ignores all of the complexity and subtlety of this magnificent organ, focusing on just enough small-scale details to help us grasp the essential features of the ECS.

At its core, the brain exists to provide us with accurate information about the outside world for the purposes of guiding our behavior, filing away accumulated experiences and knowledge in memory and anticipating the future. Take these examples: that stripy buzzing insect over there gave me a nasty sting last year; I'd best avoid it. Or, my wife's birthday is next week. I'll need to buy a nice card on the way home tonight. The basic starting point for all of these complicated perceptions and behaviors are nerve cells, or neurons, the simplest units of brains and nervous systems, whose purpose is to pass along and integrate little packets of information. This information may come from the outside world, for example, provoked by sounds or images, or from inside the brain itself in terms of memories, thoughts, and feelings. Nerve cells come in different shapes and sizes, but the typical neuron resembles a child's drawing of an octopus, with one super-long tentacle (called an axon) that ends in a tuft of branches that connect with other neurons. In the biggest neurons, the part that looks like the octopus' squishy body and contains the nucleus (named the cell body) can be as big as 1/10 of a millimeter across with the teeniest ones being almost 100 times smaller.

From the cell body, a series of mini-tentacles (named dendrites) poke out. These are the means by which the cell receives input from other neurons. The disproportionately long axon tentacle, which is the main unit to conduct nerve signals to other neurons, can be anywhere from a few millimeters to several feet in length. The billions of nerve cells in our brains, form trillions of possible connections, and communicate with one another through electrical impulses that have an underlying chemical basis. This communication happens across tiny gaps between the neurons (that are called synapses). Nerve cells are excitable, not in the Warren Zevon sense, but because they are specialized in generating, modifying, and passing along fast-traveling electrochemical pulses akin to coded signals. Neurons are primed to fire these electrical charges because little molecular pumps inside of them are constantly transporting electrically charged particles such as potassium ions, in and out, preparing the cell to discharge electrically when triggered by chemical compounds (known as neurotransmitters). The "happy hormones" dopamine (DA) and serotonin are two perhaps familiar examples of the more than 200 known neurotransmitters produced by our bodies that act as molecular messengers between nerve cells. Their release from one nerve cell into the synapse causes a wave of nerve impulses to leap to another neuron, resulting in the latter either firing, or switching off the electrical signal. When and how often, the neuron fires codes information that it passes along down the chain. If as a child you were bored by Sid the Science Kid, uninspired by Mr. Wizard's World, suffered ennui with Bill Nye the Science Guy, or became carsick on the Magic School Bus, you may want to skip the next several fact-packed paragraphs and dive straight into the ECS. Otherwise, bear with me.

Neurotransmitters are released from secure stores in tiny bag-like sacs inside the neuron on one side of the synapse, where they have usually been synthesized from amino acids. The name dopamine (dop-amine), for example, indicates that origin. This compartmentalized storage is necessary so that the transmitter chemicals don't leak out and cause neurons to fire randomly. Rather, their release is usually provoked by a prior electrical signal. The direction of this signal transmission is usually one-way from a pre-synaptic originating neuron on one side of the synapse to a post-synaptic target neuron on the other side, and so on down the chain. Numerous targets for different neurotransmitters may exist on the surface of a given neuron, allowing the cell to respond to and integrate incoming information from many sources before deciding whether or not to fire. In general there are specialized neurotransmitter receptors on the post-synaptic neuron to which specific neurotransmitter molecules bind. These docking sites are highly specific, so that the shape of a particular neurotransmitter allows it to fit into its dedicated receptor like a tiny distinct chemical key into a personalized lock. The neurotransmitter "key" is referred to as a "ligand", because it binds to the receptor. This term stems from the same Latin root word that gives us names of things that bind, including ligatures and ligaments. The receptor "locks" sit on the cell surface and are often configured in the form of so-called G-protein coupled receptors,

(GPCRs), proteins that detect specific transmitter molecules. In response they activate signaling pathways inside the cell, acting like molecular switches, (or, if you prefer, given their G-protein basis, tiny G-spots). GPCRs of one sort or another are targets for over one third of all FDA approved drugs. They perform this function by being woven snake-like through neuron cell membranes, simultaneously touching both the cell's inside and outside. When a ligand binds to its unique GCPR, the latter changes its three-dimensional shape. This temporary modification makes it less of a lock or docking site for the transmitter, and more of a mini-machine that triggers a cascade of chemical changes inside the post-synaptic neuron. A ligand that binds to a receptor and switches it on or activates it is called an "agonist" at that particular site, from the Greek word for contest. Logically, a ligand that fits into the same lock and blocks an agonist is termed an "antagonist." Likewise, a ligand that binds to the receptor and activates it weakly, in a half-assed way compared to a full agonist, is called a "partial agonist." As soon as their signaling duties are done, neurotransmitters are quickly whisked away from the scene of action at the synapse.

Once acted on by a neurotransmitter ligand, postsynaptic neurons are connected to yet other neurons, (typically via several thousand synapses) so that the electrical message gets passed on like a baton in a relay race of information across the brain. Over time, synapses that are frequently activated work faster and more efficiently so that in a very basic way they can "learn." From this simple, basic arrangement of neurons, one can imagine scaling everything up into something more brain-level in nature. Remember that these nerve cells may be metaphorical building bricks of the brain, but unlike bricks they are dynamic and ever-changing. Throughout development, groups of neurons within the brain that perform similar functions or contain specific neurotransmitters will organize to form complex circuits and systems. In turn, these circuits are themselves components of hierarchically organized meta-systems. Thus, when the brain is called on to perform a particular task such as recognizing an object or remembering a fact, the relevant modules of neuronal architecture can coordinate with each other across different regions most effectively to get the job done, with the whole arrangement being underpinned by millions of coordinated neurons acting in synchrony.

An apology is in order at this point. My use of words such as "usually" and "typically" earlier, is a prelude to confessing that the ECS, both inside and outside the brain confounds many of the rules regarding neurotransmission that I just laid out. For example, the relevant ECS neurotransmitters are built from fats, not amino acids. They are not stored long-term in tiny sacs, but synthesized on the fly as needed and their direction of transmission is backwards compared to the usual one. Also, while there are specific receptors for cannabinoids, there are no unique, specialized cannabinoid neurons like those that exist for DA or serotonin [3]. Finally, so far we've only got as far as Revelation number one:

like the eager explorers I mentioned above, there are many nuanced ways in which the ECS will reveal itself as utterly novel.

What is the endocannabinoid system?

So, what's the ECS and what's going on under the hood here? The general term "endocannabinoid system" refers to the whole kit and caboodle of cannabinoid neurotransmitters manufactured in the body, the specialized receptors that they lock onto in the brain and elsewhere in our anatomy, and the many enzymes and proteins that synthesize, transport, and break them down. Cannabinoid neurotransmitters that we will meet in a moment such as anandamide are unique in being fat-(lipid) based. Lipids are small, carbon-based fatty molecules that don't dissolve in water, but in organic solvents like acetone/nail polish remover and benzene. Lipids are used as general-purpose molecules all through biology, to build cell membranes, to store energy, and most relevant here only in the ECS, in sending signals from one part of the body to another.

In more technical terms, the ECS is an evolutionarily ancient, lipid-based signaling system and one example of the type of biological network I described earlier where neurons responding to particular neurotransmitters hang out together in a cannabinoid clique. In this case, the two main cannabinoid neurotransmitter ligands are N-arachidonoylethanolamide, thankfully abbreviated to AEA and most commonly known as anandamide, and 2-arachidonoylglycerol (2-AG). Much more 2-AG exists in the brain than anandamide, 10–100 times more in fact [4]. Together, anandamide ad 2-AG are known as "endocannabinoids," an abbreviation for "endogenous cannabinoid." This refers to their status as substances found naturally inside (endo) the body and produced there. This distinguishes them from exogenous cannabinoids (exocannabinoids) such as THC and CBD that are produced outside (exo) the body either by plants or in chemical labs. The two endocannabinoids are synthesized in the body from arachidonic acid, a substance that is named for peanuts (arachis) that contain this molecule and not for arachnids like spiders. Arachidonic acid is an omega-6 polyunsaturated fatty acid that enters our bodies through eating meat and eggs and is used for important functions like building cell membranes.

Anandamide and 2-AG are neurotransmitter ligands for two distinct types of cannabinoid receptors known unsurprisingly as cannabinoid receptors 1 and 2, abbreviated to CB_1 (cloned in 1990) and CB_2 (cloned in 1993). These two ligands glom onto specific cannabinoid receptors in the typical key-and-lock fashion. CB_1 and CB_2 are GCPRs of the type discussed earlier, located predominantly on neuron cell surfaces in the brain and peripheral nervous system, and in cells that are part of the peripheral immune system, respectively. The CB_1 receptor protein is built out of a twisted chain of 472 amino acids that weaves in and out of cell membranes 7 times like an intoxicated snake. Researchers have learned how to build mutant CB receptors by altering the chain's building bricks one amino acid at a time, and thereby determined exactly where THC and CBD bind to different locations on the CB_1 molecule.

CB_1 is one of the most abundant GCPRs in the brain and central nervous system, an impressive fact when you realize that 30 years ago nobody suspected that cannabinoid receptors even existed. These receptors are widely distributed throughout the human body. As Megan Scudellari nicely summed it up, "your body is teeming with weed receptors" [2]. Anandamide/AEA binds primarily to CB_1 receptors, where it acts as a partial agonist (refer earlier to the explanation of partial agonists). It is also a weak partial agonist at CB_2 receptors. 2-AG is the ligand for both CB_1 and CB_2, where it is a full agonist. CB_1 receptors are found in very large numbers in some brain areas such as the hippocampus and cerebellum, but outside of the brain they are also distributed in fat cells and in the pancreas, where they influence metabolism. CB_2 receptors are located mostly in the immune system, and in adaptive immune cells, macrophages and monocytes. These latter cells float around in the blood slurping up and neutralizing potentially dangerous microbes and particles. CB_2 receptors are believed to help fine-tune the body's pain responses, and protect tissues by helping balance levels of immune defenses and inflammation. CB_1 receptors are coded for by the gene CNR1 (an abbreviation for cannabis receptor type 1) and each is comprised of 472 amino acids. It is virtually identical to the rat CB_1 receptor, whereas the rodent form of the CB_2 receptor (built by the gene CNR2) is closer to 80% identical to the human one, which contains 360 amino acids.

THC, the plant-derived cannabinoid (phytocannabinoid) binds to CB_1 receptors, where it is a partial agonist, albeit one that hangs around for a long time, much longer than anandamide [1]. Anandamide and 2-AG are synthesized by a set of enzymes inside post-synaptic neurons, and are broken down by other enzymes, mono acyl glycerol lipase (MAGL) in the case of 2-AG and fatty acid amide hydrolase (FAAH) for anandamide, back into arachidonic acid and other simple chemicals. Though the abbreviations MAGL and DAGL (pronounced "maggle" and "daggle"), sound like Tolkien's discarded names for Lord of the Rings characters, they are found not in underground houses, but inside of pre-synaptic and post-synaptic neurons, respectively.

Endocannabinoid neurotransmitters are produced on demand—when certain post-synaptic neurons fire off an electric current, calcium flows into them, which ultimately triggers the enzyme diacylglycerol lipase (DAGL) to produce 2-AG, and a second, complicatedly named enzyme known as N-acyl phosphatidyl ethanolamine-hydrolyzing phospholipase D (NAPE-PLD) to synthesize anandamide in these cells. These cannabinoid compounds then travel backward to the presynaptic neuron to influence its behavior, mainly by stopping neurotransmitter release. Once they have accomplished that, they are quickly mopped up by, MAGL and FAAH. Anandamide and 2-AG unlike most plant cannabinoids are thus both unstable and short-lived. Although, the latter can have both stronger and more persistent effects at cannabinoid receptors. Interestingly there appear to be a host of endocannabinoid compounds in the human body, all of which consist of a fatty acid coupled with either an unmodified or altered amino acid. My favorite example is N-arachidonoyl DA, commonly abbreviated to NADA, fitting as we have absolutely no idea what this molecule does. That also turns out to be true for many of these lesser-known endocannabinoids, most of which are currently mystery substances.

Just as some are compounds are agonists or activators at cannabinoid receptors, there are antagonists, that is to say chemical ligands that bind to the receptor and block its effects. Most of these have been synthesized by pharmaceutical companies, with the exception of the cannabinoid Cannabigerol, synthesized naturally in the cannabis plant as a CB_1 antagonist. Most cannabinoid antagonists bind to a CB receptor and compete with the naturally occurring endocannabinoid to block its effects. The two molecules duke it out, and the more numerous one gloms onto the receptor. Antagonists are the deadbeats of the psychopharmacology universe. They tend to hang around the receptor and get in the way of the busy agonists that just want to do their job. More recently, pharmaceutical companies have designed compounds active at CB receptors called "positive allosteric modulators," or "PAMs" (like the non-stick cooking spray). CB receptor PAMs latch onto the receptor sites and actually change the 3-D shape and properties of the GCPR, so that its usual response to binding a natural molecule of anandamide or 2-AG is magnified. "Negative allosteric modulators" have the opposite action, that is changing CB receptors so that the usual effect of agonists is diminished. To add a final complication, in this bizarro, mirror-image world of receptor antagonists, there is a newly discovered family of "inverse agonists." These block a receptor and reverse its usual activity. Thus, for example, if an inverse agonist binds to a receptor, it will have the exact opposite effect of a regular agonist such as anandamide, rather than merely neutralizing it as an antagonist would. If you find that your head is spinning with all this talk of NAMs and PAMs, inverse agonists and receptor blockers, you'll be gratified to know that there are whole teams of pharmacologists working for commercial drug companies whose entire careers are focused on designing these ECS-modulating compounds as important new therapeutic agents. I have more to say about them and the drugs that they are discovering, a little later in this chapter.

How was the ECS discovered?

Various cannabinoids including THC and CBD were first extracted from cannabis back in the 1940s, but it was unclear which of these chemicals was responsible for cannabis' effects. Early studies suggested that THC could alter the characteristics of artificial cell membranes, so that scientists clung to the false hypothesis that the drug worked non-specifically on the cell membrane to alter cellular characteristics in a broad, unfocused way. People held to that view for a surprisingly long time, until THC's primary intoxicant role was discovered by Rafael Mechoulam at the Weizmann Institute in Rehovot, in 1964. Born in Sofia, Bulgaria in 1930 and forced out of his hometown by the

Nazis, Mechoulam emigrated to Israel as a teen in 1949. Trained as a biochemist interested in plant-derived medicines, he figured logically that if morphine and cocaine could be extracted from plants and shown to act on specific brain systems, there might be a parallel story with cannabis. He sweet-talked the local police into giving him 5 kg of confiscated Turkish hashish. (Advice to readers: if you want to score some weed, this is definitely not a recommended tactic). There is an often-told story of Mechoulam subsequently traveling on the bus back from the police station to his lab with his cannabis-laden shopping bag, surrounded by suspicious passengers curious about the characteristic skunky odor. Together with his colleague Yehiel Gaoni, he was able to extract the component chemicals from this stash, including THC and CBD. One at a time he then tested them carefully and systematically on animals. Thus the two colleagues determined that THC was the main intoxicant present in the plant. Subsequently they identified the 3-D chemical structure of both CBD and THC, and ultimately synthesized both chemicals. Mechoulam, now aged almost 90, still appears at cannabis conferences and his research group actively pursues multiple questions regarding the ECS and the drugs that interact with it. Despite being referred to affectionately as the "OG of cannabis chemistry," he has never used the drug himself [5,6].

The questions still remained, however, as to how THC exerted its effects. Solomon Snyder and Candace Pert had shown in 1973 that brains harbored opiate receptors, but the idea that there might be a specific receptor for cannabinoid drugs was widely discounted. After all, scientists "knew" that it did something or other to cell membranes. Then, in 1988, over 20 years after Mechoulam's key paper was published, US scientist Allyn Howlett and her colleague Dr. William Devane conclusively demonstrated that specific receptors for cannabinoids could be detected in rat brains. They first showed that a number of pharmaceutical company-developed analogs of THC had all of the properties of G-protein-coupled receptor agonists. Next, they attached radioactive labels to these drugs and demonstrated that the binding sites to which they docked were different than any previously known to pharmacologists, and that the cannabinoid ligands bound to them very avidly and selectively. A few years later, Bill Devane traveled to Jerusalem where Mechoulam now had his lab. Working together with a visiting Czech scientist, he used pig brains to track down the naturally-occurring nervous system chemical that bound to the newly discovered cannabinoid receptor. Incidentally, knowing that pigs possess ECS has been exploited recently. Farmers in Washington State have been feeding leftover cannabis stems, roots, and plant trimmings to hogs in order to increase their appetites and enhance their meat flavor [7].

As Mechoulam said at the 13th European Congress on Epileptology, "Receptors don't exist because there's a plant out there. They exist because we, through compounds made in our body, activate them. So we went looking for the endogenous compounds that activate the cannabinoid receptors." In March 1992, his team isolated the first of these chemicals, a cannabinoid-like substance naturally present in the body. Devane, who had an intense interest in Hindu religion named it "anandamide," after the Sanskrit word "ananda" for bliss. A little later in 1995, Mechoulam's team discovered another endogenous brain cannabinoid (2-AG) by processing rat brains and dog intestines. By then, Devane had returned to the United States to work in the National Institutes of Health laboratory of Julius Axelrod, where a team of scientists helped show more specifically how anandamide bound to the CB_1 receptor, which was first cloned in rats and then in humans in 1990. A cloned version of the human CB_2 receptor followed next in 1993. These were truly exciting times for cannabis scientists, and one discovery followed another. Much of this receptor work had been triggered in Axelrod's lab in 1990 when Lisa Matsuda found the DNA sequence that coded for CB_1 receptors. The next logical step was for a lab to breed genetically altered mice completely lacking CB_1 receptors, and to see whether THC would still get them high. Unsurprisingly, it didn't. Because there was nothing specific in their brains for the THC to bind to, there were zero psychoactive effects from the drug. Instead, the lonely partial agonist wandered around seeking receptors in vain. But all of this novel molecular research raised much broader questions—when did endogenous cannabinoids and cannabinoid receptors first start appearing in animals, why are they present in the body to begin with, and what's the overall purpose and function of this molecular signaling network? Next, let's tackle some of these big questions.

How early did cannabinoid receptors originate in evolution?

The genes that code for CB_1 receptors (*Cnr1*) are found across the animal kingdom in fish, frogs and other amphibians, birds, and mammals. Similarly CB_2 receptor genes (*Cnr2*) have been found in the spiky puffer fish from which we obtain the sometimes deadly fugu sushi. Incredibly however, something that is essentially a mash-up between the CB_1 and CB_2 receptor genes exists in the primitive undersea invertebrate *Ciona Intestinalis*. This beast (known technically as a sea squirt or vase tunicate) gets its intestinal name not from the fact that it inhabits somebody's gut, but because its body's hollow, pillar-like form is reminiscent of an intestine. A colony of *Ciona* resembles a mass of guts anchored on an undersea rock. Up close, each pillar of *Ciona* is a few inches long and resembles a 5-year-old's attempt at making rigatoni, or a tiny condom with its end snipped off. For such a primitive beast, parts of its biology are surprisingly complex. The genome of this little creature is tiny—barely 5% of yours or mine, but almost every gene family in our human DNA is represented there. *Ciona* doesn't have a brain or even a proper nervous system, but it does have axons, so that the available evidence suggests that cannabinoid receptors first evolved inside of those primitive nerve cells to help regulate and coordinate them in performing one of the few activities in which sea squirts excel, squirting. That behavior consists of sucking seawater into their bodies through a mouth siphon and spritzing it out of the other end of its digestive tract [8]. The aptly named, (in this context) US Department of Energy's Joint Genome Institute [9] documents much of the relevant cannabinoid genetic information on its website.

In addition to the unexpected occurrence of cannabinoid receptors in a primitive undersea creature, Italian scientists were also startled to find not only the presence of anandamide but also of the major metabolic enzymes of the ECS such as NAPE-PLD and FAAH, inside the delicious and pricey winter black truffle, *Tuber* melanosporum [10]. These truffles do not express the endocannabinoid-binding receptors CB_1 or CB_2, or contain 2-AG (despite having all the RNA for producing DAGL and MAGL). The researchers concluded that anandamide and ECS metabolic enzymes evolved much earlier than endocannabinoid-binding receptors, and speculated that anandamide might be an ancient attractant to truffle-eating creatures. Putting the relevant dates together, we learned in the prior chapter that the cannabis plant emerged about 28 million years ago; truffles and their kin were around for about 128 million years prior to that, at the end of the Jurassic period, (and sea squirts evolved about 550 million years ago). It appears that parts of the ECS are truly an evolutionarily ancient part of our biological history.

What are some broad properties of ECS components?

When they bind to their respective CB receptors, both anandamide and 2-AG are agonists, that is they switch on (activate) the receptor to produce a cellular response. While brain anandamide generally has a diffuse, modulating effect on multiple neurons, it also has highly specific effects in particular brain regions. Some researchers have summarized one of the ECS's primary purposes as homeostasis, that is, maintaining key systems in the body on an even keel [11]. Meanwhile, 2-AG is more straightforward in its activity, backtracking within the synapse from a post-synaptic nerve cell to influence presynaptic neurons, (which is why it is sometimes termed a "retrograde messenger"). Unlike anandamide it has a more precise point-to-point profile of action. It's easy to see how this type of activity might act as a feedback mechanism, from post-to presynaptic neurons.

As well as their primary effects in the ECS, both anandamide and 2-AG have secondary actions on other neurotransmitter systems, including circuits ruled by glutamate and DA. This activity enables endocannabinoids to balance the overall stirring up versus calming down of neural inputs across the entire nervous system, in both the short- or long-term. In a tit-for-tat system of regulation, release of those other neurotransmitters in turn can trigger neurons to squirt endocannabinoids that then act on presynaptic CB receptors to regulate neurotransmission. You can think of this round-robin of activity as a kind of self-regulating feedback loop, consistent with the hypothesis that the ECS may have a role in homeostasis.

Thus, on the one hand there are straightforward, tissue-specific effects of endocannabinoids; this compound on that receptor in this organ to elicits this

precise effect. But from another, more general perspective, the body's cannabinoids also act as non-specific cross-system modulators, tweaking numerous
general molecular processes including immune function, pain perception, appetite, metabolism, cognition, and motor coordination, like a team of busy office
managers or ringmasters. Moreover, cannabinoids initiate multiple downstream
molecular signaling mechanisms inside of cells, resulting in altered cell functions. Here they act more like online influencers. Later in the book I will discuss
these multiple ECS effects in the body in more detail.

How did people figure out what the ECS and its components do?

Over the years, scientists have gradually chipped away at the varied complex
functions of the ECS in order to understand better how each works. To do so,
they have used many different approaches, including genetic tools to create
experimental animals completely lacking CB_1 or CB_2 receptors, or having reduced numbers of them, in order to study the modified animals' behavior and
biology. With such genetically modified beasts, they determined the effects of
cannabinoid molecules that are not mediated by cannabinoid receptors. More
explicitly, if they dosed a CB receptor-deficient animal with a cannabinoid such
as THC, and something important changed in its behavior or physiology, that
effect could not have been driven by CB receptors, as this creature had none.
Other research has focused on pharmacological engineering to alter the shape
and function of cannabinoid receptors by tinkering with the genes coding for
them. This strategy might employ synthetic cannabinoid molecules similar to
anandamide or THC, but designed in the lab to bind many times more strongly
to CB receptors. Some of these designer drugs have escaped the research lab to
resurface as synthetic cannabinoids such as "spice" and "K2." More recently researchers have discovered drugs that influence (e.g., block or rev up), the effects
of the enzymes that synthesize or break down endocannabinoids. They've then
used these drugs to build up or deplete those ligands to much higher or lower
levels than normal. They can then study the effects on health and behavior of the
modified lab animals. These ongoing experiments, have clarified that the ECS is
a unique, complicated, vast-ranging cell-signaling system whose functions are
only beginning to be unraveled. Now let's take a closer look at what some of
those functions might be.

Molecular studies of receptor structures

What do CB_1 and CB_2 receptors look like? In layperson's terms, these GCPRs
are present on cell surfaces as tiny, precise, perfectly designed pieces of machinery that resemble a mash-up between a nanobot and a series of Thomas
Joynes vortex sculptures tossed together with a handful of Slinky springs. If
you want more help in visualizing a CB receptor protein, then you can borrow
this idea from Richard Powers novel *The Gold Bug Variations* and turn it into

your very own science project. First, take the cardboard tube from the center of a roll of kitchen towels and unpeel it so that it becomes a loose cardboard spiral. Repeat this process 6 times, so that you have seven cardboard rolls, representing the seven trans-membrane domains of the receptor. Then spray paint one of the rolls yellow, two green, one orange, one teal, and one navy blue. Stack them in a vertical row from left to right, rearranging them slightly so that the greens cross behind the red and the teal hides unsuccessfully between the red and blue. Then bend the bottom of the red so that it crosses all the way over to the right side. Now you have your very own artist's model of a CB_1 receptor.

Alexandros Makriyannis, a pharmaceutical chemist and acknowledged expert in the field of cannabinoid compounds, formerly directed a research lab at the University of Connecticut in the early 2000s. At that time, he would attend lectures at my home institution and we'd chat about his latest cannabinoid discoveries. Several years ago, he moved his lab to Northeastern University in Boston, where over the last few years he has helped decipher the active crystal structure of CB_1 and CB_2 receptors. His work is the type of complicated, finicky, and exact science that requires a large-scale team effort. Indeed, Alex has been part of several international collaborations, teaming up with research scientists from Moscow, Shanghai Tech University, and Beijing as well as other US universities. In a series of fascinating publications, these investigators have helped unravel the structural intricacies of the cannabinoid receptors, how to build them inside of bacteria, how their 3D shapes are altered when occupied by agonists and antagonists, and where novel molecules can bind in various nooks and crannies on the receptors to alter their properties. In addition, they have genetically engineered mutations in the receptors, and altered their protein to make easier to study without altering its structure or function [12–15]. Theoretically, one could use THC to probe some of these questions on CB_1 receptors, but because it's a partial agonist, what's really needed are to understand how these receptors work are specially engineered "super-agonist" and antagonist molecules that fit precisely and decisively into the molecular binding pockets in the receptor. When bound by these specially designed cannabinoid ligands the CB_1 receptor morphs its 3D shape like a Transformer. The molecular binding pocket, where these ligands dock can change its volume by more than 50% to accommodate active molecules of various sizes, shapes, and functions. On the surface of all of these twisted, pretzeled-together Slinky protein structures are additional niches and crevices separate from the originally discovered binding pocket. Into these molecular niches within the proteins a variety of novel cannabinoid compounds can insinuate themselves and alter the properties of the receptors. And because the cannabinoid receptors are fancy molecular toggle switches, the different resulting 3D shapes will pass along different molecular messages with new downstream effects on cellular targets. These "unofficial" binding sites turn out to be very different functionally depending on whether they belong to CB_1 or CB_2 receptors, despite the latter's superficial resemblance and structural overlap of 40%. Consequently, substances that activate one receptor can actually weaken or inhibit signaling on the other, in what has been described as a "yin and yang" relationship between CB_1 and CB_2. All of this work has practical consequences in helping us understand basic principles of rational endocannabinoid drug design.

Measuring human CB_1 receptors with PET scans

The inverse agonist drug compound MK-9470 binds strongly to CB_1 receptors in the human brain. So strongly in fact that tiny doses can be given in a radioactively labeled form that can produce an accurate map of receptors in the brain and enable us to count their numbers. The fact that minuscule doses are sufficient for such mapping means that there are no undesirable drug effects associated with the studies. Thus, the dose of administered radioactivity can be small. Through this and similar approaches, we can identify where the highest concentrations of CB_1 receptors are located in the brain, and also see whether they are changed in cannabis users compared to non-users, and as we'll see later, what such measurements reveal in the brains of stoners and in patients with schizophrenia.

What body functions is the ECS responsible for?

Comprehensive reviews of this topic can be found in a series of recent articles [1,3,4,16,17]. A list of the activities that the ECS participates in is seemingly exhaustive and wide-ranging over almost every known body system. As well as neural signaling, it plays an important role in blood pressure regulation, immunity, inflammation, sleep, memory, gut motility, fat metabolism, insulin sensitivity, bone development, pain control, and fertility. The cannabinoid system is with us from our very first moments. The early developing embryo is covered in cannabinoid receptors that help it implant in the wall of its mother's womb.

The ECS, although biologically ancient, is so recently discovered and novel that much still remains to be uncovered and understood about it. It is also a relatively hot area for pharmaceutical companies to investigate. Because cannabinoid receptors and endocannabinoids are widely disseminated through multiple systems in the body, this suggests they play an important role in many physiological processes and as promising targets for novel drugs. Some of these prospects are already starting to pay off. For example, some cannabinoids delay tumor progression [18], while others help treat kidney fibrosis [19]; however many of these compounds have yet to make it into routine clinical practice. One factor that makes studying the ECS so complex is that at least at a brain level, many of the effects of cannabis and of THC are biphasic, that is, a low dose may have one effect (such as anxiety reduction), whereas a higher dose has exactly the opposite outcome [20]. Consequently, nothing about the ECS can be taken for granted and everything requires careful exploration by both consumers and scientists.

Thus, we need to think beyond the cannabis plant more broadly, in terms of how to manipulate the ECS for therapeutic purposes, something we will review

in more detail in Chapter 10. In addition, we need to understand the intricacies of molecular structure of cannabinoid receptors in order to custom-design
novel drugs that interact with them in predictable ways. But first, let's try and
take one argument off the table. A common assertion raised by individuals opposed to medical cannabis, is that one drug could not plausibly be effective (as
has been suggested for cannabis), in treating among other disorders, arthritis,
depression, glaucoma, chronic pain, anorexia, epilepsy, dementia, Parkinson's
disease, dystonia in multiple sclerosis, anorexia in AIDS and cancer, and inflammatory bowel disease. But to review what we've just learned, because of
its seemingly ubiquitous distribution, its at least feasible that drugs acting on
the endocannabinoid system really could have far-flung and multiple beneficial
effects on the brain and body through a variety of different mechanisms. With
that mental math in mind, let's review some of what's been discovered, in terms
of the normal role of the ECS, and how it interacts with chemicals in cannabis.

Brain development, brain maintenance, and beyond

Clues to what the ECS does can be garnered from its distribution in the body.
Though its reach may be far and wide, nevertheless it is especially concentrated
in some areas and cell types. For example, large numbers of CB_1 receptors are
present in impressive quantities in particular parts of the brain. I list a few here,
followed by an overly simplified summary of some of the processes that they
are likely involved in, (in parentheses). Their highest density is in the prefrontal
cortex (various cognitive functions), hippocampus (memory), cerebellum (balance), cingulate gyrus (choice and decision making), and basal ganglia. More
modest quantities exist in basal forebrain (sleep; anandamide injection into rats
produces increases in slow wave and REM sleep), amygdala (emotional regulation and fear processing), nucleus accumbens (reward), peri-aqueductal gray
(pain processing), and the hypothalamus (appetite). Notably there are few or no
CB receptors in the brain stem or primary motor region. The former probably
accounts for why cannabis even at high doses has almost no damaging effects
on breathing or cardiac function, unlike alcohol and opioids.

The ECS is crucial to normal fetal brain growth and continues to play an
important part in regulating multiple aspects of normal brain development and
maintenance, through adult life. Anandamide, for example, functions in early
brain development to influence the formation of new synapses, fosters the growth
of just-born neurons, and helps guide these developing fetal nerve cells to their
correct targets in the busily expanding brain. 2-AG also plays an important role
in the growth and guidance of neurons. Both cannabinoids also help axons to
grow, provoke new neurons to develop and mature, and beyond individual cells,
aid in the development of more complex neural circuits. These important brain
sculpting and guidance functions continue through infancy, adolescence and
early adulthood. CB_1 receptors in particular, are also required for the growth of
axons [21]. Later, in adolescence, anandamide contributes to pruning or paring

down existing synapses to help brain connections become fewer in number but more efficient in function [22]. Some researchers have argued that these persisting ECS activities in the brain may be one reason that activating the CB_1 receptor with THC in a way that exceeds normal physiological levels can produce ill-effects on the developing brain when cannabis is used in adolescence.

Reward circuits

As mentioned, the distribution of CB_1 receptors in the brain allows us to infer some of their physiologic effects. The moderate concentration of receptors in nucleus accumbens (NAcc), that is connected to the ventral tegmental area (VTA), brain regions containing high concentrations of dopamine neurons and their receptors, plays a role in how we respond to behavioral rewards, predict environmental events, motivate us and allow us to calculate the pluses, and minuses of potential behaviors. This topic is nicely reviewed in several articles [23,24]. Natural, life-sustaining rewards such as tasty food, sexual activity, and social interaction feel good (hedonic response) and lead to changes in motivation that increase the odds of repeating such behaviors in the future. Artificial and abstract rewards such as money or viewing pictures of sexually attractive individuals activate the same system. This type of positive reinforcement has its mirror image in our avoiding nasty or unpleasant stimuli and circumstances. These responses are mediated through a dopamine reward system underpinned by VTA/NAcc circuitry and can go haywire if exposed to super-rewarding stimuli. This is most evident in individuals exposed to drugs. In the short-term, drugs of abuse activate the same reward pathways, but over the long haul can hijack them in the service of using more drugs. Having accomplished that, the zombie reward circuit focuses increasingly on pharmaceutical rewards with diminishing attention devoted to everyday sources of satisfaction. The ECS interacts with the reward system on multiple levels. Brain endocannabinoids dial up or down the positive reinforcing effects both of natural rewards and of abused substances. So highs are higher (or lower) as a result of ECS intervention. Second, this ECS tweaking of reinforcement and reward occurs through cannabinoids interacting with the DA, and (to a lesser degree) the mu opioid neurotransmitter systems. More on the latter is discussed later. The primary ECS site for these effects seems to be the CB_1 receptor. Activity here significantly modulates how DA acts in the circuit. In the VTA, these effects are indirect, because CB_1 receptors do not occur on these DA neurons.

Animals show a complex and unpredictable set of responses to CB_1 agonists such as THC. These compounds reinforce behaviors under some conditions or doses but not others. They can be rewarding in particular circumstances, or anxiety-provoking or unpleasant in others [4]. Neither anandamide nor THC injected directly in small amounts into this brain area in animals elicits a classic "drug of abuse profile" as do stimulants like cocaine, or opioids like heroin. In fact some cannabinoid-like medications can reduce dependence effects of other drugs of abuse including nicotine, alcohol, opioids, and cocaine. With enough effort researchers can coax animals to self-administer cannabinoid agonists, but it's difficult to get them to do so. This seems counterintuitive, since cannabis dependency is well-established in humans, so the natural assumption is that lab rats and monkeys would enjoy getting high. The fact that they usually don't is a reminder of important behavioral and brain differences between people and animal models. Rewarding effects in lab animals are also very dose sensitive. The injection of larger doses of cannabinoids is quite negatively reinforcing. The tendency of animals to actively avoid these larger doses is mediated through CB_1 receptors [23]. The role of cannabinoids in the reward system is further complicated, because recreational cannabinoids can both reinforce and negate effects of other abused substances like opioids.

Memory

The hippocampus is a curved, finger-sized piece of brain shaped a little like a seahorse. It swims deep on the inside surface of the temporal lobe in each hemisphere. As was shown in the disastrous neurosurgical removal of this area from the patient HM, the hippocampus is essential for forming so-called episodic memories, best described as information about what you experienced, where it happened, and when it occurred. An example might be where you were when you heard about 9/11. Complex electrophysiological patterns known as long-term potentiation (LTP) in the hippocampus are essential to retaining these types of facts. Interfering with the process leads to problems recalling past events or remembering new ones. THC suppresses LTP and this phenomenon in part explains the well-known memory-impairing properties of cannabis use on short-term recall. As the old joke goes, "if smoking marijuana causes short-term memory loss, what does smoking marijuana do?" Mutant mice bred to lack CB_1 receptors have altered short-term memory, spatial memory, and motor learning function [25,26]. An extension of this memory story comes from a recent study [27]. Two major information highways from the cerebral cortex into the hippocampus carry information about "what happened" and "where did it happen" that are part of the recipe for memory material. The time-stamping of memories occurs in the hippocampus via LTP, with unique labels when it occurs in the "what" circuit. This "what" signal involves the local synthesis of endocannabinoids, and their binding to CB_1 receptors. Once that binding occurs, something unexpected happens; a sustained increase in neurotransmitter release at the site, strengthening the synapse. Interfering with this mechanism damages the encoding of "what" data very specifically. So cannabinoids are clearly crucial in helping create memories in the hippocampus about what happened but not where or when it occurred. This activity in the hippocampus is not the only exception to the rule that cannabinoids act as simple neuron "circuit breakers" [28]. For example, CB_1 receptors are found on multiple cell types throughout the brain, and activating these receptors has radically different outcomes, for example, on cognition, depending on their location.

When we think about cannabinoids and the ECS, the first thing that comes to mind (so to speak) is the brain. But the ECS extends throughout the entire body, and only some of its many distant extensions even communicate with the brain. For instance, both endocannabinoids and phytocannabinoids have well-established effects on appetite. THC acts directly on the hypothalamus to increase appetite and hunger (the munchies) and in the nucleus accumbens to re-balance food-related reward as well as taste perception. The endocannabinoid system helps regulate food intake, appetite, and metabolism. These effects relate to the therapeutic relevance of medical cannabis to stimulate appetite in medically underweight patients (e.g., those with HIV/AIDS or wasting due to cancerous conditions). Again, the role of the ECS in pain relief extends beyond the brain to the spinal cord and peripheral nerves. There seems to be a primary role for cannabinoids as pain-relieving agents through direct activity on CB_1 receptors, in addition to the more complicated interactions with the opioid system discussed later.

Stress, anxiety, and mood

Endocannabinoids and CB_1 receptors have important functions in dealing with emotions, mediating stress response and release of stress-induced neurotransmitters. They also consolidate emotional memories, process threatening stimuli (fear regulation), and manage anxiety [28–30]. Although the ECS clearly plays an important role in regulating these emotional behaviors, manipulating cannabinoid signaling in humans results in complicated and sometimes contradictory effects unpredictable different effects depending on the brain circuits and cell types targeted. In the Medical cannabis section, we will discuss the role of plant cannabinoids as potential treatments for a variety of disorders.

Cortical CB_1 receptors have also been implicated in other neuropsychiatric disorders. For example, the ECS seems to be significantly altered in Parkinson's disease, multiple sclerosis, seizure disorders, Alzheimer's disease, ALS, major depression, and schizophrenia. Abnormal anandamide levels have been detected in the cerebrospinal fluid of schizophrenia patients. Intriguingly, THC seems to promote psychosis while CBD may have antipsychotic properties. Those schizophrenia-related issues will be discussed in Chapter 8.

Other organ systems

The relationship of the ECS to appetite and the gut goes way beyond the munchies. Both CB_1 and CB_2 receptors are present in the intestines in large numbers; the former senses fat as part of its role in regulating appetite, separate from brain effects. The ECS also regulates intestinal movement and controls both how quickly food moves through the gut, and how the body's energy balance is regulated. In this latter role, CB_1 receptors help control glucose metabolism, fat breakdown, and energy balance, as well as modulating effects of steroids. Blocking these receptors tends to slow down weight gain. The role of cannabinoids and their receptors on modulating symptoms of nausea and vomiting is likely mediated both through the gut and the brain. Notably, ECS components seem out of balance in many gastrointestinal (GI) disorders, such as inflammatory bowel disease, and somewhat surprisingly chronic liver disease. CB_1 and CB_2 receptors in the liver seem to have opposite roles. CB_1 over-stimulation lards the liver with fat, promotes the formation of scar tissue, and causes liver cells to multiply and then die off prematurely [17]. Conversely, CB_2 receptors have opposite effects on the organ and thus seem to have therapeutic potential for treating chronic liver disorders.

CB_1 receptors are located in the *heart and blood vessels*. Their presence explains why smoking marijuana causes people to get bloodshot eyes (through blood vessel dilation) and a speeded-up heart rate. In addition, CB_2 receptors are found in these same locations, where they likely play a role in reducing the inflammation associated with fatty deposit (plaque) formation in coronary artery disease. It is now believed that CB receptors have an important function in the development of common, chronic heart and blood vessel disorders [17], and that overactivity and dysregulation of the ECS underpin injury and inflammation in this system. But there is no real evidence that marijuana use is associated epidemiologically with the development of coronary artery disease or heart attacks, unlike the clear link seen with tobacco smoking.

CB receptors are also activated by external, energy-related factors. For instance, aerobic exercise causes release of anandamide that may contribute to "runners high" as well as mobilizing energy stores and reducing exercise-related muscle pain [31]. Athletes tend to favor the use of CBD to ease such pain, but no convincing clinical trials have yet been conducted to investigate this possibility. The ECS plays an important role in muscle formation; CB_1 receptors are well represented in muscle cells and play an important role there in managing how muscles budget energy. Cannabinoids also influence bone growth and repair [17], so that it will be important to figure out whether there are effects of THC use in adolescence on height. Maybe pot use gets one less "high". Research in Israel has begun investigating cannabinoids in bone repair following fractures.

In the *reproductive system*, CB_1 receptors seem to play a role in fertility, are present in the placenta and necessary for embryos to implant normally in the wall of the uterus. This explains some of the concerns regarding deleterious effects of cannabis smoking associated with use immediately before and during pregnancy. CB receptors are also present in the testicles, where they influence sperm production and help orchestrate effects of sex steroid hormones in the reproductive tract [17,32].

The immune system and inflammation

The human immune system consists of two distinct parts, the innate and adaptive divisions. The innate immune system is a very general purpose, short-term frontline defense apparatus consisting of a collection of chemicals and proteins that block and destroy bodily invaders. The adaptive immune system is much more specific, recognizing and remembering proteins that don't belong to our own bodies and neutralizing them to prevent infection and disease. This latter, more sophisticated system operates through specialized cells classes including B lymphocytes, whose surfaces are studded with numerous CB_2 receptors. Endocannabinoids and their receptors alter the functional properties of immune cells and 2-AG helps control when and where these cells are deployed in their miniature search and destroy missions. The fact that the ECS seems to be so bound up with immunity has caught the attention of many researchers. Designing novel drugs that are potentially powerful anti-inflammatory agents or can modulate the immune system through ECS-related activity is a very hot area in the pharmaceutical community.

CB_2 receptors—scientists thought initially that these receptors were only present only outside of the brain and restricted to the immune system. Both of these assumptions turned out to be wrong. It is now known that CB_2 receptors have an extremely wide distribution in multiple body systems, including the brain [33]. That panoply of locations might help explain why cannabis' claim to have so many diverse effects on health and to be effective in treating a wide array of very different types of diseases occurring in different tissues, may have some underlying basis. CB_2 receptors are present in bones, the gut, and the immune system, (including the spleen, thymus, tonsils, and the surface of immune cells). To the surprise of researchers, these receptors are also present in the brain, on specialized non-neuron brain cells known as microglia, as well as on dopamine neurons in the midbrain. Preliminary experiments suggest that activating these DA reward neuron-related CB_2 sites changes behavioral responses to both cocaine and alcohol, at least in mice. CB_2 receptors in their brains are involved in anxiety reduction and in modifying the animals' response to stress. Their role in the human brain remains to be fully elucidated. The recently identified cannabinoid receptor GPR55 (see later) is also located in the brain.

Cannabinoid activity via receptors outside of the conventional ECS (and in Scotland)

We have mentioned in passing that cannabinoids can have effects on other well-known neurotransmitters such as the DA and opioid systems. In addition, outside of their primary actions on CB_1 and CB_2 receptors, the endogenous cannabinoids anandamide and 2-AG influence brain function through a variety of other mechanisms. One important function that falls into this non-classic role seems to be related to regulating one of the most important human experiences: pain.

To understand how, let's take a quick trip to Scotland to meet a remarkable woman whose peculiarities of perceiving pain trace back directly to her mutated ECS. Seventy-two year old Jo Cameron, who lives next to the gloomy Loch Ness Lake in Scotland, was born with a unique combination of two genetic differences that render her insensitive to pain. The first alteration is a small mutation in a recently identified gene located on chromosome 1 named FAAH-OUT, (presumably named by a geneticist with a memory of 60s hippies), that is normally expressed in the brain and in large, pain-processing nerve clusters located next to the spinal cord. The second difference is a common functional variant in the FAAH gene that reduces its effectiveness and activity. Together, these two distinct genetic hits significantly disable her body's FAAH activity and subsequently her ability to experience pain from everyday scrapes, bumps, and bruises. Jo's blood tests reveal significantly higher than normal concentrations of anandamide (increased by 70%) and related cannabinoid compounds (some of which are triple normal levels) that are normally broken down by the FAAH enzyme. Because her MAGL enzyme is unaffected, Jo's concentrations of 2-AG are entirely normal. The result of Ms. Cameron's combined genetic differences is that her endocannabinoid signaling is unleashed and thus boosted, by her relatively ineffective FAAH, and she lives virtually pain-free. This newly recognized and likely unique condition has its upsides—she experienced pain-free childbirth without an epidural. But there are reciprocal downsides. For example, a severely degenerating hip joint that required replacement had generated no warning pain signals. she barely batted an eye after being involved in two car accidents and can eat scorchingly hot chili peppers, that produce only a "pleasant glow." Interestingly, Ms. Cameron has other differences from the average person. One is her personality. She is unflappable and nearly always in a good mood: (she scores 0 on standard anxiety disorder and depression questionnaires). A second is her memory capacity. Given what we know about anandamide's function in the hippocampus, it's not surprising that she also reports lifelong short-term memory problems, such as frequently forgetting words mid-sentence. The neurologists who tested her did not perform detailed memory testing however, to quantify this problem more precisely. So that we don't know whether she has deficits in "what" versus "where" memories. And although she inadvertently burns and cuts herself frequently because of a lack of warning pain signals, she has few scars, so perhaps her wound healing is speeded up [34,35]. Jo Cameron's case is fascinating, not only because of the extraordinary way her condition has affected her life, but also how it clearly implicates the role of the ECS in pain control.

As well as complex activities in the dopamine signaling network, the ECS also has effects on opioid receptors that are an integral component in our ability to tolerate pain. Cannabinoid receptor agonists are associated with pain relief in animal models of both chronic and acute pain. Part of this effect definitely occurs through the opioid system. Indeed, the specific type of opioid receptors in the brain and spinal cord to which pain-relieving opioid drugs such as morphine, heroin, and OxyContin bind are often located right next to CB_1 receptors, suggesting that the two influence one another (what scientists term "cross-modulation"). Under some circumstances, opioid and cannabinoid receptors will actually link up functionally in a kind of temporary hookup, to alter each other's biological properties. Thus there is at least some evidence that cannabis' effect on pain reduction is not just due to release of opioids but involves these sorts of complicated cross-receptor interactions. In addition, peripheral CB_2 receptors may modulate pain outside of the central nervous system, not only through inhibiting inflammatory responses, but surprisingly by causing release of endogenous opioid peptides. Cannabis-derived agonists such as THC likely relieve pain by encouraging the body to synthesize or release endogenous opioid chemicals such as endorphins in the brain and spinal cord. Recent studies from the University of New Mexico show that substantial pain relief was reported by patients who used high-THC cannabis flower across a variety of medical conditions [36]. Relatedly, individuals who use medical or recreational marijuana and then need to undergo major surgery rate their post-surgical pain as significantly more severe than non-users [37]. A similar phenomenon occurs with individuals who are taking pre-operative prescribed opioid pain medications and have them discontinued in the hospital prior to surgery. In both cases, an effective pain-relieving medication is stopped as part of routine care, and due to tolerance to the drug, the body's pain sensitivity rebounds. Logic here leads us to the conclusion that cannabis is an effective pain reliever, as might be intuited from the case of Ms. Cameron in Scotland.

Understanding these interactions between cannabinoids and the opioid system may seem abstruse, but if we want to figure out whether medical cannabis might help reduce opioid consumption in individuals battling chronic pain, dissecting the underlying neuroscience might help to guide clinical practice. Compared to opioids, cannabis has a more benign side effect profile and less physically addictive. Thus, a harm-reduction argument is often made for cannabis's role as an opioid substitute in managing severe chronic pain, a claim that we will examine in more detail in Chapter 10.

ECS/opioid interactions turn out to be only part of why cannabis has been known for millennia as an effective pain reliever. A newly discovered and equally important role in pain reduction occurs through interactions with $TRPV_1$ receptors. In brief, any nasty stimulus, from a physically hot object to juice from a spicy hot pepper (such as a Scotch bonnet) produces a characteristic hot, painful sensation. This process occurs by activating so-called vanilloid receptors (known as TRPV1 sites) that are found mainly inside of the pain-sensing neurons of the peripheral nervous system, but are also located in the brain. Nobody would make the intuitive connection without examining their molecular structures, but vanilla flavoring and the key hot pepper ingredient capsaicin are relatives. In another example of chemicals produced by plants interacting with the nervous system, capsaicin from chili peppers binds highly specifically to $TRPV_1$ receptors. So along with cannabis, tobacco, opium poppies, and cocaine, red hot chili peppers are yet another example of a plant whose natural chemicals alter body function by binding to specialized receptors. $TRPV_1$ is involved in transmitting pain signals, by boosting, lowering or merging them together. These receptors also regulate body temperature. It was previously thought that only heat or capsaicin could activate $TRPV_1$ receptors, but the sites are also molecular targets not only of anandamide, but also of CBD [38,39]. This explains part of CBD's role in modulating pain. $TRPV_1$ receptors are also located on the surface of blood vessels, where their interactions with anandamide causes them to dilate. Interestingly, chili peppers and anandamide synergize to reduce gut inflammation and reduce the severity of type I diabetes, in part through their joint action on $TRPV_1$ [40]. Perhaps in the future, patients with inflammatory bowel disease may shun Tums in favor of jalapeno-laced cannabis edibles.

Cannabinoids also have an important involvement with yet another obscure receptor entity, known as the GPR55 receptor. This structure is a type of GCPR whose functions were unknown for a long period of time, during which it was classified as one member of a biological ragbag of so-called "orphan receptors." This is another way of saying "this molecule is clearly a GCPR, but we have no idea what its function in the body might be." Ultimately it was discovered that GPR55 was activated by not only anandamide and 2-AG, but also by THC [41]. The receptor, (like $TRPV_1$) processes stimuli related to pain and inflammation, and is found all over the brain, especially in the cerebellum. It has also been discovered in sensory integration sites next to the spinal cord, where information on pain and sensation is blended and organized, and in the gut. Although GPR55 bears scant molecular resemblance to either CB_1 or CB_2 receptors, recently many have suggested that it should be re-categorized as a third member of the cannabinoid receptor family, and hence renamed the CB_3 receptor. This would be the molecular equivalent of finding a sibling that you never knew existed via a consumer genotyping outfit such as 23andme. This reclassification is not yet scientifically official however. What we can infer though, is that GPR55 is yet another mechanism through which the ECS controls pain signals in the nervous system. I'm sure that's not the end of the story. At least in animal models, cannabinoids synergize with this former orphan receptor to kill off pancreatic cancer cells suggesting that its role in health seems to be broader than initially suspected.

I have already discussed some of the primary functions of the ECS in the brain as related to its key neurotransmitters and receptors. But that's only the beginning of the story. As hinted at in our discussion of the role of CB_2 receptors in the brain's reward system, the ECS in general has important and broad interactions with the DA system that influence alertness, reward, motivation and possibly risk for addiction and schizophrenia. The DA system imbues incentive motivational value into stimuli that we encounter in the world. That's a fancy way of saying that one vital function of brain networks that use this neurotransmitter is to prioritize what's truly important in the environment and needs attending to, and that DA plays a key role in these value judgments. The ECS influences this monitoring and judging process. Acute stimulation of CB_1 receptors in rats and mice leads indirectly (i.e., via other neurotransmitters such as glutamate), to DA release in reward regions, similar to other recreational drugs such as cocaine. But compared to other addictive substances this effect is not especially impressive. Seemingly every behavioral science teacher shows students grainy videos of rats pressing a lever to deliver electrical stimulation to these brain reward regions, something they (the rats not the students) will perform to the point of exhaustion. Similarly rats can be trained to lever-press to deliver micro doses of potentially abusable substances to appropriate brain reward locations. Joseph Brady at Hopkins and his fellow researchers produced hundreds of papers examining just how hard various animals would work for particular drugs as models of their abuse potential. In the case of rodents given access to cocaine and opioids, it's easy to elicit lots of repeated lever pressing, showing that rats find these substances rewarding and presumably pleasurable. For THC the equivalent effects are mixed and not very impressive. Rats are reluctant to self-administer THC, although squirrel monkeys are not. Consistent with this, the DA release provoked by THC is significantly less than that provoked by amphetamine or cocaine in reward-relevant brain regions. Rats will show some behaviors (e.g., favoring a particular location where they have been previously rewarded with a given drug) to low doses of THC, indicating that they find the substance rewarding, but higher doses are aversive and they will actively avoid receiving them. Chronic administration of cannabinoid agonists may reduce DA levels in parts of the prefrontal cortex in rodents, with higher doses having relatively more marked effects [16]. Where rewarding responses can be elicited at a particular drug dose, animals ultimately develop tolerance to them and need higher amounts of drug to produce the same effect.

The question of how chronic cannabis use might trigger later psychosis has focused to a large extent on possible alterations in the DA system, as we go into more detail in another chapter. One obvious key experiment to test that link is to let volunteers smoke cannabis, or to inject them with intravenous THC. Then, one can use brain imaging techniques such as positron emission tomography (PET scans) that employ short-lived, mildly radioactive isotopes to probe DA receptors, and to see whether cannabis actually triggers DA release compared to placebo. Several labs have tried exactly this approach to address the cannabis/DA question. Here, humans differ from rats. The results in human volunteers are somewhat confusing and contradictory, but at present the evidence isn't clear that acute exposure to THC causes significant brain dopamine release. Thus, the answer to a simple, obvious hypothesis, (cannabis causes excessive DA release, which leads to psychosis) is neither simple nor obvious. On the other hand, there is more convincing evidence that individuals with cannabis dependence have a chronically sluggish brain DA system. There is also a reduced ability to synthesize new dopamine in those chronic cannabis users who develop psychotic symptoms, despite having few or no specific changes in their numbers of brain DA receptors compared to those users who do not. So, while long-term cannabis use causes changes in the DA system, there are not necessarily the ones we might have predicted.

The neuroscience of cannabinoid dependence is still in its infancy. At Yale, Deepak Cyril D'Souza [42], used PET scanning techniques similar to those described earlier, this time using a radioactive labeled chemical that binds very tightly to CB_1 receptors, enabling their numbers to be tallied. Daily cannabis users, who had been briefly abstinent, showed a 15% reduction in these receptor numbers, suggesting that chronic exposure to cannabis had decreased (strictly "down-regulated") them, exactly as occurs in animals treated chronically with CB_1 receptor agonists. D'Souza then followed his subjects as outpatients over the next 4 weeks, checking their urine regularly to ensure that they were not relapsing to cannabis use. What he found was that the cannabinoid system "re-booted," with receptor numbers returning to normal levels after about 2 weeks of abstinence. The two-week time period corresponds with what clinicians have observed is the typical time span for cannabis withdrawal symptoms to last. Symptoms generally emerge after a day or two, (unless they are dramatically provoked by giving an antagonist drug such as Rimonabant), and are usually gone in a fortnight.

How do THC, CBD, and other plant cannabinoids interact with the endocannabinoid system?

THC is a lipid and partial agonist at the CB_1 receptor, that binds to this docking site and switches it on, albeit rather weakly. Although its binding to the receptors is less specific than that of endogenous cannabinoids, it hangs around for much longer at the synapse. It persists there because it is not broken down by enzymes such as FAAH or MAGL, but metabolized elsewhere over several hours. In addition THC and some of its metabolites are stored in the bodies fat cells for days to weeks. Therefore its partial agonist effects are much more persistent than those of natural endocannabinoids such as 2-AG, even though those are full agonists. In addition to its actions at CB_1, THC is a weak CB_2 agonist, activates GPR55 and TRPV2 receptors, as well as yet other receptors of greater obscurity including the nuclear receptor, peroxisome proliferator-activated receptor, gamma (PPAR-g) [43]. When THC is given to lab monkeys they literally chill out and drop their body temperature by a degree or two, likely due to its interaction with $TRPV_1$ sites. Similarly when THC is given to rats they develop couch lock, or at least significantly drop their spontaneous activity [44]. Rats also become more pain tolerant after being dosed with THC, as we might expect from cannabinoid interactions with both opioid and $TRPV_1$ systems as well as primary endocannabinoid effects. Preliminary human clinical trials suggest that THC is an effective pain reliever [36].

What are synthetic cannabinoids?

Synthetically designed cannabinoids brewed in the lab, such as the active agents in "Spice" and "K2" are super-agonists. As such, they bind super-tightly to the subtype of cannabinoid receptor (CB_1) found exclusively in the brain. And once these designer drugs have latched onto the receptors, they keep them switched on for hours or days, rather than the milliseconds of activation normally produced by the body's own chemicals such as anandamide. To use the light switch metaphor we've employed earlier, these synthetic drugs not only switch on the light and keep it on, but they replace the bulbs with those you'd ordinarily find in football stadium floodlights. In the 1970's the Pfizer drug company designed an experimental compound named CP55940 with similar super-agonist effects. Most synthetic cannabinoids were synthesized in the 1980s as lab tools, to help scientists study the cannabinoid system by binding tightly to its receptors. Most first-generation synthetic cannabinoids bear prefixes to designate the chemist who first designed or synthesized them, or the company of origin. For example "AM" compounds such as AM-2201 were designed by the cannabinoid researcher Alexandros Makriyannis, whom we met a few pages ago. Similarly, "JWH" compounds are named after perhaps the most well-known synthetic cannabinoid chemist, Clemson University's John W Huffman. In the 1990s, Huffman designed and produced hundreds of designer cannabinoid compounds for research and medicinal purposes, little suspecting that many of his super-agonist compounds would later manifest as toxic street drugs. Instead, he had crafted them as molecular probes of cannabinoid receptors, specialist compounds that would act as tools to unravel the ECS. Subsequently, homebrew chemists seized on Huffman's original scientific publications as recipe books, and found in them the means of making a quick buck by synthesizing and dealing these compounds. The resulting products were first reported as being sold illicitly in the United States in November 2008. Currently these synthetic cannabinoid drugs are mainly synthesized in and imported from China, sprayed onto neutral dried herbs in the United States and sold relatively cheaply in rest stops and gas station convenience stores as "air fresheners" and "potpourri" branded as Spice and K2. Most packets bear the disingenuous disclaimer "not for human consumption" as a kind of legal liability CYA. They are also sold in e-cigarette cartridges as "c-liquid" under brand names including Kronic. Their effects on hapless consumers are unpredictable. The products often contaminated with unknown chemicals, and their health consequences can be dire. After hitting the market, emergency calls to poison-control centers related to these compounds began to skyrocket. In 2019 in New Haven, for example, there were numerous hospitalizations resulting from their use. First responders attended to more than 50 very sick people in a matter of several hours in the central New Haven Green area alone. And the next day, several of the same individuals were back in the same emergency department, either not having learned their lesson the first time around, or being unwilling to abandon a product that they had paid good money for.

Related abused synthetic cannabinoids bind not only to CB_1 receptors but have other unwanted effects on other metabolic systems in the body. Among other varieties of medical mischief, they can produce states of prolonged delirium, psychosis, huge spikes in blood pressure, dangerous increases in heart rate, seizures, coma, and in some cases even death.

There is a cat-and-mouse game with regard to synthesizing and selling these substances. Beginning in 2011, some of the original compounds were banned by the DEA, including a handful of Huffman's progeny. But as soon as a particular synthetic cannabinoid is analyzed, identified in government labs and declared illegal, the manufacturer rapidly changes the recipe to a related, still-legal chemical structure, temporarily avoiding its detection and identification as an illicit substance, until the novel compound too is reclassified as illegal. At which point the rogue chemists merely switch to the next synthetic cannabinoid and so on. Many of the compounds are fairly quick and easy to synthesize, so that the bait and switch or whack-a-mole process is a rapid one. Thus unwary users can never be certain what they have purchased. Huffman, still alive and productive as of this writing, labels synthetic cannabinoid users as "idiots" for putting substances never properly tested in humans into their bodies and brain receptors [45]. So far, over 150 of these ultra-potent compounds have been identified. Synthetic cannabinoids known as "mojo" first appeared in US prisons in 2010. They proved irresistible to some inmates because the substances were powerful and took up much less space than traditional cannabis, thus being easy to conceal. Moreover, they had no odor and were not detected by drug-sniffing dogs. Urine toxicology screens designed to pick up THC came up negative, so that routine testing failed to detect use, and the compounds were hard to identify through analysis due to continual cycling into new synthetic formulas. To the terminally bored, long-term incarcerated individuals these designer cannabinoids were like catnip. Some mojo-using prisoners "started having seizures and aneurysms and some people were freaking out and getting paranoid and scared...... They were going crazy on it. But they loved it" [46]. Such synthetic cannabinoids are so toxic because compared to cannabis and THC, they have hugely greater affinity for CB_1 receptors, greater activity at the receptors since they are full-on versus partial agonists (such as THC), and much longer-lasting in their activity there through their own effects and those of their metabolites. Also, they lack possible offsetting effects of CBD. In addition to ECS activity, these synthetics have multiple, unpredictable targets outside of the CB_1 system. In my opinion, the sale of these drugs would likely dwindle significantly if recreational marijuana were legal and affordable.

As an aside, chemical wizardry is not restricted to cannabinoids but also extends to terpenes, compounds manufactured in the cannabis plant that we will encounter in Chapter 9. The terpene beta-caryophyllene, found in not only in cannabis, but also other plants (such as black pepper) is a natural CB_2 agonist. By tinkering with this drug, chemists created an FAAH-specific COX-2 inhibitor drug that resembles non-steroidal anti-inflammatory compounds like ibuprofen or naproxen, but lacking any of the typical GI side effects of drugs in this class (indigestion, ulcers, GI bleeds). Surprisingly, some of these non-cannabinoid drugs have significant effects on the ECS. For example, acetaminophen, (marketed as Tylenol among other brands) has metabolites that indirectly activate CB_1 receptors and modulate $TRPV_1$ sites [4].

Cannabidiol, the "Cure-All."

CBD is found in both non-intoxicating hemp and THC-containing cannabis plants. CBD was present at the level of a few percent in 1960s-style street cannabis, but this amount is now much lower as a result of selective breeding to drive up the percentage of THC to obtain more intoxicating cannabis. The logic of the biologic pathways in cannabis trichomes is such that breeding plants for higher THC levels necessarily drives down the amount of CBD and vice versa. When it comes to THC's activity at the CB_1 receptor, the combination of modern high-potency THC and low or almost absent CBD is thought to be the equivalent of pressing on the gas pedal while taking the foot off the brake, with resulting consequences for psychosis risk. This hypothesis is discussed further in Chapter 7. CBD has been described as pharmacologically promiscuous because of its wide range of activities and interactions with multiple other neurotransmitter systems [44]. Conventional wisdom holds that CBD is not psychoactive (or at least not intoxicating like THC), and that it opposes THC's effects. All of these claims turn out to be more complicated than they appear [47–49], as we discuss elsewhere.

Is CBD an effective drug, and if so for what conditions? And what about all those other health claims?

CBD is a substance that on the one hand has demonstrated therapeutic potential in controlling pain, reducing symptoms of schizophrenia, and diminishing seizure activity in young people with specific types of intractable epilepsy. We will review the evidence for these therapeutic effects in Chapter 10. CBD is also widely touted as being beneficial to one's mind and body in other ways, but the evidence for claims such as its ability to cure insomnia and significantly reduce anxiety are almost entirely anecdotal at this point, particularly so, given the very low doses contained in most consumer CBD products [50]. This hasn't stopped the compound being included in a multiplicity of products from hummus, sodas, and coffee to lip balms and body lotions. Gwyneth Paltrow's company Goop famously sells a variety of CBD-infused health and beauty products, and the unlikely combination of Snoop Dogg and Martha Stewart hosts an Emmy-nominated cooking show on VH1, "Martha and Snoop's Potluck Dinner Party" that features the substance. As mentioned in Forbes magazine, Canopy Growth, the Canadian cannabis behemoth, appointed Stewart their official company advisor for a new line of CBD products for people and their pets, and helps market Snoop Dogg's 'Leafs by Snoop' branded cannabis products. Canopy will begin marketing official Martha Stewart-branded, hemp-derived CBD products to the US market by the end of the 2020 fiscal year [51]. Chelsea Handler is also about to launch a self-branded line of cannabis products. And Kim Kardashian, ever eager to get in on the act, hosted a CBD-themed baby shower in April 2019. Thanks to all of this publicity, CBD oil and CBD-containing products are seen as "hip" and sales for them are skyrocketing. "Wellness with an edge" is how Eleanor Morgan describes CBD consumer health products in the Guardian [50]. These highly-touted CBD-containing wares have become all the rage, and their soaring sales suggest their reign is just beginning.

Alas, there is little quality control over CBD products. Consequently some contain vanishingly small amounts of the substance, whereas others derived from drug cannabis plants rather than non-THC-containing hemp, are found on analysis to contain significant amounts of THC. Thus unwary consumers of "pure CBD oil" have tested as THC-positive on routine employment drug testing. The UK Center for Medical Cannabis blind-tested 30 purportedly CBD-containing products that they purchased online or in retail stores. Almost half of them had significant detectable THC, making them illegal. Fewer than 40% of the products had CBD levels within 10% of the advertised amount. FDA-approved pharmaceutical CBD (see later) is subject to rigorous purity testing. But when marketed as a food additive or supplement there is no legal requirement for testing. As we discuss elsewhere, the US FDA has recently cracked down on companies for making unwarranted claims regarding the health benefits of CBD. Essentially, the hype has run way ahead of the evidence and there is a serious lack of properly controlled clinical trials for most of the purported health benefits claimed for CBD [52].

This is not to pooh-pooh CBD's genuine therapeutic promise and already-demonstrated efficacy in certain conditions. The substance has shown promising antipsychotic properties, as reviewed in detail in Chapter 8 [53,54]. And as we will see in the Medical Cannabis section, such orally administered CBD is very effective at reducing the frequency of seizures in certain rare, severe childhood epilepsies such as Dravet syndrome [55–60]. In 2018, the CBD drug Epidiolex was approved for treatment of these two neurologic conditions by the US Food and Drug Administration. CBD also boosts endocannabinoid tone indirectly through mechanisms that remain somewhat obscure, and counteracts the tendency of the glutamate neurotransmitter system to become overly excited and kill neurons by overexertion. This and other evidence that CBD has antioxidant and anti-inflammatory properties, suggests that it may be neuro-protective, (i.e., that it prevents damage to compromised neurons that may be affected by trauma or too low a concentration of oxygen). The compound also has probable anti-anxiety properties [53], although as mentioned better-powered clinical trials are needed to nail this down.

Pharmaceutical companies are currently screening allosteric modulators of CB_1 and CB_2 to find novel molecules that replicate desirable medical effects of cannabis without getting patients high. In the long term, they hope that their novel chemicals will supplant botanical cannabis in the medical marketplace.

CBD's mechanism of action

Despite barely binding to cannabinoid receptors, CBD is a weak, low-affinity non-competitive CB_1/CB_2 receptor antagonist [61]. Neuroscientists have speculated that CBD may either have indirect inverse agonist or negative al-losteric modulator properties at the CB_1 receptor [11,62]. The latter property may reduce the ability of CB_1 agonists such as THC to bind to that receptor, providing a simple explanation of how CBD modulates THC activity. How might this molecular binding work? The THC molecule locks directly into a special molecular pouch known as the orthosteric or primary binding site on the CB_1 receptor (the same site where anandamide and 2-AG dock) and switches it on (rather weakly) like a key in a lock, to begin a cellular cascade. Imagine a molecular pinball ricocheting through the neural arcade machine and in the case of CBD switching off a whole series of other neurotransmitters. CBD does not bind to this primary site within the CB_1 receptor complex, but to another part of the CB_1 protein called the allosteric binding site (ABS). When CBD gloms onto the ABS, the entire CB_1 receptor changes shape in subtle ways, that in this case make it harder for CB_1 agonists such as THC to bind there, and temporarily blunts their effect. Basically CBD is a molecular dimmer switch at CB_1 recep-tors [63,64]. Interestingly CBD is also a positive allosteric modulator at opioid receptors [65].

Despite its efficacy, the drug's underlying pharmacological mechanism of action in epilepsy and psychosis is yet unknown. There is no shortage of hy-potheses regarding receptor systems outside of the traditional endocannabinoid universe via which CBD exerts these effects. Scientists have speculated that the drug may be an antagonist of the orphan receptor GPR55, [27,41], an allosteric modulator of the μ- and δ-opioid receptors [65], or a modulator of electrical cur-rents within the $TRPV_1$ receptor. Any of these is plausible, but none is proven.

Cannabidiol and pain

In addition to the terpene beta caryophyllene, discussed in Chapter 9, CBD is another marijuana constituent that may hold promise as a pain reliever [66]. We have already referred to CBD's a complicated chemical activity profile as a CB_1 negative allosteric modulator, with ECS receptor-independent nervous system actions, and significant activity at other GCPR's, including the 5-HT1A, serotonin, TRPV1, and GPR55 receptors [64,67–71]. There is also evidence that CBD is an allosteric modulator of mu opioid receptors. How CBD is bro-ken down in the body and how its levels in the blood change over time are well established [72–75]. When given by mouth, only about 13%–19% of the CBD swallowed is actually absorbed. Most of it passes out of your body un-changed. Think about that the next time you pay $300 for 10 mL of 40% CBD oil. Like THC, CBD is a highly fat-soluble compound. Its elimination half-life is 18–32 hours; in other words if you ingest a dose of the drug, it takes that long to get rid of 50% of the part that's absorbed, from your body. CBD is me-tabolized in liver and gut by enzymes. There is now considerable support from animal studies that CBD has pain relieving properties, and strongly suggestive evidence for the same effect from the smallish number of human clinical trials that have measured CBD effects [76]. This analgesia appears due to action at peripheral CB_1 cannabinoid receptors that sit cheek-by-jowl with specialized, pain-sensing peripheral receptors [77]. Recent information, using convergent evidence from rat models, revealed that repeated low-dose CBD treatment induced pain relief in a neuropathic pain model, predominantly through $TRPV_1$ activation, reduced anxiety through 5-HT1A receptor activation, and rescued impaired 5-HT neurotransmission [78]. Among the phytocannabinoids, CBD is joined by THC as a pain reliever. Both central and peripheral cannabinoid receptors appear to be implicated in THC's effects on muscular and other types of pain [79–88].

It's easy to forget that the cannabis plant contains scads of other cannabinoid compounds, most of which are unexplored, both in terms of potential therapeutic possibilities and biological activity in the ECS and other systems. Preliminary work shows that delta-9 tetrahydrocannabivarin (THCV), another naturally occurring cannabinoid in the marijuana plant, modulates the effects of THC via direct blockade of cannabinoid CB_1 receptors, behaving like a first-generation CB_1 receptor inverse agonist, such as rimonabant, the abandoned ECS-acting appetite suppressant drug. In test tubes, THCV is also a potent, high-affinity CB_1 receptor antagonist. But when given to live animals it mysteriously has very few if any of the behavioral effects typical of CB_1 receptor antagonism. THCV also has high affinity for CB_2 receptors and acts there as a partial agonist, differing from both CBD and rimonabant [62]. Cannabigerol, another plant cannabinoid acts as a weak CB_2 antagonist. The point of discussing this complexity is to illustrate that of the over 100 cannabinoids synthesized by the cannabis plant, the few that biochemists and neuroscientists have examined thus far exert a dizzying complexity of effects inside and outside of the ECS. Many of these compounds have opposite modes of action, so that how the body reacts to them in combination is pretty much unknown. The same holds true for their medicinal and subjective effects. Just as we would never think to draw conclusions regarding the beliefs, opinions, and customs of people in the entire United States by interviewing individuals only in New York and LA, these non-CBD/ non-THC cannabinoid compounds require independent and careful study.

CBD interactions with THC

An up-to-date review of this important topic can be found in a recent publication [44]. Briefly, the popular belief is that CBD initiates brain changes that are opposite to those resulting from THC, thus in some sense protecting it from the harmful effects of THC, for example, on cognitive performance and psychosis-proneness. However this interrelationship is much more complex than was initially portrayed, as I'll explain. For one thing, CBD may change THC levels in the blood before the latter even gets to the brain, by altering THC's distribution in the body, breakdown to simpler chemicals and its excretion [44]. In addition, the order of drug administration may be important. CBD given prior to THC clearly has different effects than when the two compounds are given at exactly the same time. Third, the ratio of the two compounds seems to make a big difference: I will have a lot more to say about that in a little while. Finally, the range of CBD and THC doses that have been explored in human studies is considerably more limited than the range that has been administered to lab animals. Another part of what makes the THC/CBD interrelationship even more complicated, is that beyond a simple tug-of-war, the two compounds have complicated and separate effects of their own that blend and combine unpredictably when they are administered together. Sometimes the two drug's effects on a particular function are in the same direction in one study and opposite in another. For example, both compounds are effective in reducing pain, but probably through different mechanisms. THC effects on pain in mice and rats are amplified by CBD, so here the two effects apparently are additive. But that is the exception to the rule, and even in this case the picture is likely more complicated. Although the ratios of THC and CBD in marijuana flower are relatively constrained, vaporizer cartridges now allow people to self-administer a wide range of ratios of the two compounds, (for example, equal ratios of each compound that don't occur in nature). Different CBD and THC doses may have varying effects, in a so-called "biphasic" pattern. In some experiments low doses of CBD reduce anxiety but higher doses don't. This further complicates the well-known fact that low doses of THC reduce anxiety but higher doses make people increasingly anxious. In other experiments, neither low nor high doses of CBD were effective in reducing the anxiety associated with public speaking, but medium doses were. Nobody has yet explored in detail the question of how different doses of CBD modulate those of different doses of THC in humans to produce particular psychological effects such as anxiety or paranoia.

Along these lines research using relatively limited dose combinations has shown that CBD can modulates particular THC-related effects in humans. For example, pre-treatment with oral CBD reduced cognitive impairment and paranoia produced by subsequent intravenous THC administration. CBD at particular doses also protects against THC's tendency to raise anxiety [89–91]. When administered to cannabis users at 200 mg a day for 10 weeks, CBD improved cognitive and psychological status and (on MRI scans) increased volumes of the hippocampus, part of the brain crucial in memory [92]. On the other hand, THC impairs working memory in rhesus monkeys. When CBD was given alongside the THC doses, it not only failed to help the memory problems but may have worsened them [93], although it did reduce other cognitive problems caused by THC. Acute exposure to THC in humans also causes short-lived impairments in decision-making and abstraction. It is particularly impairing for short-term memory and the ability to learn lists of words, comparable to the kinds of cognitive abnormalities seen in rats and monkeys [44]. CBD acutely administered by itself, does not seem to be associated with cognitive problems and may revoke those provoked by THC. In a human study of recognizing different emotions, orally dosed THC caused impairments that were significantly reduced when oral CBD was administered alongside it [94]. In conclusion, the relatively small amount of human research to date suggests that CBD may offer some protection against some acute learning deficits provoked by THC at certain doses and under particular circumstances. These types of considerations are ultimately important not only for teens using cannabis, but for users of medical marijuana in such conditions as multiple sclerosis, where cannabis may improve muscle spasms and pain but at the potential cost of worsening any underlying cognitive deficits.

Other important interactions of THC and CBD occur in and around psychosis. Outside of its subjective effects such as feeling "'stoned," "chilled out," or merely "altered," cannabis causes some people to feel paranoid, suspicious, conceptually disorganized, and susceptible to unpleasant sensory distortions [44]. These effects can be easily demonstrated on clinical rating scales that measure psychotic symptoms in individuals with schizophrenia and related disorders. CBD does not produce these psychotic-like symptoms, but may have antipsychotic effects, as we will explore in Chapter 10. Epidemiologic surveys have explored how ratios of CBD to THC influence various symptoms, and pharmacologic studies have administered CBD as an antipsychotic to people with schizophrenia. In addition, a handful of human studies have compared the acute effects of THC and CBD interactions in the realm of psychosis. These data are fairly consistent in suggesting that CBD dilutes the tendency of THC to cause psychosis-like symptoms in some people [44]. However this claim needs much more study for several reasons. First, chronic psychosis risk is associated with repeated, chronic exposure to marijuana, and most human lab experiments are necessarily limited to acute dosing. Second, in epidemiologic studies out in the community much of the dosing data relies on participants self-reported cannabis use, which is of variable accuracy. And both the total amounts of cannabinoids consumed as well as THC/CBD ratios are often unknown. Third, despite the increasing popularity of concentrates and vaporizers of various sorts that enable blending of CBD with THC there are remarkably few lab studies of acutely inhaled CBD. Overall, as summarized recently [47], the bottom line of how THC and CBD modulate psychotic symptoms likely depends on the ratio of the two drugs, how they are administered (e.g., orally, by injection or by inhalation), how far apart in time they given (and in what order) and the absolute amounts of each in the doses.

Can CBD get you high?

Recently, experiments carried out in New South Wales Australia, by Nadia Solowij and colleagues [47] highlight the potential value of carefully conducted lab studies in puncturing conventional wisdom regarding effects of cannabinoids. She examined interactions of various doses of THC and CBD alone and in combination administered to subjects by a vaporizer, compared to placebo. The two substances were provided in sessions 1 week apart to

36 occasional and regular cannabis users, most of whom were men. The different dose combinations were administered in apparent random order to each subject. Following acute dosing the volunteers were then observed for 3 hours and their behavior scored by researchers. The subjects also self-rated their level of subjective intoxication on a regular basis. The actual amounts were high-dose CBD (400 mg), high dose THC alone (8 mg), a combination of high dose THC (8 mg) *plus* low dose (4 mg) CBD (that resembled typical cannabis flower), and finally combined, high doses of both THC (8 mg) *plus* CBD (400 mg). The study was generally well-designed, as it was both double-blind (neither the subjects nor the researchers actually administering the drug to subjects were aware of what was in a particular dose) and also placebo-controlled. The experiment's main results are very interesting. Surprise number one was that contrary to the expectation that CBD would have no psychoactive effects, the high CBD dose subjects reported "distinct feelings of depersonalization, derealization and altered internal and external perceptions" compared to placebo, and these were especially marked soon after dosing, and persisted for a couple of hours. In other words, CBD made them stoned. Their ratings of intoxication with high-dose CBD were however much lower at every time point sampled than their highness scores resulting from the high dose of THC. Subjects who received the high dose of CBD also appeared objectively intoxicated to the trained research observers, who were unaware of what volunteers had received at each session. This effect of CBD was only apparent in the infrequent cannabis users. To put numbers on these results, on a 10-point subjective scale of feeling stoned (where 10 is the most buzzed), subjects rated themselves as a steady 1 on placebo, (i.e., close to zero) on placebo (no drug), between 2 and 3 on the high-dose CBD alone, 6 on the THC by itself or THC with low-dose CBD, and between 4.5 and 5 for the combined THC plus high dose CBD.

More intriguingly, given the numbers we just reviewed, when subjects inhaled the THC/low CBD combination, they self-rated their highness level as being ever-so-slightly greater than when using the same amount of THC by itself, rather than lower as might be expected. Blood levels of THC also appeared to be increased by the low dose of CBD, suggesting that the small amount of CBD was boosting, (albeit slightly), rather than diminishing THC's psychoactive effects. At the high dose of CBD, the expected effects of reducing THC's intoxication were observed. The high dose of CBD was also not associated with subjects rating themselves as drowsy, contrary to popular lore that it is mildly sedative and an effective sleep aid. Thus CBD administered at different doses appears to have opposite effects on the high produced by THC. This suggests that medical marijuana patients seeking to offset THC effects (such as intoxication or memory loss) with CBD need to pick their relative doses judiciously, or they may be inadvertently increasing rather than decreasing these side effects. And CBD, at least when inhaled from a vaporizer at relatively high doses, does seem to have intoxicating effects compared to placebo, at least in occasional cannabis users. Another important issue that we are still unsure about is how CBD oil taken orally influences THC effects from smoked marijuana.

Revelation number two: why the ECS turns out to be different than our initial expectations

At the beginning of this chapter, I claimed that recent research has contradicted some of our earlier ideas about how the ECS functioned. Let's revisit that assertion here. From humble beginnings as research curiosities, seemingly restricted to understanding how we get high from marijuana, endocannabinoids have been described as "among the most widespread and versatile signaling molecules ever discovered" [17]. Given its biological roots in CB receptors functioning as a primitive neural coordination system in sea squirts, and with endocannabinoids mysteriously stuffed inside of truffles, it is incredible that the human ECS has evolved into a vastly complicated, ubiquitously distributed coordinating system inside our bodies. It is responsible for so many different kinds of activities in different organs and systems that it is impossible to sum up its many functions in any straightforward way. So, lets begin picking it apart one piece at a time, beginning with the receptor portion of the ECS. As was nicely summarized in a recent comprehensive review, [95] CB_1 receptors have fewer predictable "intrinsic" signaling properties than "typical" GCPRs such as opioid or dopamine receptors; instead their effects largely "emerge" in a time-dependent manner from where the cannabinoid receptors are located. Researchers initially believed that local concentrations of CB_1 sites correlated with the extent of their role in that region, as is typically the case for other neurotransmitter receptors. But, the overall number of CB_1 receptors in a particular location turns out to be an unreliable guide to their local importance. For example, in the hypothalamus receptor numbers are low yet their function is critical, as illustrated by important behavioral consequences and unexpectedly high levels of cannabinoid-dependent signaling. Brain geography also determines what function endocannabinoids subserve; as we saw in the hippocampus, there are opposite consequences of their activity in adjacent territories. Cannabinoid receptors located on different types of brain cells not only interact differently with other neurotransmitter chemicals, but also evince multiple behavioral consequences. The more we learn about the ECS, the more it sounds like it makes up the rules as it goes along depending on the location and the biological task at hand. There are other surprises. Unexpectedly, CB_1 receptors can buddy up with other brain neurotransmitter receptors. When they do so, stimulating these joint receptor couples elicits a signal very different than that derived from either receptor alone.

Our original, simple-minded concept of the ECS was that it acted as a type of modulating "off," or "circuit breaker" switch at neural synapses. This view turns out to be overly simplistic. Typically, when we give a large dose of a particular drug, we will see more of the same effect that was produced by a smaller dose. But, this is not true for behavioral responses resulting from either stimulating CB_1 receptors, or dosing people with exogenous cannabinoids such as THC. In both cases there can be biphasic, opposite effects on anxiety, novelty seeking, fear responses, and food intake. There's a fair amount of evidence that the cannabinoid system interacts mainly with brain excitation systems at one dose range and with inhibition systems at another, in ways that are consistent for a particular behavior but differ from one behavior to another. Finally, numerous endogenous molecules similar in structure to endocannabinoids are floating about the brain; although they seem to be both biologically active and an integral part of the ECS, their function is essentially unknown. All of these molecular features are uncharacteristic of classic neurotransmitter systems. So far we have a very limited understanding of their purpose, necessity, or biological advantages. These many unexpected findings challenge many dogmatic views in neuroscience in a broad manner that extends well beyond the ECS, and serves to remind us that much work lies ahead before we can fully comprehend this complex, multi-functional system.

How are psychopharmacologist and pharmaceutical companies trying to build new endocannabinoid-related drugs?

A psychopharmacologist friend of mine likes to say that all drugs are poisons, whose properties humans can harness to work for or against us. An extrapolation of that line of thinking is that when we discover how a drug works we can design compounds that achieve a similar end result, but faster, bigger, and safer with fewer side effects. Alternatively, we can craft compounds with converse actions. For example, if THC gives us the munchies, a drug with opposite effects can theoretically be used as an appetite suppressant. Similarly a CB_1 antagonist drug could be used to treat individuals acutely who have over-indulged in edibles and have become acutely anxious or psychotic.

There is a dynamic tension here. On the one hand, numerous medicines in common use today (such as digitalis for heart failure or morphine for pain) originated as plant extracts, and began with crude plant preparations such as foxglove leaves and opium poppy resin. But in every case, the ultimate path has always been from the plant to a specific, biologically active molecule such as digoxin or morphine that can be purified, measured, and prescribed in precise doses. From a pharmaceutical company's point of view, the problem with plants is that they were not designed by pharmacologists. For every desirable compound that they contain, the plant is often synthesizing a related compound with an effect opposite to the one they are looking for, requiring complex and perhaps expensive separation. And besides, natural products are hard to patent. So from the business standpoint, it's ultimately more straightforward and profitable to custom-design a compound that resembles something originally found in the botanical world but is more powerful weight-for-weight, has a more specific effect on the target, and manifests fewer undesirable side effects. If you can imagine a face-off between pharmaceutical company executives and medical marijuana dispensary owners, the executives will object that cannabis contains multiple ingredients—literally many hundreds of cannabinoids, flavonoids, and terpenes that vary in their relative proportions from not only from chemical variety (chemovar) to chemovar, but even within a particular chemovar from batch to batch. Not to mention significant variation from the top of the individual bud to its bottom. In contrast (they would say), that if you go to your local drugstore and purchase a Tylenol, it has a single active constituent and one tablet always contains precisely the same number of milligrams of that ingredient. Besides that, no one has performed the proper large-scale, randomized, double-blind placebo-controlled studies to show that your favorite medical cannabis chemovar is effective for your illness. The federal government probably won't ever allow us to license it, they will continue. We can't patent it. And it has major, undesirable side effects—it makes people intoxicated and alters their behavior. From our point of view, (they would continue) getting high is a bug not a feature. Given all of that, what we need to do is to invest millions of dollars to create single-molecule drugs that act on the cannabinoid system in precise and subtle ways, not by not mimicking natural cannabinoids but by tweaking the enzymes that synthesize or break them down.

On the contrary! dispensary owners would exclaim in protest. Cannabis' multiple ingredients are a help not a hindrance, because they synergize in useful entourage effects that are medically useful in ways above and beyond the utility of individual plant constituents. And of course there are very few large-scale clinical trials; cannabis is illegal at the federal level, and cannabinoids are federal Schedule 1 drugs, making it almost impossible to conduct clinical lab studies using plant strains typically consumed in the real world. Consequently, the best evidence we can show is necessarily limited to anecdotal reports. And what we are selling in our dispensaries is always going to be far less expensive than anything the pharmaceutical industry can create in your giant factories. Unlike you, we have no need to recoup the millions of dollars spent in drug development.

So while budtenders ply their trade, drug company pharmacologists have plowed ahead exploring compounds that act at cannabinoid receptors and modify the action of the relevant enzymes, that is, MAGL, FAAH, DAGL, and NAPE-PLD. They have also explored compounds that don't cross the blood-brain barrier and act only peripherally, without any behavioral effects. (Remember that to them, getting stoned is a bug not a feature). Positive allosteric modulators of CB_1 receptors have been suggested as potentially effective in treating post traumatic stress disorder [96]. Separating medical benefits from psychotropic effects can be difficult however. Pharmacologists have tried for years with opium poppy-derived compounds and their synthetic analogues to create pain relief without addictive potential, but in the long journey from opium to morphine to heroin to oxycodone to fentanyl, better pain relief has always proceeded hand-in-hand with greater abuse liability and increased lethality. So far, progress with the pharmaceutical industry's endocannabinoid drugs has been mostly discouraging [1], but nevertheless remains full of promise [4,11]. In a complex and widespread system such as the ECS, progress is potentially booby-trapped with unexpected drug effects on multiple off-target systems. Let's review a few examples.

One of the first such prescription compounds to reach the market was the pharmaceutical company Sanofi Aventis' CB_1 inverse agonist drug rimonabant. Marketed under the trade names of Accomplia and Zimulti, the drug was approved for human use in Europe in 2006. The idea was that the medication would have the opposite effect at the receptor than agonists or partial agonist such as THC, and thus provoke the "anti-munchies," resulting in reduced appetite, weight loss, and improvement in the metabolic syndrome (obesity/hypertension/type II diabetes/high blood fats). From that perspective, the drug was effective, leading to significant gradual weight loss compared to placebo. Unfortunately, it had to be withdrawn due to severe side effects of anxiety, depression, and suicidal thoughts that occurred in a small proportion of patients who took it. The risk of notable psychiatric disorders in people taking the drug was about double that expected to occur in the general population. For that reason, the drug had to be removed from the market worldwide in 2008 before it could be approved in the United States, as the risks were judged to outweigh the benefits. In retrospect, this side effect profile is utterly predictable. If for most people THC provokes effects of mild euphoria, calmness, increased appetite, sleepiness, improvement in nausea, and calming of the GI tract, then the most commonly reported side effects of rimonabant are low mood, anxiety and crankiness, decreased appetite, insomnia, nausea/vomiting, and diarrhea.

In the beginning of 2016, a stage I clinical trial in France of a drug designed to inhibit FAAH also went disastrously wrong. The drug, BIA 10-2474, was designed by a Portuguese pharmaceutical company to treat various brain-based disorders. When trial doses were being given to humans for the first time in order to test the compound's safety, one of the healthy volunteers being given the medication sank into a coma and was ultimately pronounced brain dead. Another five participants needed to be hospitalized, two of them with severe neurological damage, that fortunately reversed within a few days. Because of these severe adverse events the drug trial was immediately stopped. It's currently thought most likely that the drug's effects outside of the ECS (what drug companies call "off-target effects") were most likely responsible for the damage, by inhibiting multiple fat-metabolizing enzymes essential for brain function.

Currently, Pfizer is examining the experimental FAAH inhibitor compound PF 04457845, one of a class of drugs expected to boost brain anandamide levels. The medication was designed to ameliorate symptoms of cannabis withdrawal in individuals with cannabis use disorder, that is, to lessen the usual withdrawal symptoms of insomnia, irritability, anxiety, gastrointestinal distress. Based on the results of a preliminary clinical trial run at Yale by Deepak Cyril D'Souza's group, this compound seems to reduce those symptoms. It produces about 97% inhibition of the FAAH enzyme's activity for 2 weeks. It is not psychoactive or subjectively rewarding and produced no withdrawal, tolerance or dependence. So, stay tuned for more news on this particular drug. Pharmaceutical companies are actively exploring ECS-based drugs to treat a wide range of disorders. Some of the more prominent areas of investigation include chronic pain control, slowing cardiovascular inflammation, and addressing cardio-metabolic disorders [17]. The future hope is that ECS-active drugs will also be used for effective treatment of psychiatric disorders including schizophrenia, anxiety and substance use disorders including PTSD, as well as for epilepsy, irritable bowel syndrome, and various cancers. To reach these goals, a number of potential hurdles remain, among them designing a series of ECS-active compounds that do not reach the brain, avoid off-target effects, (a particularly difficult demand in such a widespread signaling system), and acquiring a more thorough understanding of how plant cannabinoids exert their effects inside the body.

In summary, this chapter has summarized the discovery of the ECS, our realization of its growing medical and behavioral importance, and our changing view of its many functions. This whole area of cannabinoid-related research is of intense interest not only to scientists of various stripes, but also to pharmaceutical companies, who glimpse the possibilities of designing novel drugs to manipulate this widespread, little-known but vital system in order to help treat a wide range of illnesses.

References

[1] Chanda D, Neumann D, Glatz JFC. The endocannabinoid system: overview of an emerging multi-faceted therapeutic target. Prostaglandins Leukot Essent Fatty Acids 2019;140:51–6.

[2] Scudellari M. Your body is teeming with weed receptors, The Scientist. 2017. Available from: https://www.the-scientist.com/features/your-body-is-teeming-with-weed-receptors-31233.

[3] Nicolussi S, Gertsch J. Endocannabinoid transport revisited. Vitam Horm 2015;98:441–85.

[4] Scherma M, et al. Brain activity of anandamide: a rewarding bliss? Acta Pharmacol Sin 2019;40(3):309–23.

[5] Mechoulam R. The most famous cannabis scientist that you've probably never heard of. Available from: https://gbsciences.com/2018/03/01/raphael-mechoulam-cannabis-scientist/.

[6] Wing N. Godfather Of marijuana research says he's never even tried the stuff, Huffington Post. 2017. Available from: https://www.huffpost.com/entry/raphael-mechoulam-marijuana-research_n_58de7ac1e4b0c777f786e94a.

[7] Lighter side: This little piggy ate pot, Pork Business 2016. Available from: https://www.pork-business.com/article/lighter-side-little-piggy-ate-pot.

[8] Elphick MR, Satou Y, Satoh N. The invertebrate ancestry of endocannabinoid signalling: an orthologue of vertebrate cannabinoid receptors in the urochordate Ciona intestinalis. Gene 2003;302(1–2):95–101.

[9] Joint Genome Institute. Available from: https://jgi.doe.gov/.

[10] Pacioni G, et al. Truffles contain endocannabinoid metabolic enzymes and anandamide. Phytochemistry 2015;110:104–10.

[11] Di Marzo V. New approaches and challenges to targeting the endocannabinoid system. Nat Rev Drug Discov 2018;17(9):623–39.

[12] Hua T, et al. Crystal structures of agonist-bound human cannabinoid receptor CB1. Nature 2017;547(7664):468–71.

[13] Hua T, et al. Crystal structure of the human cannabinoid receptor CB1. Cell 2016;167(3):750–62. e14.

[14] Mallipeddi S, Zvonok N, Makriyannis A. Expression, purification and characterization of the human cannabinoid 1 receptor. Sci Rep 2018;8(1):2935.

[15] Li X, et al. Crystal structure of the human cannabinoid receptor CB2. Cell 2019;176(3):459–67. e13.

[16] Cohen K, Weizman A, Weinstein A. Modulatory effects of cannabinoids on brain neurotransmission. Eur J Neurosci 2019;50(3):2322–45.

[17] Maccarone M, et al. Endocannabinoid signaling at the periphery: 50 years after THC. Trends Pharmacol Sci 2015;36(5):277–96.

[18] Xiang W, et al. Monoacylglycerol lipase regulates cannabinoid receptor 2-dependent macrophage activation and cancer progression. Nat Commun 2018;9(1):2574.

[19] Zhou L, et al. Targeted inhibition of the type 2 cannabinoid receptor is a novel approach to reduce renal fibrosis. Kidney Int 2018;94(4):756–72.

[20] Mechoulam R, Parker L. Towards a better cannabis drug. Br J Pharmacol 2013;170(7):1363–4.

[21] Malone DT, Hill MN, Rubino T. Adolescent cannabis use and psychosis: epidemiology and neurodevelopmental models. Br J Pharmacol 2010;160(3):511–22.

[22] Basavarajappa BS, Nixon RA, Arancio O. Endocannabinoid system: emerging role from neurodevelopment to neurodegeneration. Mini Rev Med Chem 2009;9(4):448–62.

[23] Parsons LH, Hurd YL. Endocannabinoid signalling in reward and addiction. Nat Rev Neurosci 2015;16(10):579–94.

[24] Wenzel JM, Cheer JF. Endocannabinoid regulation of reward and reinforcement through interaction with dopamine and endogenous opioid signaling. Neuropsychopharmacology 2018;43(1):103–15.

[25] Varvel SA, Lichtman AH. Evaluation of CB1 receptor knockout mice in the Morris water maze. J Pharmacol Exp Ther 2002;301(3):915–24.

[26] Niyuhire F, et al. The disruptive effects of the CB1 receptor antagonist rimonabant on extinction learning in mice are task-specific. Psychopharmacology (Berl) 2007;191(2):223–31.

[27] Wang W, et al. A primary cortical input to hippocampus expresses a pathway-specific and endocannabinoid-dependent form of long-term potentiation. eNeuro 2016;3(4).

[28] Haney M, Hill MN. Cannabis and cannabinoids: from synapse to society. Neuropsychopharmacology 2018;43(1):1–3.

[29] Morena M, Hill MN. Buzzkill: the consequences of depleting anandamide in the hippocampus. Neuropsychopharmacology 2019;44(8):1347–8.

[30] Zimmermann T, et al. Impaired anandamide/palmitoylethanolamide signaling in hippocampal glutamatergic neurons alters synaptic plasticity, learning, and emotional responses. Neuropsychopharmacology 2019;44(8):1377–88.

[31] Hillard CJ. Circulating endocannabinoids: from whence do they come and where are they going? Neuropsychopharmacology 2018;43(1):155–72.

[32] Wang H, Dey SK, Maccarone M. Jekyll and Hyde: two faces of cannabinoid signaling in male and female fertility. Endocr Rev 2006;27(5):427–48.

[33] Mechoulam R, Parker LA. The endocannabinoid system and the brain. Annu Rev Psychol 2013;64:21–47.

[34] Murphy H. At 71, she's never felt pain or anxiety. Now scientists know why, NY Times. 2019. Available from: https://www.nytimes.com/2019/03/28/health/woman-pain-anxiety.html.

[35] Habib AM, et al. Microdeletion in a FAAH pseudogene identified in a patient with high anandamide concentrations and pain insensitivity. Br J Anaesth 2019;123(2):e249–53.

[36] Li X, et al. The effectiveness of self-directed medical cannabis treatment for pain. Complement Ther Med 2019;46:123–30.

[37] Liu CW, et al. Weeding out the problem: the impact of preoperative cannabinoid use on pain in the perioperative period. Anesth Analg 2019;129(3):874–81.

[38] Costa B, et al. Vanilloid TRPV1 receptor mediates the antihyperalgesic effect of the nonpsychoactive cannabinoid, cannabidiol, in a rat model of acute inflammation. Br J Pharmacol 2004;143(2):247–50.

[39] Lowin T, Straub RH. Cannabinoid-based drugs targeting CB1 and TRPV1, the sympathetic nervous system, and arthritis. Arthritis Res Ther 2015;17:226.

[40] Acharya N, et al. Endocannabinoid system acts as a regulator of immune homeostasis in the gut. Proc Natl Acad Sci USA 2017;114(19):5005–10.

[41] Ryberg E, et al. The orphan receptor GPR55 is a novel cannabinoid receptor. Br J Pharmacol 2007;152(7):1092–101.

[42] D'Souza DC, et al. Rapid changes in CB1 receptor availability in cannabis dependent males after abstinence from cannabis. Biol Psychiatry Cogn Neurosci Neuroimaging 2016;1(1):60–7.

[43] Mechoulam R, et al. Early phytocannabinoid chemistry to endocannabinoids and beyond. Nat Rev Neurosci 2014;15(11):757–64.

[44] Boggs DL, et al. Clinical and preclinical evidence for functional interactions of cannabidiol and delta(9)-tetrahydrocannabinol. Neuropsychopharmacology 2018;43(1):142–54.

[45] McCoy T. How this chemist unwittingly helped spawn the synthetic drug industry, Washington Post. 2015. Available from: https://www.washingtonpost.com/local/social-issues/how-a-chemist-unwittingly-helped-spawn-the-synthetic-drug-epidemic/2015/08/09/94454824-3633-11e5-9739-170df8af8eb9_story.html.

[46] Davies R. From pecan pralines to 'dots' as currency: how the prison economy works, The Guardian. 2019. Available from: https://www.theguardian.com/us-news/2019/aug/30/prison-economy-informal-markets-alternative-currencies.

[47] Solowij N, et al. A randomised controlled trial of vaporised Delta(9)-tetrahydrocannabinol and cannabidiol alone and in combination in frequent and infrequent cannabis users: acute intoxication effects. Eur Arch Psychiatry Clin Neurosci 2019;269(1):17–35.

[48] Bhattacharyya S, et al. Opposite effects of delta-9-tetrahydrocannabinol and cannabidiol on human brain function and psychopathology. Neuropsychopharmacology 2010;35(3):764–74.

[49] Leweke FM, et al. Cannabidiol enhances anandamide signaling and alleviates psychotic symptoms of schizophrenia. Transl Psychiatry 2012;2:pe94.

[50] Morgan E. CBD lubricant is a bestseller: cannabis oil products are booming – but does the science stack up? The Guardian. 2019. Available from: https://www.theguardian.com/lifeand-style/2019/sep/14/cbd-lubricant-bestseller-cannabis-oil-products-booming.

[51] Shapiro K. Martha Stewart on how Snoop Dogg got her into the cannabis business, her new CBD line, and aging well, 2019. Forbes. Available from: https://www.forbes.com/sites/katie-shapiro/2019/06/24/martha-stewart-on-how-snoop-dogg-got-her-into-the-cannabis-business-her-new-cbd-line-and-aging-well/.

[52] Schraer R. CBD oil: Have the benefits been overstated? BBC News. 2019. Available from: https://www.bbc.com/news/health-48950483.

[53] Lewis MA, Russo EB, Smith KM. Pharmacological foundations of cannabis chemovars. Planta Med 2018;84(4):225–33.

[54] Ranganathan M, et al. Highs and lows of cannabinoid-dopamine interactions: effects of genetic variability and pharmacological modulation of catechol-O-methyl transferase on the acute response to delta-9-tetrahydrocannabinol in humans. Psychopharmacology (Berl) 2019;236(11):3209–19.

[55] Billakota S, Devinsky O, Marsh E. Cannabinoid therapy in epilepsy. Curr Opin Neurol 2019;32(2):220–6.

[56] Thiele E, et al. Cannabidiol in patients with Lennox-Gastaut syndrome: interim analysis of an open-label extension study. Epilepsia 2019;60(3):419–28.

[57] Devinsky O, et al. Long-term cannabidiol treatment in patients with Dravet syndrome: an open-label extension trial. Epilepsia 2019;60(2):294–302.

[58] Szaflarski JP, et al. Long-term safety and treatment effects of cannabidiol in children and adults with treatment-resistant epilepsies: expanded access program results. Epilepsia 2018;59(8):1540–8.

[59] Devinsky O, et al. Randomized, dose-ranging safety trial of cannabidiol in Dravet syndrome. Neurology 2018;90(14):pe1204–e1211.

[60] Devinsky O, et al. Open-label use of highly purified CBD (Epidiolex(R)) in patients with CDKL5 deficiency disorder and Aicardi, Dup15q, and Doose syndromes. Epilepsy Behav 2018;86:131–7.

[61] Pertwee RG. The diverse CB1 and CB2 receptor pharmacology of three plant cannabinoids: delta9-tetrahydrocannabinol, cannabidiol and delta9-tetrahydrocannabivarin. Br J Pharmacol 2008;153(2):199–215.

[62] McPartland JM, et al. Are cannabidiol and delta(9) -tetrahydrocannabivarin negative modulators of the endocannabinoid system? A systematic review. Br J Pharmacol 2015;172(3):737–53.

[63] Marcu J. Is CBD really non-psychoactive? Project CBD. 2016. Available from: https://www.projectcbd.org/science/cbd-really-non-psychoactive.

[64] Laprairie RB, et al. Cannabidiol is a negative allosteric modulator of the cannabinoid CB1 receptor. Br J Pharmacol 2015;172(20):4790–805.

[65] Kathmann M, et al. Cannabidiol is an allosteric modulator at mu- and delta-opioid receptors. Naunyn Schmiedebergs Arch Pharmacol 2006;372(5):354–61.

[66] Cunetti L, et al. Chronic pain treatment with cannabidiol in kidney transplant patients in Uruguay. Transplant Proc 2018;50(2):461–4.

[67] Abood ME. Allosteric modulators: a side door. J Med Chem 2016;59(1):42–3.

[68] Ignatowska-Jankowska BM, et al. A Cannabinoid CB1 receptor-positive allosteric modulator reduces neuropathic pain in the mouse with no psychoactive effects. Neuropsychopharmacology 2015;40(13):2948–59.

[69] Khajehali E, et al. Biased agonism and biased allosteric modulation at the CB1 cannabinoid receptor. Mol Pharmacol 2015;88(2):368–79.

[70] Kulkarni PM, et al. Novel electrophilic and photoaffinity covalent probes for mapping the cannabinoid 1 receptor allosteric site(s). J Med Chem 2016;59(1):44–60.

[71] Straiker A, et al. Aiming for allosterism: evaluation of allosteric modulators of CB1 in a neuronal model. Pharmacol Res 2015;99:370–6.

[72] FDA. Epidiolex: Full Prescribing Information. 2018, Greenwich Biosciences, Inc Carlsbad, CA, USA.

[73] Scuderi C, et al. Cannabidiol in medicine: a review of its therapeutic potential in CNS disorders. Phytother Res 2009;23(5):597–602.

[74] Mechoulam R. Discovery of endocannabinoids and some random thoughts on their possible roles in neuroprotection and aggression. Prostaglandins Leukot Essent Fatty Acids 2002;66(2–3):93–9.

[75] Devinsky O, et al. Cannabidiol: pharmacology and potential therapeutic role in epilepsy and other neuropsychiatric disorders. Epilepsia 2014;55(6):791–802.

[76] Bagues A, Martin MI, Sanchez-Robles EM. Involvement of central and peripheral cannabinoid receptors on antinociceptive effect of tetrahydrocannabinol in muscle pain. Eur J Pharmacol 2014;745:69–75.

[77] Agarwal N, et al. Cannabinoids mediate analgesia largely via peripheral type 1 cannabinoid receptors in nociceptors. Nat Neurosci 2007;10(7):870–9.

[78] De Gregorio D, et al. Cannabidiol modulates serotonergic transmission and reverses both allodynia and anxiety-like behavior in a model of neuropathic pain. Pain 2019;160(1):136–50.

[79] Aviram J, Samuelly-Leichtag G. Efficacy of cannabis-based medicines for pain management: a systematic review and meta-analysis of randomized controlled trials. Pain Physician 2017;20(6):E755–96.

[80] Hill KP, Palastro MD. Medical cannabis for the treatment of chronic pain and other disorders: misconceptions and facts. Pol Arch Intern Med 2017;127(11):785–9.

[81] Russo EB, Marcu J. Cannabis pharmacology: the usual suspects and a few promising leads. Adv Pharmacol 2017;80:67–134.

[82] Genaro K, et al. Cannabidiol is a potential therapeutic for the affective-motivational dimension of incision pain in rats. Front Pharmacol 2017;8:391.

[83] Neelakantan H, et al. Distinct interactions of cannabidiol and morphine in three nociceptive behavioral models in mice. Behav Pharmacol 2015;26(3):304–14.

[84] Goncalves TC, et al. Cannabidiol and endogenous opioid peptide-mediated mechanisms modulate antinociception induced by transcutaneous electrostimulation of the peripheral nervous system. J Neurol Sci 2014;347(1–2):82–9.

[85] Serpell M, et al. A double-blind, randomized, placebo-controlled, parallel group study of THC/CBD spray in peripheral neuropathic pain treatment. Eur J Pain 2014;18(7):999–1012.

[86] Ward SJ, et al. Cannabidiol inhibits paclitaxel-induced neuropathic pain through 5-HT(1A) receptors without diminishing nervous system function or chemotherapy efficacy. Br J Pharmacol 2014;171(3):636–45.

[87] Ward SJ, et al. Cannabidiol prevents the development of cold and mechanical allodynia in paclitaxel-treated female C57Bl6 mice. Anesth Analg 2011;113(4):947–50.

[88] Russo EB, Guy GW, Robson PJ. Cannabis, pain, and sleep: lessons from therapeutic clinical trials of Sativex, a cannabis-based medicine. Chem Biodivers 2007;4(8):1729–43.

[89] Morgan CJ, et al. Impact of cannabidiol on the acute memory and psychotomimetic effects of smoked cannabis: naturalistic study: naturalistic study [corrected]. Br J Psychiatry 2010;197(4):285–90.

[90] Morgan CJ, Curran HV. Effects of cannabidiol on schizophrenia-like symptoms in people who use cannabis. Br J Psychiatry 2008;192(4):306–7.

[91] Schubart CD, et al. Cannabis with high cannabidiol content is associated with fewer psychotic experiences. Schizophr Res 2011;130(1–3):216–21.

[92] Solowij N, et al. Therapeutic effects of prolonged cannabidiol treatment on psychological symptoms and cognitive function in regular cannabis users: a pragmatic open-label clinical trial. Cannabis Cannabinoid Res 2018;3(1):21–34.

[93] Wright MJ Jr, Vandewater SA, Taffe MA. Cannabidiol attenuates deficits of visuospatial associative memory induced by delta(9) tetrahydrocannabinol. Br J Pharmacol 2013;170(7):1365–73.

[94] Hindocha C, et al. Acute effects of delta-9-tetrahydrocannabinol, cannabidiol and their combination on facial emotion recognition: a randomised, double-blind, placebo-controlled study in cannabis users. Eur Neuropsychopharmacol 2015;25(3):325–34.

[95] Busquets-Garcia A, Bains J, Marsicano G. CB1 receptor signaling in the brain: extracting specificity from ubiquity. Neuropsychopharmacology 2018;43(1):4–20.

[96] Shekhar A, Thakur GA. Cannabinoid receptor 1 positive allosteric modulators for posttraumatic stress disorder. Neuropsychopharmacology 2018;43(1):226–7.

Chapter 6

Psychology + human behavior

"And I saw anew many of nature's panoramas & landscapes that I'd stared at blind-ly without even noticing before; through the use of marijuana, awe & detail were made conscious. These perceptions are permanent—any deep aesthetic experience leaves a trace, and an idea of what to look for that can be checked back later"

Allen Ginsberg, *The Great Marijuana Hoax*. The Atlantic Magazine [1].

"The brain on marijuana will never deviate from its destined disposition, nor be driven to madness. Marijuana is a mirror reflecting man's deepest thoughts, a magnifying mirror. It's true, but only ever a mirror"

Charles Baudelaire, Les Paradis Artificiels 1860 [2].

This chapter deals with some essential questions—what's it like to be stoned? Why do different individuals have different experiences when they use the drug? What are some of the reasons that people use cannabis? Does cannabis enhance creativity? How does cannabis affect sexuality? What's the story on cannabis' effects on memory, cognition, and driving ability? How are these experiences related to dose and method of administration? How does what we learned about the endocannabinoid system and brain function help explain some of these experiences? Related issues, including cannabis use and risk for psychosis, the drug's long-term effects on IQ and the modification of THC-related effects by terpenes are dealt with in other chapters, specifically Toxicology (Chapter 8), Chemical Analysis (Chapter 9), and Epidemiology (Chapter 7).

An introduction to intoxication

Let's begin with a big, broad, general question. Before we even focus on can-nabis, it's important to ask why human beings choose to tinker voluntarily with substances that alter their mental states for recreational purposes. These discussions of cannabis raise much wider issues that apply to the myriad of recreational drugs affecting mood and consciousness from legal everyday examples, such as caffeine, tobacco, and alcohol; through the illegal, including ecstasy, marijuana, opioids, and psychedelics. But alcohol Prohibition in the United States illustrates that a drug that is legal one day can become illegal the next. Many of these issues are addressed at length in Richard Davenport-Hines' masterful and comprehensive book *The Pursuit of Oblivion* [3].

Weed Science. http://dx.doi.org/10.1016/B978-0-12-818174-4.00006-9

The tendency to intoxicate ourselves for pleasure is far from unique to humans and bound up with the desire for amusement and diversion. Anthropomorphism be damned. If you don't believe that crows like to entertain themselves, then Google the video of a Scandinavian crow sliding down a roof repeatedly on an aluminum plate. Rats divert themselves by playing hide-and-seek [4]. Also, for such spontaneous amusement, many creatures incorporate chemical intoxicants. Elephants seek out and ingest fermented fruit to get drunk, and similarly many birds consume fermented berries that render them so squiffy that they fly into windows or stagger around. Despite the many liquor brands named after their kin, they don't actually drink Old Crow or Wild Turkey. As anyone who has smoked really poor quality cannabis flower can attest, inadvertently included cannabis seeds contain no THC and serve no apparent purpose for the would-be stoner other than to make interesting popping sounds when heated in a "joint." Hemp seeds are included in many store-bought wild bird food mixes, but because of the near-zero available THC content, no birds or animals (such as the bears who raid backyard bird food containers) are known to become intoxicated by consuming the seeds or indeed marijuana plants themselves in the wild. Crows, lemurs, and sundry other beasts share the general desire to alter their states of consciousness. We already encountered the opium poppy-intoxicated parrots in Chapter 3. Crows rub ants that secrete psychedelic substances into their feathers, while in Madagascar lemurs bite and sniff millipedes that contain toxic mind-bending chemicals. For both bird and mammal, alongside the fact that these chemicals often usefully repel pests, getting buzzed is the goal. And in case you're wondering, given the British expression "stone the crows," it is possible for crows to be intoxicated by cannabis. This experiment was carried out by Sir William Brooke O'Shaughnessy in 1843 [5], who fed cannabis preparations to crows and observed their subsequent insobriety.

Like the birds and beasts, but provided with something as inherently fragile, uniquely beautiful, and highly-evolved as human consciousness, we mortals share a highly developed urge to put the boot in once in a while and to wallop our minds with one of a variety of chemical sledgehammers. Given the multiplicity of neurotransmitter systems that underpin the myriad complex functions of our humanoid brains, there are hundreds of ways to effect such consciousness change. These approaches range from the subtle and feather-light (microdosing) through the full-on cranial battering ram all the way to brain-disrupting chemicals with the relative wallop of a limited-yield thermonuclear device.

Many of these relevant enabling substances occur in the natural world, the majority within plants and fungi, with a few in more recondite locations such as amphibian secretions. But humans bring two unique and novel aspects to chemical consciousness alteration. The first is a conscious, intentional ability to tinker with these natural substances to boost their effectiveness and to get more pharmaceutical bang for our buck. Given a combination of human ingenuity and a strong push to escape temporarily from our nasty, short, and brutish lives, we've both extracted the key molecules from what the Earth provides and more recently have synthesized novel molecular structures that improve on or multiply the effects of whatever Mother Nature created originally. This human chemical ingenuity is a distinct step up from the humble crow, who naturally feels less pressure to escape from the drudgery of 9–5 jobs or mortgage applications. The second uniquely human aspect is that cultural blessing or its opposite, often alongside legal penalties, attaches to use of these mind-altering substances. Historically, different societies have made the choice to celebrate or to ban alcohol consumption, to glorify or criminalize cannabis use, and to either build religions around or to lock people up for use of hallucinogenic cacti. I would like to focus on just one of these many substances, derived from the familiar cannabis plant. Although of course we will discuss synthetic cannabinoids, these exist mostly on the thermonuclear end of the drug spectrum.

What are motivations for peoples' use of cannabis?

As with all recreational drugs, motivations are numerous and varied: different people choose to use cannabis for widely disparate purposes and intentions [6]. As Randall Jarrell reminds us, the same water can run a prayer-wheel or a turbine. As a parallel with alcohol, a Catholic priest using a tiny amount of communion wine as part of the sacrament, a first year college student getting wasted on vodka as part of a fraternity initiation and a skid row alcoholic with terminal liver disease and DTs are as different from one another as a sadhu using cannabis to contemplate the divine, a wake-and-bake stoner grabbing his first bong hit of the day, and a 75-year-old grandmother delicately nibbling an edible to ease her arthritis pain. Listed further are some typical motivations, a list that is not intended to be exhaustive and with my acknowledgment that reasons for cannabis use can vary from one session to another within the same person.

Purposeful controlled recreational intoxication

An example here is provided by Devendra in Chapter 3. Individuals employ cannabis in this manner to get pleasantly stoned, experience mild euphoria, heighten perception, tweak their creativity, better appreciate music or lovemaking, see the pretty colors and so on. Consistent with this, a very small scale survey of cannabis use in Canada [7] reported that many subjects use marijuana recreationally "to enhance relaxation and concentration while engaged in leisure activities."

Recreational use taken to an extreme

Examples include getting trashed as a rite of passage, for example, during spring break in the United States. Employing cannabis in this manner offers chance to become intoxicated, shed inhibitions, behave outrageously, celebrate, and hook up. Entangled with these motivations are the urge to rebel, equally applicable to teen tobacco smoking and binge drinking alcohol. Also bound up with this drive can be a desire to push things to the limit and beyond as part of a personal quest or dangerous journey, in other words to prove oneself and return safely (hopefully) from the other side. But the uncertainty in that word "hopefully" is part of the thrill for some people.

Specific use to heighten perception to explore one's internal world and to increase creativity

Vikram in Chapter 3 represents one example of this motivation. Reaching higher states of consciousness, achieving transcendence and/or attaining religious insight probably represents one of the early uses of cannabis as we explored in Chapter 4. Although THC, some other cannabinoids and probably some terpenes undoubtedly possess psychoactive properties, cannabis occupies a kind of middle ground when it comes to defining drugs as psychedelics (significantly altering everyday perception and conscious experience), or as entheogens (psychedelics used to induce religious/spiritual states or for sacred use). Mescaline/peyote DMT, LSD, psilocybin/magic mushrooms, and *Salvia divinorum* fall into both these categories. Psychedelics are defined variously, both by what they do, (see earlier) and how they do it (sometimes defined as their acting as agonists at serotonin $5HT_{2A}$ receptor sites). Cannabis partly meets the first criterion of being experience-altering, depending on the dose used, set and setting, but does not meet the second definition because it acts primarily via cannabinoid, not serotonin receptors. A recent literature review from the University of Buenos Aires tried to shed light on this question by analyzing reports on the online Erowid database of psychoactive substance reports [8]. Examining a total of 165 substances, and using the metric of similarity to lucid dreams, unsurprisingly LSD led the pack of psychedelics. However, cannabis came in at a respectable number 5, and was the only substance in the top 20 that was not classified as a classic hallucinogen. On the other hand, these results are a little puzzling in that hashish and cannabis were very differently classified in these ratings.

Whether or not cannabis increases one's creativity is a hotly disputed and contentious area, in part because measuring creativity in an objective manner is an elusive goal, so that most reports are merely anecdotal. Individuals who have used cannabis to spur their creative process include William Burroughs (specifically for *Naked Lunch*), Modigliani, the musicians Mezz Mezzrow, Billie Holiday, Snoop Dogg, Bob Marley, Louis Armstrong, plus Apple's Steve Jobs, and the filmmaker Benjamin Dickinson. The general idea is that in the context of amplified emotions and modified thinking patterns cannabis enables individuals to make new and unexpected connections between ideas more fluidly, to take a looser and more open approach to artistic creation that fosters building bridges between ordinarily unrelated ideas. All of these in turn boost creative originality. One study that tried to probe the interaction between creativity, divergent thinking and cannabis use [9], had individuals smoke a "spliff" of their own cannabis compared to their own non-intoxicated drug-abstinent baseline. On both occasions, the authors measured performance in a number of psychological domains, including verbal fluency, (e.g., "How many words can you think of in one minute that begin with the letter 'B'?"), and the Remotes Associate Test, (where subjects were given three words at a time and had to generate a single word related to the original triplet in four minutes.) An example here would be being presented with the probe words *night, wrist,* and *stop*, with the correct answer being *watch*. The only significant effect was found for verbal fluency, where the experimenters concluded that acute use of cannabis boosted divergent thinking only in those individuals who were not particularly creative at baseline.

The association of cannabis with musical creativity either in composition or performance has been much-discussed particularly in the realm of jazz, although we should note that Melissa Etheridge, Snoop Dogg and Sigur Ro's are all engaged in marketing their own cannabis products or are part of related "joint business ventures." One person who explored marijuana and music 20 years ago was Peter Webster [10]. He suggested that a general loosening of associations, de-synchronization of thinking, mild interference with memory and altered time sense are particularly suited in jazz to the need to re-improvise a given piece over time.

Medical use

As a reminder, the majority of 1000 customers surveyed at 2 adult use dispensaries in Colorado used cannabis to relieve pain (65%) and to promote sleep (74%) [11] (see Chapter 10).

Misguided self-prescribed psychotherapy

Examples here would be using cannabis to deal with personal problems, frustrations, anxiety, and depression. Users in this category are likely to be more troubled before they use the drug, and this pre-existing psychopathology contributes at least some part of their motivation to indulge. When cannabis effects wear off, the prior unpleasantness returns, so that this population is more likely to use the drug more frequently, and in turn this type of habitual use enhances the risk of developing new psychopathological problems, and of cannabis dependence, as explored in Chapter 8.

Use of cannabis by individuals in the context of pre-existing serious mental illness, so-called "dual diagnosis" is a separate phenomenon that we will discuss in Chapter 8. This entire last category of individuals is the one that most concerns policy experts. Their worry is that cannabis will be sold to individuals who will be psychologically harmed by it.

The subjective experience of being stoned

Let's move next to the typical experiences of people intoxicated with cannabis. Immediately we run into several problems. The first is that with all drugs, individual subjective effects are as unique as individuals themselves with their different brain chemistries, genetics, personalities, expectations, and prior exposures [12,13]. Going on a roller coaster ride inside your brain is even more varied than an actual roller coaster, where there is a limited range of experiences from thrills to incipient nausea, and your brain is under your control. In contrast, nobody customarily meets God on a roller coaster or has metaphysical out-of-body experiences there. For cannabis, there is a significantly wider spectrum of "typical" encounters with the drug.

A second problem is nicely summarized by Michael Pollan in his essay *Smoking the Toad* [14], and involves the difficulties in trying to describe the inherently indescribable, that is psychedelic drug-provoked encounters of all types. The paradox is that the person describing their altered state is by definition in an altered state and thus no longer an objective observer. Because of their intoxication, they may not be their usual articulate self. Once they are no longer under the influence of the drug, details of the altered state may be hard to recollect accurately. This is partly due to a phenomenon known as "state-dependent memory," where details and memories are most accurately and efficiently recalled when their brain is in the same state as it was when the memory was laid down. An often-cited example is that of Charlie Chaplin in the movie *City Lights*. Chaplin, in his tramp persona, saves a drunken millionaire from jumping to his death, thereby earning his gratitude. Once sober, the millionaire completely fails to recognize Chaplin, until the former is again inebriated, when he immediately recollects the identity of his bosom buddy. So, it is with the complex mental states evoked by cannabis intoxication. As soon as a person is stoned, he or she often immediately recognizes a wealth of thoughts, emotions, and perceptions that elude capture and description when not under the drug's influence. Standard clinical scales that purport to measure marijuana intoxication do a remarkably bad job of capturing its salient characteristics. They rely disproportionately on self-reports of physical phenomena such as "my mouth is dry right now" or "my heart is beating fast."

Marijuana intoxication is one particular and characteristic example from a diversity of altered states of consciousness that we can potentially sample. Humans are not lab rats and the complexity of the human brain and the marked individual differences between different people argue against there being a single, modal one-size-fits-all cannabis experience, (in fact, as we see below there are many varieties). Nevertheless, despite these caveats the task of describing what people feel when they are stoned is feasible, albeit necessarily hedged with many provisos. Some of these potentially distracting influences to take into account include the following:

Expectations and placebo effects

One of the most vivid demonstrations I had of these effects occurred in medical school. In the class on alcohol use everybody was given a fair-sized glass of a vodka/fruit juice mixture and encouraged to swallow the contents in one gulp. The glasses were wrapped in a vodka-soaked napkin secured with a rubber band. We were told that the majority of the tumblers contained two shots of vodka, but that some consisted only of juice with a small amount of vodka floated on the top. Between that and the boozy napkin, in the act of swallowing everybody experienced the smell and taste of alcohol, so that it was not obvious who had quaffed the placebo beverage. We then had to rate each other's behavior. The person who was most obviously and stereotypically alcohol-intoxicated staggered around the room, giggling, making inappropriate jokes, slapping everyone on the back and generally being loud and obnoxious. This was a vast departure from his customary restrained baseline behavior. When we opened the envelopes at the end of the one-hour study, it was revealed that this seemingly alcohol-inebriated young medical student had received placebo. He was familiar with the idea of how people act when drunk, and believed that he had received alcohol, so he behaved accordingly. For precisely this reason most drug administration studies include a placebo condition and are also performed under double-blind conditions so that any expectations on the part of the experimenter are not conveyed to the subject.

Set and setting. The 1960s LSD-guru Timothy Leary first drew attention to these considerations. The term *"set"* (an abbreviation for mindset) refers to the person who is receiving the drug's physical and mental condition at the time they take it, in other words what they are bringing to the drug rather than what the drug is bringing to them. Mental set includes the individual's personality characteristics, cultural background, mood at the time of dosing, recent experiences, beliefs, current preoccupations and expectations regarding the substance. Someone who is depressed, anxious, and upset at the time of cannabis use, or concerned about the risk will have a different set of experiences than another individual who is calm and relaxed. I tried to incorporate some of these principles into the description of the three revellers and their experiences of cannabis at Holi. Prior experience with the drug likely falls into this category too, as do a series of learning processes subsuming such things as knowing how to titrate one's drug intake to maximize pleasant effects and to minimize any unpleasant ones associated with excessive dosing.

The term *"setting"* describes the environment in which the person is consuming the drug. This accounts for variables such as physical location, comfort, familiarity, prior associations, who else is there and the degree of social support, sociocultural factors (is the drug legal, does the culture disapprove of it), even room temperature may all affect the user's experience. Different social circumstances call for different behaviors when one is intoxicated. Alcohol offers an obvious example. A bunch of businessmen out carousing, celebrating a colleague's birthday by drinking liquor in a stripper bar will behave differently than a gaggle of academics sipping sherry with an academic Dean at a faculty get-together, although their blood alcohol concentrations may be identical.

Inter-individual biological differences

We are all familiar with individuals who can "hold their liquor" or conversely are "cheap drunks" who become obviously intoxicated after receiving small doses of alcohol. Presumably what is accounting for these differences and for parallel effects with cannabis is a host of biological factors, from liver enzymes that break down the drug, relevant brain receptors, one's sex, body weight and percentage of fat (given the fat-solubility of cannabinoids). Some of these metrics are genetically driven, and results of a small experiment examining effects of variation in CB1 receptor and FAAH genes offer some suggestions that reward responses to marijuana cues are biologically mediated [12]. Additional biology-related factors, including perhaps what you had to eat prior to using the drug (that might affect absorption of edibles) and configuration of non-cannabinoid brain neurotransmitter systems (e.g., dopamine that might influence experiences of paranoia or reward) may all be relevant. I've known personally individuals who find it impossible to experience the high of marijuana despite trying various doses and methods of administration, and others who have experienced panic attacks and episodes of mild paranoia on first use of the drug that discouraged them from ever trying it again. It seems reasonable that individual differences in biology are determining some of this experiential diversity. Many people seem to have their preferred recreational drug; for some this is cannabis whereas for others the same substance is at the very bottom of their list. One important biological difference seems to exist for sex/gender; as reviewed recently, women may be at greater risk of adverse outcomes from cannabis use. Men tend to use cannabis more, but in the United States, the sex gap is shrinking; while cannabis use disorder is more prevalent in men, women transition faster from first using the drug to cannabis use disorder. Addiction researchers term this effect "telescoping." Overall though, sex/gender effects are understudied for marijuana use, and this area is ripe for exploration [15].

Dose: With virtually any drug including alcohol, effects are dose-related, but with cannabis there is an extra layer of complexity. From what we learned earlier in Chapter 5, some cannabinoid effects are biphasic, for example, low doses of THC may be anxiety relieving, whereas higher doses are anxiety-provoking. Dosing also assumes a standard cannabis chemovar, but different ratios of THC/CBD and differential terpene content are important to know about in this context. As we'll see later, with cannabis, as with many drugs, new behaviors can emerge at high doses that are not seen at lower ones. After all, a spring shower gently watering the flowers and a monsoon flattening fields and washing away hillsides represent dose effects.

Method of administration: To perform their basic function of extracting oxygen from the air and eliminating carbon dioxide, the inner spaces of the lungs are in intimate contact with the body's blood supply. So, inhaling vaporized or combusted cannabis, or the contents of vape cartridges, is an extremely efficient way of quickly absorbing those chemicals into the body and have them circulate rapidly to the brain. Absorption from the gut, on the other hand, is slower, less efficient, and results in a significantly longer time between taking the drug and feeling its effects. Behavioral consequences of taking edibles are significantly both later-appearing and longer-lasting than those following inhalation. For inhalation, onset is within a few minutes and offset within 3–4 hours. For edibles, onset is typically an hour or more, with effects lasting anywhere from 6 to 10 hours. Recent studies suggest that dose-for-dose, use of a vaporizer produces a significantly greater high than smoking the same amount in a "joint." There are also additional considerations in administering cannabis to human volunteers under laboratory conditions. If you want to give precise doses to individuals, and ensure that every research participant receives exactly the same amount without some unknown quantity disappearing into the air as occurs with smoking, then intravenous doses of pure THC may seem appealing. But the method of administration interacts here with both your subjects' prior experience and the laboratory setting in which the drug is being administered. The average cannabis-using volunteer is most unlikely to have previously received IV THC, and probably isn't sure what to expect. Thus, their altered mental set will affect how the drug is experienced. Therefore, the aim of precise dosing has to be balanced against numerous other design factors.

Experience with the drug

There are two separate major effects at work here. The first is that people who are inexperienced with the drug, surprised by its effects or unsure how to titrate and modulate intake may become overwhelmed by the experience of intoxication or anxious about what's happening. The second is a metabolically-based phenomenon, that of diminishing response or short-term desensitization to successive drug doses; one technical name for this is tachyphylaxis. This effect alongside the longer-term one of drug tolerance is due to a variety of biological mechanisms, including desensitization of brain cannabinoid receptors to the drug, and the liver's homeostatic response of producing more enzymes to break down a substance more effectively when presented with it frequently. Tolerance is one aspect of physical dependence, (i.e. needing to use higher and higher doses of the drug to get the same effect), but many people exhibit cannabis tolerance without other symptoms of dependence such as drug withdrawal. These mechanisms explain why the frequent dab users in the Colorado social club that we meet in Chapter 8 can inhale enormous amounts of pure THC from concentrates without being massively intoxicated.

Something else that is worth mentioning because it is so characteristic of subjective cannabis intoxication, (although its precise mechanism is unknown) is its wave-like nature, where the "high" feeling will fade away periodically, only to well up again unexpectedly, as mentioned by Lester Grinspoon in his well-known book *Marihuana Reconsidered* [16]. One possible explanation for this effect is based purely on the metabolism of THC. The initial hydroxy-metabolite of THC is psychoactive; 11-hydroxy-THC in fact is even more powerful in this regard than the parent Delta-9 THC, and more easily gains entry

into the brain. So as THC circulates dynamically about the body and is metabolized, one can imagine waves of different psychoactive chemicals periodically
cresting and swirling into the brain.

One of the few places where anecdotes perform better than statistics is in
exploring subjective drug effects. The next section of this chapter aims to try
and capture what "being high" is all about. The researcher who literally wrote
the book on cannabis-altered states of consciousness back in 1971 was Charles
Tart. His book titled naturally enough, *On Being Stoned* [17] is available online
and is a classic in the genre. Tart elicited information from 150 marijuana
users, mostly West Coast college students, an idealistic, serious, and religiously
unconventional group who as Tart says "overcame their fear that the study was
a police trap, and so gave of their time and experience." Cumulatively they
had used marijuana about 37,000 times for a total of 421 years of experience.
It's worth spending some time reviewing Tart's findings, because of his fairly
thorough research methods, his attempt to probe a wide variety of possible drug
induced experiences despite the inherent difficulty of capturing complex subjective journeys. He wanted to catalog information that went beyond the limits
of traditional laboratory experiments on the one hand and captured individual,
unstructured subjective reports on the other. His approach was straightforward
and relied on asking users who had smoked marijuana more than a dozen times
to report on cannabis intoxication that they had experienced over the last 6
months, using standardized questionnaire items that ranged across almost
20 different domains of experience. This allowed him to collect a wealth of
detailed information that he then boiled down to extract commonalities. He also
explored his subjects' backgrounds, encounters with the law related to cannabis
use, alcohol consumption, age, educational level, political beliefs, marital status, occupation, and gender. Finally, he documented for each type of subjective
experience how intoxicated subjects thought they were at the time. Because
that's difficult, subjective, and unreliable to track retrospectively, I've paid significantly less attention to those aspects of his data. So here follows the range of
topics that were explored and the major findings: I admit to rearranging a few
components within categories in minor ways so that at least from my point of
view, like sorts with like a bit more logically. I've done my best here to summarize the key points of a 300-page book in a couple of pages, sometimes using
Tart's terminology and sometimes paraphrasing, while interspersing my own
observations and quotes from both Grinspoon and from online Erowid users
[18]. The Erowid website [19] provides access to a wealth of curated information regarding psychoactive substances, including very detailed user reports.
After we review what the cannabis-intoxicated think, feel, and experience, we
will backtrack into the world of neuroscience to try and explain those findings
in terms of how cannabinoids affect the brain.

Perception of the external world: Tart hypothesizes that cannabis alters
aspects of perception and distorts interpretation of patterns in the environment
that reach the brain through the senses, a phenomenon that Grinspoon [16]

believes can lead to a subjective feeling of unreality or of the external world being altered or skewed in some way. Tart embeds this phenomenon within a general "reorganization of mental functioning," which to me is a nice encapsulation of what a psychedelic drug does. In the realm of *visual perception*, participants reported seeing objects in sharper contrast, and some experienced people's faces morphing and changing identity as they watched them. They perceived novel hues, and could appreciate more subtle shades of color. Visual depth perception altered, so that near objects appeared closer and far objects seemed much further away (an effect that one might be concerned about if driving when intoxicated). "The patterns look awesome. The textures seem richer. The colors seem purer - more vivid - delicious - almost edible! I feel like I'm David Bowman at the end of 2001 - Jupiter and beyond. A very cosmic motif. ...... Also, the artificial synesthetic properties seem to be enhanced - I notice the relationship of the patterns to the music much more than I used to" [20]. Some subjects reported "Alice in Wonderland" experiences of objects appearing suddenly huge or tiny. Correspondents told of discerning distinct patterns in normally ambiguous or amorphous visual material such as bumpy wallpaper or cracks in the wall. Some remarked on a sensual quality of their vision, as if their eyes were actually touching the object they observed in some manner. Visual imagery in the mind's eye became more vivid; with their eyes closed some subjects saw little dancing cartoon figures or richly detailed 3-D scenes. With their eyes open, fringes of colored light might surround objects, and some individuals experienced stroboscopic flickering effects, afterimages, or perceived apparent movement in static objects, especially those in their peripheral vision. "Every time I'd move my eyes, it lagged. Like watching frames of a movie slowly, like flipping through a slideshow. This was amazing, my vision had changed, and my brain was interpreting movement of my eyes differently" [21]. At higher levels of intoxication, inanimate material in the environment such as a heap of blankets might seem to become animated. Frank visual hallucinations unprovoked by anything in the environment (i.e., vividly seeing things with their eyes open that were not there and appeared real), were extremely rare and only occurred at high doses of the drug. To me these latter experiences seem more reminiscent of delirious states.

Hearing: Tart's students reported being able to perceive more subtle qualities within sounds. They provided examples of musical notes seeming purer and more distinct, rhythm being more prominent, and being able to distinguish words of songs that were unclear or inaudible when they were not high. Along the same lines, spatial separation between various instruments sounded magnified and more 3-D, as if they were physically further apart. These phenomena were experienced as positive and pleasurable. With their eyes closed, sounds seemed arrayed in auditory space according to their audible characteristics, more complex, and in some cases provoked visual images. At high levels of intoxication synesthesia occurred, so that different notes appeared in various colors or provoked distinct tastes, and also serious or complex music was experienced as markedly "sensual and profound." On a personal level, I recall listening to Karlheinz Stockhausen's *Stimmung* when stoned at age 20 and being able to understand the piece and to feel it intuitively for the first time.

The whole issue of marijuana's improving the appreciation and enjoyment of music is much discussed. Anyone who has attended Coachella, a Grateful Dead concert or a Sigur Ro's music performance (where CBD/THC edibles are an intrinsic part of their Icelandic music experience) will appreciate that many cannabis users employ the drug in a general way to "enhance relaxation and concentration while engaged in leisure activities" [7]. Music offers a particularly striking example of this general tendency. When stoned, many people are absorbed in the here and now, receptive to sounds in the environment, have altered time perception, are in a euphoric, relaxed state where emotions can be exaggerated, are relatively free of mental chatter about the past or future, somewhat disinhibited. They are thus free to explore novel implications of events in the environment. The combination of these various cannabis-induced experiences seems pretty much designed to increase musical appreciation and enjoyment. In addition, we know from Chapter 4 that harps were discovered in the same ancient Chinese grave sites as cannabis braziers, strongly suggesting use of both simultaneously during funeral processions and ceremonies. I discuss the separate issue of whether or not cannabis increases musical creativity in the section later in this chapter that deals with motivations for people's use of the drug.

Touch, temperature, and taste

"Touch becomes more entertaining and far more acute. My ears are more sensitive, sounds become full bodied and subtleties become easier to distinguish. Music takes on another dimension, it flows from the air, from the walls, the ceiling, from within. Taste, well, I need not say much about that--we all know about the munchies. In short, I become a creature adept-at-feeling everything to its full capacity" [22]. "I opened up my Snapple and took the most glorious sip of my life. The flavor was exquisite. Raspberry and sugar were orgasmic and going down, the tea was the most refreshing drink I've had to date. My friend lent me some of his sour candies and upon eating them, flashes of light started dancing around in my head" [23]. Surfaces were reported to feel rougher or smoother or more irregular, often in an interesting or entertaining way. Objects felt heavier or lighter, tastes embodied novel qualities and the experience of eating was more pleasurable. Food not only tasted and smelled better, but these sensations became enhanced and changed for the better in unique and unfamiliar ways. "I first noticed that my senses had changed. I was suddenly aware of my socks when I walked. They felt like they had a wrong shape or maybe they were too thick. Anyway walking became something I constantly would notice. The feeling in my fingers also changed, it was as if I had Band-Aids on the finger tips" [24].

Space and time

Time distortion is one of the most characteristic of cannabis' effects. A majority of Tart's subjects remarked that when intoxicated, time passed unusually slowly. For example, a CD seemed to play for hours, or time even seemed to come to a complete halt, so that experiences were felt as going on "outside of time, or timeless." Related to this, events seemed located completely in the "here and now," with intense immediacy and no thought given to the future. Other subjects reported the opposite phenomenon, that is, time seemed abnormally speeded up so that events were over as soon as they had barely begun. Some of his respondents related either that events flowed more smoothly one into the other, or alternatively segued jerkily and changed suddenly.

Around 1999, when I worked at Johns Hopkins, our research group conducted an experiment that recruited cannabis-using volunteer subjects to participate in a study consisting of two sessions, a week apart, to record how THC alters time perception. The experiment included going into a PET scanner and using an injection of mildly radioactive isotope-labeled water to measure brain blood flow. In a double-blind manner, each subject received either dronabinol (brand name Marinol), a synthetic form of Delta-9 THC, or an identical-appearing placebo capsule. All of our experiments use placebo controls to deal with expectancy effects. Dronabinol is a controlled prescription drug approved by the FDA to increase appetite and weight in individuals who've lost significant amounts of weight, for example, due to HIV/AIDS-associated disease, or to combat nausea and vomiting caused by cancer chemotherapy. Common sense would tell us that eating a dronabinol capsule should be no different than ingesting an edible such as a marijuana brownie. In fact when we un-blinded our data to reveal which subject received which drug on which day, the subjective effects of the active drug were very different from standard descriptions of somebody who had eaten (or smoked) an equivalent amount of cannabis. On standard scales to rate a marijuana "high" our subjects reported variously that the dronabinol made them sleepy, interfered with their ability to think clearly, and made them dizzy or slightly anxious, but definitely not "high", "buzzed", "mellow", or "feeling good." (This might represent a consequence of missing entourage effects; pure THC in the absence of other cannabinoids and or terpenes, produces significantly different psychoactive effects. Similarly, Deepak Cyril D'Souza's subjects at Yale dosed with pure intravenous THC, also found the experience very different than smoking marijuana and generally not pleasant).

The crucial time estimation part of our Johns Hopkins experiment had subjects listen to two tones from a pitch pipe that were sounded by an experimenter a few seconds apart and then to estimate the time between the notes. This procedure was repeated on multiple occasions throughout the experiment. In some of our subjects, as expected, compared to placebo, the drug caused errors in these time estimates. Other participants however were completely unimpaired (in part because dronabinol given by mouth is absorbed very variably from the gut). The most striking temporal distortion effect was in one subject, whose first response to the time estimation question was a lengthy pause, followed by the answer "A billion, trillion, quadrillion milliseconds." This was significantly different than the correct answer, which was 3.5 seconds.

Experiences might be difficult to communicate because they seemed disconnected from what Tart terms the everyday physical "space/time matrix." Distances were skewed, appearing longer or shorter than they were objectively, especially when walking. Some individuals reported temporary loss of spatial orientation, for example becoming absorbed into fantasies and forgetting where they were. Others lost bodily awareness, so that they felt their bodies floating in space or actively accelerating through it. Many of the experiences recalled Rod Serling's evocation of realms "as vast as space and as timeless as infinity." Some respondents reported that the space or air around them took on a solid quality and thus was no longer "empty." Space sometimes became defined purely by auditory experiences. At high doses a small number of subjects described déjà vu feelings, where everything they were experiencing seemed a "rehash" of prior events. Such experiences lead naturally to the next category.

Paranormal phenomena such as ESP

Tart mentions drug-induced credulousness or suspension of disbelief as possibly explaining some of these experiences. It was the 1960s. A full 76% of his population believed in the reality of ESP, for example, so that their collective mental set may have predisposed them to perceiving such phenomena. When intoxicated, they reported feeling very aware of what other people in the environment were thinking, not just in the sense of being more sensitive to subtle cues in their behavior, but actually feeling what other people were experiencing with enhanced intuitive and empathic understanding. "I'm tripping on an oblique angle to her mind, so I figure it's about 92% telepathy. I become convinced that what I hear her say is really my mind reconstructing her words into a language I can only understand at a peripheral level. I feel if I could fully understand this language, I could read her mind perfectly. I could gain ultimate insight into the meaning of her speech" [20]. Some individuals experienced magical or psychokinetic abilities, believing that they were capable of performing supernatural operations resulting in direct effects on neighboring objects or people while stoned. Instances include beliefs that they could, with 100% accuracy, identify other individual's astrological signs, or predict those people's body movements before they occurred.

Bodily experiences

Some changes in whole-body proprioception seem to be dependent on distraction or inattention, for example, being so absorbed in inner experiences that subjects failed to notice their own body or where their limbs were located in space. Tart's respondents stated that their bodies felt abnormally heavy, light, larger or smaller, stronger or weaker, or differed somehow from their true body shape. Many subjects described their body feeling unusually light, or sensations of floating in limitless space, feeling tinglings of energy and power flowing in one's body. These are experiences that would nowadays be described as a "body high." Out-of-body experiences, perceiving oneself as located outside one's actual physical self from a vantage point in the same room, or far removed in space, or via the mind visiting another dimension leaving the body behind, generally sum up this type of report. Subjects reported altered pain tolerance, likely related to analgesic effects of both THC and CBD. In addition, some mentioned either being able to focus on and make pain or other sensations more intense, or to concentrate on other matters and make pain disappear altogether. Touch in some cases became more exciting. A number of subjects reported greatly increased relaxation, or what would now be termed "couch lock." They felt extremely mellow and disinclined to move. In Québec the slang for stoned is "gelé", which translates as "frozen." Occasional restlessness was reported. Some respondents reported changes in coordination—occasional improvement but more likely a sense of awkwardness and impaired balance. "Rather than just doing things and saying things, it was like I was ordering my body to say and do things. I could literally feel the delay in reaction and the messages coursing through my nervous system. Like I was controlling an organic robot from inside" [25]. Some individuals reported that their body shape and location of their "self" was "messed up." Interior perceptions changed in the sense that people became much more conscious of parts of their bodies that they were normally unaware of, such as their internal organs, and related processes such as heartbeat and breathing, as well as bodily accompaniments of emotion. A handful of individuals have mentioned to me that practicing biofeedback while mildly cannabis-intoxicated seemed to be significantly easier for them, both because they were more aware of internal bodily processes and because time was subjectively altered.

Social interaction

Some reports in this category are clearly due to cognitive alterations, for example, the experience of forgetting the start of a sentence or the beginning of the conversational thread. Other subjects reported that despite being at noisy and boisterous social gatherings they were taciturn, while others became more sociable, wanting to be with and interact more with others. A number reported that they were disinclined to play ordinary "social games", for example, feeling that everyday chit-chat seemed hollow and worthless. On the other hand, a number of people reported liking to play elaborate or silly games when stoned. While some individuals characteristically became more distant and objective in their feelings of insight into other people and their social strategies and what motivated them, or "saw them more clearly," others became more empathic. Some subjects became more aware of their value judgments about others. A few became preoccupied by their awareness of how stoned other people thought they were. Tart also reported that while some subjects experienced a greater sense of cohesion and unity with the group or with the world, others felt isolated. These latter individuals reported a "kind of barrier or glass wall between me and the world" along with suspiciousness about companions, or the feeling that others seemed "dead or lifeless" or that their statements were ambiguous. Many also reported a heightening of the overall significance of conversations and events, perceiving that things that they and others uttered were more profound, subtle, appropriate, important, and interesting than usual.

Sexuality

Many of Tart's subjects reported that their sexual drive increased, not in a general sense, but amplified only in situations where they would usually be sexually aroused. Numerous individuals noted that because of a heightened sense of touch that incorporated new sensual qualities, with greater focus on awareness of bodily feelings, focus on the here and now and feeling closer to others, sex was significantly more enjoyable. However, at very high levels of intoxication people retreated into themselves and sexual desire diminished. Several described that when making love they felt in close mental contact with their partner, sharing the experience more, "a union of souls as well as bodies." A rare few felt more isolated or distracted by synesthesia. Subjects reported new qualities to orgasm, either more prolonged (possibly an effect of time slowing), ecstatic feelings of energy flowing or exploding in the entire body, energy being exchanged with their sexual partner, or total undistracted immersion in the experience. Others commented on feeling that they were a better lover than when straight, using descriptions (I'm paraphrasing in some cases here) such as "less inhibited ....more gentle and giving....more sensual... more here and now, more confident."

Cognitive processes: memory

The universal Middle Eastern word for stoned is "mastool." It's Arabic origin hints at someone with a bucket over their head, but the general meaning is befuddled and confused. Within this overall cognitive impairment, memory seems particularly affected. Among Tart's subjects, there were reports of altered long-term memory, for example, long-ago forgotten events would pop up in consciousness, or subjects remembered spontaneously events that they hadn't recalled in many years. Short-term memory seemed generally impaired. Conversation span was shortened, and conversations lost track of, unless effort was exerted. However, respondents reported being able to carry on an intelligent conversation (subjectively) despite not being able to remember how the exchange had started. Many recorded that thoughts would slip away before they could be grasped, with subjective confusion between thinking of saying something and actually having said it. There were many experiences

of state-dependent memory, that is during the period of intoxication, recalling familiar thoughts or feelings specific to being stoned, that did not pop up
in an ordinary state of consciousness, (and were unrecalled until next time the
person was intoxicated). Some subjects described their memory problems as
being due to attention difficulties, that is, being easily sidetracked and distract-
ible. Others reported that when sober, they were unable to remember what they
had read when stoned. A recent book chapter provides a detailed overall review
of short and long-term effects of cannabis on cognition [26]. A major concern
among public health workers has been the fact that adolescence is the period of
statistically greatest risk for beginning regular cannabis use, and coincides with
significant brain-shaping processes that are theoretically especially vulnerable
to neurodevelopmental disruption. A review of the literature on cognition in
adolescent cannabis users provides some evidence that earlier and more regular
use of the drug is significantly more likely to have adverse cognitive effects on
memory, attention and decision-making, with partial recovery if people reduce
or stop use. Effects on cognition appeared to be dose-dependent [27].

These reports are generally borne out by laboratory experiments that we
will explore in more detail in Chapter 8. Briefly, cannabis along with other
CB_1 receptor agonists produces quite marked and frequently replicated acute
effects on impairing short-term and working memory in humans. This topic
has been reviewed by Deepak Cyril D'Souza and his coworkers at Yale, who
have a wealth of experience in this arena [28]. To sum up a large and complex
literature, THC impairs all component stages of memory. It has a short-term
effect in scrambling an individuals' ability both to repeat back (recall) immediately or provide later, information on such material as stories or lists of words
presented to them to remember after they were stoned. The drug didn't affect
memory for similar material that was presented before they were given the drug.
In addition, the higher the cannabinoid dose, and the longer the wait between
administering the material to be remembered and asking the subjects to repeat it
back, the worse was the memory problem. Not only were there issues recalling
stimuli that had actually made it into the brain, but there were also difficulties
getting information in there (what psychologists refer to as "encoding."). This
type of memory problem seemed to be dependent on several factors, including
subjects being distracted. Another memory process, consolidation, that turns
labile, transient short-term memories and transforms them into stronger, longer-
term, easily recalled memories is also clearly impaired by cannabis. So not only
was short-term recall impaired, but longer-term memories were also harder to
access. Finally, when subjects given cannabis tried recalling what they had been
asked to learn (retrieval) there was intrusion of irrelevant material, or items
that were incorrect but conceptually related to the stimuli being encoded. For
example, if after smoking marijuana you were given a list of tools to remember
(hammer, saw, screwdriver, etc.), then you would tend to falsely recall tools
that were not on the original list, but belonged to the same general category,
for example, wrench or crowbar. Cannabinoid effects on memory are almost

certainly mediated through the hippocampus in a major way, and dependent on CB_1 receptors, since they are reversed by rimonabant [26,29].

Cognitive processes: thought

Tart's subjects recorded that they became so caught up in thoughts and fantasies that they failed to notice what was going on around them or perceive that another person was talking to them. Their thoughts would drift somewhat aimlessly so that they would experience blank periods, where they would become aware that nothing seemed to have been happening for an extended time period. Some individuals reported experiencing spontaneous insights about themselves or how people in the environment conducted themselves socially. "Thus, the brain has everything it's ever learned written down and recorded on task-specific discs. That way it doesn't have to keep re-learning the stuff it has to deal with every day. It learns and it stores the information on these strange, organic discs. I can see them, clear as day. The brain has learned these behaviours and somehow it has them all stored away in sections, just like the library has its books stored in sections, from horror fiction through to romance, from science fiction through to mysteries. The brain is exactly the same. It's just a clever learning machine, nothing more" [30].

Many stated that thinking was entirely limited to the here and now with no thought to the future. Sometimes this lack of distraction gave rise to a sense of additional energy, efficiency, or increased absorption in a task. Thinking itself seemed to be altered; individuals would reflect on a topic in a manner that seemed intuitively correct but didn't follow the usual rules of logic or proceed through the usual intermediate steps of thought. Critical logic seemed weakened, with individuals more willing to accept contradictions or things that didn't entirely make sense, sometimes in creative ways. Subjective feelings about the workings of one's own mind varied considerably, with some individuals feeling that their mind was processing thoughts more efficiently than usual, and grasping new insights, while others reported muddy thinking or thoughts eluding capture and uncontrollably slipping away before the person could grasp them. The latter phenomenon seemed dose-related. "I get this feeling that my brain constantly tries to predict what it is going to happen, and then when something else happens, I become aware of it, slightly confused, because I thought the thing my brain predicted had actually happened" [24]. Salience and meaning were altered, less in the area of logic or correctness, but more in the realm of depth and subtlety. Commonplace sayings or conversational topics suddenly seemed to have new meaning or significance, or to be more important and portentous.

Emotions

A more positive mood is almost universal in cannabis intoxication, sometimes to the point of euphoria, although this may diminish at higher drug levels. As an example of set, one can see amplification of any pre-existing mood or preference. Obvious examples are that the emotional accompaniments of sensory phenomena are turned up all the way to 11 on the dial. If you like chocolate, then it tastes like manna when stoned. If you love Bach, then joyful passages on cannabis are redolent with bliss, and sad dramatic music plumbs the depths of wretched despondency. On another dimension of happiness, being silly or child-like, with characteristic giggling are all frequent occurrences when stoned, even when the situation is not that intrinsically funny. Many subjects reported smiling a lot and being more aware of the bodily components of emotion. Depth of emotions increased, so that not only were sentiments felt more strongly, but they affected the person more profoundly. Some individuals reported feelings of love and compassion toward others. Others felt more powerful, capable, or intelligent when stoned. Other individuals had strongly negative experiences. Acute emotional crises might occur in emotionally unstable people, or in previously unremarkable individuals harboring worries or doubts, when using marijuana. Such people tend to have their problems amplified by the drug, as can inexperienced users who inadvertently smoke or consume more edibles than they know how to handle, and are temporarily frightened by overwhelming experiences of disorientation, anxiety, paranoia, unreality, or fears of losing control. "Waves of paranoia washed over me with sickening frequency. Who were those strange entities I sensed but couldn't see, lurking in the shadows, beyond the shadows? And what did they want with me? I felt as though it were only a matter of time before one of them might choose to manifest itself and begin the inevitable process of reducing me to a gibbering wreck. What could I do to thwart them? Was I losing my mind?" [30].

Control of intoxication

To some extent, because inhibitions are lowered, some individuals welcome their ability when intoxicated to engage in activities that are ordinarily unavailable to them due to anxiety or shyness. Similarly, a tendency to let go and allow fantasies to flow spontaneously, or allow pleasant emotional states to take the user where they will, can be tempered by a conscious push to amplify pleasant or ecstatic experiences. Almost 40% of users indicated they had special mental techniques for getting higher including focusing on current activity, contact with intoxicated companions, meditation, directly willing self to get higher, breathing techniques, music, fantasy, inducing positive emotions, and hypnosis. A very small proportion of subjects used alcohol to heighten marijuana effects. Loss of control, as mentioned earlier, can sometimes be frightening if the drug experience takes the user into unpleasant territory. One such direction can involve being confronted by or dwelling on realities about oneself or one's situation that are normally suppressed. Another is the eruption of frightening phenomena from the unconscious. Some of Tart's subjects, (albeit rarely) reported psychosis-like experiences. "I have lost control and been taken over by an outside force or will which is hostile or evil in intent for a while" was one item endorsed. Also rare was the opposite occurrence of being taken over by an "outside force or will which is good or divine, for a while." As opposed to heightening the drug experience, many individuals who wanted to lessen it and "come down" felt that they could accomplish this less via willing it directly, but more by inducing negative emotions, focusing on the situation at hand, or pushing themselves to act normally. Obviously, this was less effective at high levels of intoxication, where cannabis had the upper hand and drug effects were harder to combat. When profoundly intoxicated, some individuals use the strategy of attempting to amplify their willingness to trust the situation and "let things happen."

Identity and spiritual experiences

Tart separates these encounters two categories, but so many of the events he reports seem to cross boundaries between the two, that I collapse them here. Thinking back to Chapter 3, whether one achieves spiritual experiences through the use of cannabis likely depends both on the dose and on the user's motivation (see later). Given the well-recorded use of cannabis during ancient religious ceremonies, for example, by the Scythians, and current explicit use of cannabis as an essential part of spiritual activities by the Jamaican Rastafari, the fact that the drug evokes psychedelic and spiritual experiences comes as no surprise. Many of Tart's subjects reported changed feelings of self-identity, or "I-ness." Subjects would come out with statements such as "my personality changes so that I'm a different person for a while." They related becoming increasingly introspective, "more like the quintessential me" or their "true self." This alteration seemed to accompany emotional states of openness, such as feeling more childlike, being open to experiences of all kinds, and more filled with wonder and awe at the nature of things. Females had more of these experiences at moderate to strong levels of intoxication. Subjects reported feeling more at one with the world, or being able to focus so strongly on, or become so absorbed in contemplating an object or a person's feelings that they became that object or person. One individual reported that he was incorporated into the universal; "Some events become archetypal – part of the basic way man has always done things. That is, instead of me doing something it is just man doing what man has always done."

Other psychedelic-like experiences included inner voyages, and phenomena reported in earlier sections such as suspension of time, out of body experiences, total loss of consciousness of one's body during fantasy trips, vivid synesthesia, plus precognition, telepathy or the belief that one could perform magic. More straightforward religious experiences included feeling in touch with a higher power or divine being and "feeling more in contact with the spiritual side of things" occurred primarily at strong to very strong levels of intoxication. Some subjects used the drug to perform activities or enter states that in turn led to religious experiences. These included meditating more effectively, being less ego-focused or preoccupied with mundane concerns or more able to center oneself. Others use the drug directly to achieve religious experiences. "To me getting stoned is a communion of sorts with the Godhead", rare contact with divine beings, or experiencing feelings of unity, oneness with the universe, part of God's overall plan of things, or becoming "mystically one with the all-knowing." Accompanying such experiences were reports of feeling a high degree of spiritual empathy with others present, stimulation of long-term interest in religion, or deep peace and joy. Others experienced "fantasy being as real as reality", out of body experiences and seeing or experiencing other universes. Finally there were reports of mystical sexual experiences such as the dramatic and palpable fusion of souls as well as that of bodies.

Tart looked for effects that were fairly frequent when using more powerful psychedelics, but relatively infrequent or rare with marijuana, in the almost 75% of the sample who had used more powerful psychedelics such as LSD at least once. About one in three of his sample were 'heavy users' of psychedelics defined as having used more powerful drugs at least a half-dozen times. Experiences that were more common in these psychedelic users included visual hallucinations, losing touch with one's body and floating in limitless space, seeing another's face transform, and feeling in touch with a higher power. More than half of these users related that since having used LSD or another psychedelic drug, they were able to get much higher on cannabis than they were able to previously. They explained that their psychedelic drug experiences had let them know that certain types of psychedelic experiences were accessible and thereafter used marijuana to attain them. This type of experiential cross-drug transfer is an interesting but hard to explain phenomenon.

So the answer to the question "is cannabis a psychedelic"? is yes, it can be, especially when taken at higher doses, and via edibles, more so in individuals familiar with effects of more typical psychedelics.

Sleep and dreams

Subjects reported that it was easy to go to sleep following cannabis intoxication that drowsiness occurred earlier than usual, and that sleep was particularly refreshing with dreams being more vivid than usual. This jibes with the fact that large numbers of medical marijuana consumers choose particular chemovars to aid with insomnia.

After-effects and miscellaneous effects

Tart's participants tended to complain infrequently that it could be hard to get organized or accomplish anything the day after smoking. Some disliked that they had poor memory retrospectively for periods of intoxication. And yes, "weed hangovers" are definitely a thing, as shown by early laboratory research [31].

Dose-related effects

As well as asking individuals what quantities of marijuana they use typically, (and where reports are notoriously inaccurate as we will see in Chapter 8), retrospective inquiries regarding subjective intoxication are important to pursue, yet often yield dubious data. As implicit in Tart's account, assigning values to states of cannabis intoxication is important. It would be useful if researchers were able to allot relative quantitative measures to their subjects' levels of intoxication, as well as to compare such results meaningfully with one another. For example, "how intoxicated did your subjects recall they were when they believed that they could read other people's minds?" Such subjective scales of cannabis intoxication can range from the very simple, e.g. visual or verbal analog scales to the complex, and include both scales of current intoxication and recollections of prior such events. Typically simpler research instruments ask questions such as "on a scale from 0 to 10, where 0 is not high at all, 5 is how high you ordinarily get and 10 is the highest on cannabis you've ever been, what number indicates how high you are right now?." One of my favorite non-scientific, but nevertheless informative cannabis "highness" scales is to be found at [32]. On this "Old Hippie's Levels of Consciousness Scale," 0 is "sober," 1–2 is "buzzed"—"you can tell you're slightly high, things are more interesting...," 3–4 is "high"—you tend to stare at objects and into space often, you feel relaxed but somewhat in a daze..., 5–6 is "really high"—"...a slight vibration in your body, focusing is hard...very random thoughts. Putting together sentences is a challenge," 7–8 "blazed/stoned"—...very high, vibrations and waves of euphoria rush through your body... mild closed-eye hallucinations... everything is brighter, 9–10 "in space"—"completely lost in thought, amazing pleasure, you feel like a god... sparkling textures everywhere.. If you're not standing up... don't even try," 11–12—this ranges from "on the verge of tripping ..." to "you are now tripping. Congratulations and try to hold it together," 13, "you are tripping and losing control."—"If you don't manage to keep control at this level you may well.... find yourself in an emergency room or mental institution."

The important lesson here is that being stoned is not a unitary phenomenon. Neither is it on a smooth continuum, where more of the same experiences occur in a smoothly-graded and completely dose-dependent fashion. It's more of a roller coaster, where qualitatively different and sometimes unexpected phenomena emerge with escalating doses.

Can neuroscience explain cannabis acute effects?

What might be some explanations at a brain level for the earlier subjective phenomena associated with cannabis intoxication? This is a question that Charles T. Tart PhD would have clearly wondered about, but answering it almost

50 years later has taken a leap forward in neuroscience that was not conceived of in his day. In Chapter 5, we learned where endocannabinoid receptors were distributed in the human brain. But another important question to answer related to the physiological underpinnings of typical cannabis-related experiences and perceptions is which brain parts are switched on or off by acute doses of THC and other cannabinoids. Let's consider observations from both of these domains to help address our question of how knowledge of the drug's actions on the ECS health explains what happens when we get "stoned." Much of our understanding of this topic derives from experiments using magnetic resonance imaging (MRI) scanners.

A detour is required at this point to explain how researchers go about collecting different kinds of MRI measurements; multiple kinds of information can be gathered inside the same MRI scanner. *Blood flow* is measured using a technique known as "arterial spin labeling," that magnetically "tags" water in arterial blood. *Resting state measures* use functional magnetic resonance imaging (fMRI) to assess a combination of blood flow and metabolism, by looking for slight shifts in magnetic properties that accompany oxygen being stripped off hemoglobin in blood cells when these tiny doughnut-shaped corpuscles reach more metabolically active brain regions.

In typical resting state fMRI studies, volunteers are asked to lie quietly in the MRI scanner for 5–10 minutes, usually with their eyes open, to stay awake and to think about nothing in particular. Many people intuit that this will yield a signal analogous to a TV not tuned to any channel; a featureless low-key white noise. To most people's surprise, the brain is always busy, including being highly regionally active during this state of "just spacing out," and a number of distinct brain networks can be predictably identified thereby. One such network is most active during this "resting state," and is known as the default mode network (DMN). All cerebral networks in fMRI, including this one, are defined simply as groups of brain regions that switch on (or switch off) at the same time, and are thus implicitly connected to one other. The DMN is always "on" as a brain circuit, including when we are asleep, and even if we are unlucky enough to fall into a coma. The DMN is not unique to humans, and analogues of this circuit are found both in our close monkey relatives and even (albeit in a more primitive form) in rats and mice. The DMN is negatively correlated with other networks (i.e., its activity is high when theirs is low, and vice versa). These half-dozen other, non-DMN circuits are known as "task-positive" networks. Their function is related to various ongoing conscious cognitive processes, such as attention, or executive tasks such as working memory. When the "task -positive" brain networks are needed to participate in a particular activity (e.g., paying attention to something important such as a grizzly bear, or remembering items from a shopping list), then circuits related to the task at hand are summoned online. They immediately "wake up" more with regard to their blood flow and metabolism and move to the foreground, activity-wise, while the default mode network flips into the background and lurks there.[a] The individual regions of the DMN are active not only in the resting state during unstructured daydreaming or mind-wandering, but also during "internally focused" tasks including thinking about the future, mulling over the past, and autobiographical memories. Changes in resting state functional connectivity among circuits occur in many neuropsychiatric disorders, but altered relationships either within various circuits (including the DMN), or relationships between different circuits (e.g., cross-talk between the DMN and the executive networks) are easily provoked by any drug that impacts brain function. This of course includes cannabis. There is increasing evidence that these altered within-and between-circuit relationships might be substance-specific. Some researchers, for example, believe that every brain-changing drug leaves its own characteristic, identifiable "fingerprint" on DMN activity for a variable amount of time. All of these considerations will help us in interpreting a number of cannabis-related research reports. An important piece of additional knowledge is that one brain region may participate in multiple cognitive or behavioral activities and sometimes belong to more than one circuit. An analogy might be day laborers who have both roofing and drywall skills; one day they may participate on a tiling a roof, the next hanging plasterboard, but you won't find them welding or doing plumbing. So let's examine a few of the more important cannabis brain imaging studies to see what they discovered.

A significant paper was published in 2011 by a Dutch researcher with the Rocky Horror Show-like name of Hendrika H. van Hell, from the University of Utrecht in the Netherlands [33]. (If she turns out to be a future reviewer of one of my scientific papers or grant proposals, I mean this comment in the friendliest possible way.) van Hell used an MRI scanner to document which brain regions had greater or lesser blood flow in 23 individuals who were acutely intoxicated on THC compared to when they took placebo. She examined the size of fluctuations in their resting-state functional MRI, In addition to measuring altered blood flow and fMRI measures provoked by cannabis, she correlated these changes with the research volunteers' subjective ratings of feeling "high." Let's talk about her blood flow results first. Brain areas whose flow was most prominently shifted from placebo baseline by the drug included superior frontal regions that are involved in many complex cognitive and emotional processes, and that normally act as a "brake" to set limits on social and reward behaviors. Brain perfusion was increased in this frontal area by the drug. Blood flow was also boosted in the anterior cingulate cortex (ACC). This brain region is related to multiple psychological phenomena, including detecting errors and initiating motor responses to fix them, making choices, monitoring and resolving

---

a. The component brain regions within the DMN that constitute its "functional hubs" are the posterior cingulate cortex (PCC: located in the brain's midline towards the back end) plus the precuneus (in the midline part of the brain, towards the back, in the parietal lobe, sitting on top of the PCC), medial prefrontal cortex (also in the brain's midline, towards the front at about 11 o'clock) and finally the angular gyrus (located on the side of the brain, not in the midline, towards the back and inside the inferior parietal lobule).

conflicting situations, evaluating social phenomena, and interpreting pain phenomena. The ACC is also believed to have a role in conscious awareness and
experience, and possibly even with experiencing free will (see Francis Crick's
book [34]). Finally, the insula, associated with awareness of processes going
on inside the body as well as modeling where the body and its various parts
are in 3D space (proprioception), also showed increased perfusion following
cannabis. The insula (from the Latin word for island) is a large, complicated
brain region completely buried inside the temporal lobe. In a collection of other
brain areas, cannabis significantly decreased blood flow. These regions included
the primary somatosensory cortex, a large, well-defined gray matter region that
records and sorts out sensations of pain, touch, pressure, and temperature. As
well, it maps sensory space, (what's going on where in the body's world of sensation). Finally, blood flow also significantly decreased in the occipital gyrus,
a large swath of cortex at the back of the brain that records and interprets phenomena related to vision, all the way from the most basic (such as perceiving
light and dark) up to integrating and making sense of complex visual scenes.

Among the functional MRI measurements, cannabis increased signal fluctuations in the resting brain in the cerebellum, right insula, and substantia nigra,
areas that are packed full of cannabinoid CB1 receptors [29]. We discussed
the insula, above. The substantia nigra (Latin for "black substance") is one of
the brain's major dopamine-producing areas, and is situated in the midbrain,
underneath the main cortical areas at the back of the head. Many of its cells send
signals to the basal ganglia, large buried gray matter lumps that are related to
planning and facilitating body movements (especially eye-movement), but are
also concerned with more complicated functions including aspects of learning
and emotion, particularly reward-seeking. The substantia nigra probably has a
role in REM sleep [35], related to dreaming.

A second, separate resting state fMRI study [36] also examined the effects
of acute cannabis smoking (compared to a drug-abstinent baseline rather than
placebo) and found that during intoxication our friend the default mode network
(DMN) was transiently highly connected with parts of the cortex linked to hearing and motor movement, while simultaneously significantly negatively connected to the cerebellum and basal ganglia. The presence of this brain state was
associated both with the subjects' ratings of altered perception and blood THC
levels when they were intoxicated. We mentioned the basal ganglia in the prior
paragraph, but now need to say a word or two about the cerebellum. The structure, which resembles a large piece of pink broccoli, or a lemon-sized mini-brain
sitting in the back of the skull, deals with movement control and integration,
including learning motor movements and balancing one's posture. More recent
evidence also implicates it in cognitive domains such as attention and language
as well as modulating rewards, pleasure, and fear [37]. The cerebellum is also
intimately involved in our ability to estimate time, so-called "mental timekeeping." Evidence both from imaging and from people who've experienced brain
damage to the cerebellum suggests that the human brain possesses two separate

systems for time monitoring. One is a circuit that links the cerebellum with the frontal lobe and makes judgments regarding sub-second, fine-grained time periods. The second circuit links the frontal lobe with the basal ganglia and seems to be more concerned with time segments greater than 1 second [38]. Time stamping of memories occurs in the hippocampus, yet another region that is teeming with CB1 receptors, but was not implicated in van Hell's MRI studies.

Eight years later, and in a fresh group of research volunteers, van Hell's group repeated important aspects of their earlier study [39, 40]. They recaptured many of their earlier findings. For example, THC again increased blood flow in the insula and medial superior frontal cortex. Changes in the insula, at least on the left side were correlated with the subjects' self-rated changes in perception and in relaxation. A new finding was that of increased blood flow in the left middle orbital-frontal gyrus (MOFG), a brain region involved in making decisions about rewards in the environment, figuring out contexts, and calculating the subjective value of rewards. When subjects were stoned, the MOFG was also less connected to the default mode network. THC also increased neural activity in the brain's salience network that detects important stimuli inside the body and in the external environment in order to guide behavior and self-awareness. Some of you are likely already keeping a mental scorecard of what particular brain regions and their functions are speculatively linked to particular marijuana drug effects. We will get to that topic in a paragraph or two, so I beg your patience here.

Other research groups have conducted similar investigations and reached similar conclusions. For example, one study [40] showed that both smoked cannabis and an oral THC capsule increased connectivity between anterior prefrontal cortex, orbito-frontal cortex, and anterior cingulate cortex in direct correlation with increasing plasma THC levels. Oral CBD administration led to greater functional connectivity between the frontal lobe and the basal ganglia compared to placebo [41]. THC administration either in a joint or a THC pill reduced default mode connectivity and boosted the degree of anti-correlation between the DMN and the executive control network [42]. Other resting state studies did not use direct cannabis drug challenges, but merely made correlations with their subjects' self-reported cannabis use outside the lab, or examined cannabis users before and after a period of drug abstinence. Again the result findings are generally consistent with the other reports we've looked at earlier. For example, cannabis-dependent individuals showed connectivity between parts of the DMN and the insula that correlated with how long subjects had smoked cannabis [43] as well as different connectivity among DMN regions, middle and superior frontal cortex, cerebellum and motor cortex [44]. Other investigations reported altered functional connections between DMN and insula that correlated with amount of cannabis use [45], and increased correlation between cerebellum and inferior parietal lobule [46]. Finally, in 2012 another research group surveyed the existing literature published to date on resting state differences in functional imaging studies related to acute cannabinoid challenges, and concluded that depersonalization after cannabis administration was related to increased activity in the anterior cingulate cortex [47].

Resting state fMRI studies can be enormously informative about what's going on in the default mode network and the other brain circuits to which it's connected. However, if we want to learn more about the brain in action when engaged with cognitive tasks such as memory or reacting to different kinds of emotions, then a more appropriate experimental design is to have the research subject perform a related task paradigm, (such as learning a list of words or watching faces with different emotional expressions) when they are inside the scanner. Without getting into the complexities of how to design and interpret such paradigms, let's quickly review a handful of such task-based studies performed under the influence of cannabis to gain an additional perspective on how the drug alters brain activity in cognitive-related circuits. Healthy controls in one experiment underwent a learning task that involved reacting to threatening stimuli after being given placebo or a pill containing THC [48]. THC increased activation in the amygdala and part of the prefrontal cortex during the task. Subjects given the drug performed more poorly when tested 1 day later, and after 1 week those given THC, on the test day showed altered responses to the test cues not only in those two regions, but also significantly increased coupling between parts of the prefrontal cortex, hippocampus and dorsal anterior cingulate cortex. In other words, subjects responded differently to the threatening stimuli when acutely stoned and their brain circuits had learned to perform a task that involved recalling these cues that involved alternative brain regions. This experiment is important because it involves THC's effects on an active memory challenge, unlike the resting state studies, and clearly implicates functional alterations in the hippocampus, an area rich in CB_1 receptors and intimately concerned with memory, alongside regions that we have already seen are impacted by THC at rest. In task-related studies, the hippocampus also appears in other contexts related to emotions and to psychotic symptoms. For example, one research group of marijuana-using volunteers studied word learning under the influence of placebo or oral THC. These were individuals who were specifically selected because they usually either did or did not experience transient psychotic symptoms such as paranoia and anxiety when they used the drug [49]. All subjects activated the hippocampus during the learning task as expected, but in the placebo condition the left hippocampus was much more "switched on" when learning the list of test items in the psychosis-prone individuals. The more this brain region was activated, the more severe were the short-lived psychotic symptoms induced by THC in the drug condition. So that un-stoned baseline brain activity in the left hippocampus during memory formation was somehow linked with the tendency to experience temporary paranoia when stoned.

Now, as I promised, let's take a look at all of these brain imaging findings, especially those involving the regions most commonly implicated, and what we know about their functional relevance in relationship to what Tart and

Grinspoon had to say about the major subjective categories associated with feeling high.

- Perception of the external world. Alterations and distortions here are likely associated with changes in anterior cingulate cortex through its role in mediating perceptual errors; the insula gives information in terms of where the body is in 3D space, as well as integrating internal and external bodily information. The primary somatosensory cortex integrates information regarding spatial perception and sensory information from the external world, while the occipital/visual cortex mediates all aspects of visual perception.
- Visual perception. These changes are likely primarily mediated through the occipital/visual cortex and contributed by the anterior cingulate cortex, with its error detection function and separate effects on visual tracking through the cerebellum and basal ganglia.
- Hearing. Alterations occur in processes mediated by the auditory cortex, and this region specifically, as well as the temporal lobe in which it resides, are clearly affected by cannabis intoxication.
- Touch, temperature, and taste. The anterior cingulate cortex helps interpret pain stimuli, as does the insula in its role of dealing with awareness of events and processes occurring inside the body (including pain and temperature perception). The primary somatosensory cortex deals with pain, touch, and temperature information. Likely all of these regions contribute to altered perceptual experiences in these realms of consciousness.
- Space and time. Spatial awareness is contributed to variously by conversations among the insula, primary somatosensory cortex, cerebellum and hippocampus as well as, importantly, by the anterior cingulate cortex (that distinguishes perceptual errors). The inferior parietal lobule is a region that helps integrate the bits and bobs of incoming sensory experiences at a higher-order level to help make sense of them. The cerebellum with its role in attention and mental timekeeping likely plays a major role in subjective alterations in the passage of time. However, a portion of the substantia nigra is also activated during time reproduction, and if it is damaged mental timekeeping becomes inaccurate. So this "black substance" in the midbrain that is altered by cannabis likely plays an additional role in altered time perception. Finally, the frontal lobe is responsible for much of the ever-present mental chatter (or what Buddhists term "monkey mind") about the past and future. This is the process that mindfulness advocates urge us to suppress in order to fully experience the present. Cannabis seems to be able to stop parts of these executive processes in their tracks, thus providing a mental shove into here and now.
- Paranormal phenomena/ESP. The manner whereby the experience of these types of phenomena is coded in the brain is presumably extremely complicated and pretty much unknown. Since many people would question whether such phenomena actually exist to begin with, explaining how cannabis can convince individuals of their presence may be a fool's errand, or perhaps

an epiphenomen of 60s and 70s mystical thinking. On the other hand, paranormal phenomena may be consequences of our being tricked by our own brain processes. In that case, they are potentially amenable to study in the same manner as illusions. Cannabis' ability to reduce the anterior cingulate cortex's function of error-detection and to distinguish real from unreal phenomena, as well as the frontal lobe's role in mediating aspects of reality, likely underpin some of these experiences. When combined with some of the earlier-listed perceptual and timing distortions, all may be contributing to this category. Speculatively, if you are thinking less critically, "in the moment" and your sense of timing is disturbed, then you may be more easily persuaded, for example, that you thought of an event a short while before it occurred, rather than immediately after.

- Bodily experiences. If when stoned you are becoming increasingly aware of your heart beating in your chest and a little unsure of where your limbs are in space, then your insula and parietal lobe are likely being influenced by THC. The insula is the core brain circuit that interprets bodily awareness (such as paying attention to one's heart rate and internal sensations), with the inferior parietal lobe integrating some of these experiences with what is occurring in the environment. Impaired balance is likely mediated through disturbances in the cerebellum, which plays an important role in motor feedback. However, part of the substantia nigra also has an important (although indirect) role in dopamine-related motor feedback to the basal ganglia and in helping regulate moment-to-moment motor planning. Examples of its important role in motor regulation can be seen when it is damaged in Parkinson's disease.

- Social interaction. The anterior cingulate cortex helps evaluate social phenomena, and prefrontal regions encode many socially relevant phenomena.

- Sexuality. Increased pleasure/reward through interactions with the orbitofrontal cortex and the hypothalamus, increased tactile sensitivity (sensory cortex), as well as generalized disinhibition mediated through parts of the frontal cortex, all likely play a role here.

- Cognitive processes–memory. The hippocampus is intimately involved with memory formation, and is clearly compromised by exogenous cannabinoids. In Chapter 5, we discussed the role of cannabinoids in long-term potentiation processes related to memory construction inside the hippocampus, and their dependence on anadamide and other endocannabinoids. CB_1 agonists such as THC seem to impair this and related processes, leading to impairment of memory encoding, consolidation, and retrieval.

- Cognitive processes—thought. The frontal lobe and anterior cingulate cortex, with its connection to conscious awareness, probably play an important role in this phenomenon. If one is concentrating less on the future, and some of the brakes are off when it comes to logical linear thinking, (both effects of mild frontal lobe impairment), then these processes may account for some of the subjective reports of altered thinking.

- Moods and emotions. The amygdala handles many types of emotional stimuli, and is studded with cannabinoid receptors like cloves on a citrus pomander. Cannabis effects on this region are likely contributing to some of the drug's mood-related effects. The substantia nigra is an important hub of the brain's reward machinery through its connections with other parts of the dopamine-driven limbic system, including the nucleus accumbens and ventral tegmental area. This circuit as a whole is deeply implicated in trafficking in pleasures and rewarding situations It fires off when new events confound expectations regarding formerly predictable rewarding situations, so that updating reward-related learning is one of its more important functions. When potential goodies in the environment are judged to be important, (anything from chocolate, to dollar bills or attractive individuals) dopamine-based neurons in the substantia nigra switch on. Or, more correctly, they switch on and as a result we then judge the stimulus as important and/or rewarding. The entire circuit is also relevant to addictive behaviors. Addictive substances hijack these brain regions in the service of obtaining more drug and consequently ranking other activities as less pleasurable. When individuals with Parkinson's disease (whose fundamental pathology involves depleted dopamine levels in the substantia nigra) are treated with dopamine-boosting drugs to redress the balance, the therapy can inadvertently "overshoot" the mark. The medications goose up substantia nigra dopamine levels into the supra-normal range. In that context, there are many case reports of Parkinson patients who suddenly develop pathological gambling behavior or feel unaccustomed urges to go on shopping binges. Compulsive urges to engage in pleasurable and rewarding activities are being supercharged by figurative floods of dopamine. Parts of the orbitofrontal cortex are also clearly related to reward, and this region's more distant connections to reward-related regions are likely relevant here also, in being altered by cannabis.

Beyond the world of human neuroimaging, a recent paper from neuroscientists in London, Ontario used electrical recordings from neurons in the brains of live rats to parse anatomically rewarding versus unpleasant, avoided experiences provoked by THC [50]. The two opposite types of experiences were found to be coded in separate regions of the nucleus accumbens—nice ones in the front part and nasty types in the rear. THC's rewarding effects in these two adjacent regions were also dependent on different types of local opioid receptors with the resulting output modulating limbic dopamine levels as a kind of brain "see-saw". Translating these ratty results to the human world, the innate baseline receptor balance between the two parts of the nucleus accumbens may help explain why some individuals tilt toward bliss and others toward anxious paranoia when using cannabis. Related to this involvement of opioid receptors, at various times hashish/opium combinations have been fairly popular. Varieties of "opiated dope" circulated in Baltimore in the late 1970s and 19th century French literary types with their exaggerated descriptions of cannabis' effects also favored the combo. It's feasible that an optimal proportion of mu receptor agonists (e.g.

opioids) plus THC would tweak the latter's euphoria-inducing properties. This is a hypothesis that could be tested in a human behavior laboratory.

Control of intoxication

The frontal lobe's "brake" function is the most likely suspect when it comes to exerting temporary voluntary control and setting limits over intoxication, although this ability may be overwhelmed at higher drug doses. Many people who have used cannabis report being able to pull out of the "high" if they need suddenly to do something important that requires their immediate attention. "No officer, I have not been smoking marijuana." van Hell's functional MRI study documented that cannabis-related blood perfusion changes in the frontal cortex correlated negatively with her subjects' self-reported ratings of feeling high. She interpreted this observation as an interaction or tension between cognitive control and drug effect. In other words, this more highly developed brain region was able to figuratively step in and put a brake on the experience of cannabis intoxication when required to by circumstances. In some ways this is analogous to voluntarily waking oneself up from a state of drowsiness if necessary.

- Self-identity and spiritual experiences. These complex constructs are high-order functions of the brain that are exceptionally hard to parse, and incompletely understood, but portions of both the default mode network and the frontal lobe likely contribute. Given the possible association of depersonalization with cannabis effects on the anterior cingulate cortex, this region is also likely to be involved, in addition to its role in distinguishing real from unreal phenomena. The anterior cingulate's hypothesized role in conscious awareness and free will may be important also.
- Sleep and dreams. The substantia nigra is specifically related to sleep and dreaming, so that cannabis effects here are important.

In summary, cannabis produces a wide diversity of behavioral and psychological phenomena, some of which are explained by knowledge of the distribution of cannabinoid receptors in the brain, and experiments that indicate how drug-induced alterations in various brain circuits are associated with feelings and behaviors influenced by the drug. The "happy, hungry, horny" changes effected by the drug, along with perceptual and timekeeping alterations, seem more readily explained than the drug's effects on more complex and incompletely understood mental functions, such as sense of self and transcendence.

References

[1] Ginsberg A. The great marijuana hoax, 1966. The Atlantic. Available from: https://www.theatlantic.com/past/docs/issues/66nov/hoax.htm.
[2] Baudelaire C. Les Paradis Artificiels. France: Schoenhof Foreign Books; 1860.
[3] Davenport-Hines R. The Pursuit of Oblivion: A Global History of Narcotics. New York: WW Norton; 2004.

[4] Underwood E. Lab rats play hide-and-seek for the fun of it, new study shows, 2019. Available from: https://www.sciencemag.org/news/2019/09/lab-rats-play-hide-and-seek-fun-it-new-study-shows.

[5] On the preparations of the Indian hemp, or Gunjah (Cannabis Indica), their effects on the animal system in health, and their utility in the treatment of tetanus and other convulsive diseases. Br Foreign Med Rev 1840;10(19):225–8.

[6] Hecimovic K, et al. Cannabis use motives and personality risk factors. Addict Behav 2014;39(3):729–32.

[7] Osborne GB, Fogel C. Understanding the motivations for recreational marijuana use among adult Canadians. Subst Use Misuse 2008;43(3–4):539–72. discussion 573-9, 585-7.

[8] Sanz C, Tagliazucchi E. The experience elicited by hallucinogens presents the highest similarity to dreaming within a large database of psychoactive substance reports. Front Neurosci 2018;12:p7.

[9] Schafer G, et al. Investigating the interaction between schizotypy, divergent thinking and cannabis use. Conscious Cogn 2012;21(1):292–8.

[10] Webster P. Marijuana and music: a speculative exploration. J Cannabis Therapeutics 2001;1(2).

[11] Bachhuber M, Arnsten JH, Wurm G. Use of cannabis to relieve pain and promote sleep by customers at an adult use dispensary. J Psychoactive Drugs 2019;51(5):400–4.

[12] Filbey FM, et al. Individual and additive effects of the CNR1 and FAAH genes on brain response to marijuana cues. Neuropsychopharmacology 2010;35(4):967–75.

[13] Zhang PW, et al. Human cannabinoid receptor 1: 5' exons, candidate regulatory regions, polymorphisms, haplotypes and association with polysubstance abuse. Mol Psychiatry 2004;9(10):916–31.

[14] Pollan M. How do you put a drug trip into words? in The New York Times Book Review. The New York Times. 2018;8–9.

[15] Nia A.B., et al. Cannabis use: Neurobiological, behavioral, and sex/gender considerations. Addictions, 2018. in press.

[16] Grinspoon L. Marihuana Reconsidered. 2nd ed. Oakland, CA: Quick American Archives; 1977.

[17] Tart, CT, On Being Stoned. 1971, PaloAlto, CA.

[18] Cannabis. Available from: https://erowid.org/plants/cannabis/cannabis.shtml.

[19] Available from: https://erowid.org/.

[20] Crow HT, Peripheral visions and synthetic meaning, 2004. Available from: https://erowid.org/experiences/exp.php?ID=15293.

[21] Daniel. The experience, 2007. Available from: Https://erowid.org/experiences/exp.php?ID=55163.

[22] Das. O' bitter sweet canna, 2009. Available from: https://erowid.org/experiences/exp.php?ID=62658.

[23] G. T. Spinning while high may cause fun. 2017; Available from: https://erowid.org/experiences/exp.php?ID=70622.

[24] Dekamara. First time with powerful effect, 2019. Available from: https://erowid.org/experiences/exp.php?ID=97282.

[25] iSkank. Nostalgic panic on bong hits, 2017. Available from: https://erowid.org/experiences/exp.php?ID=87624.

[26] Castle D., Murray RM, and D'Souza DC, Marijuana and Madness. Cambridge University Press. Second ed. 2018.

[27] Solowij N, et al. Reflection impulsivity in adolescent cannabis users: a comparison with alcohol-using and non-substance-using adolescents. Psychopharmacology (Berl) 2012;219(2):575–86.

[28] Ranganathan M, D'Souza DC. The acute effects of cannabinoids on memory in humans: a review. Psychopharmacology (Berl) 2006;188(4):425–44.

[29] Iversen L. The Science of Marijuana. Second ed. New York, NY: Oxford Press; 2018.

[30] London J. Images outside dreams, 2005. Available from: https://erowid.org/experiences/exp.php?ID=19472.

[31] Chait LD, Fischman MW, Schuster CR. 'Hangover' effects the morning after marijuana smoking. Drug Alcohol Depend 1985;15(3):229–38.

[32] Old hippie's levels of consciousness scale, 2010. Available from: https://beyondchronic.com/2010/06/oldhippies-levels-of-consciousness-scale.

[33] van Hell HH, et al. Evidence for involvement of the insula in the psychotropic effects of THC in humans: a double-blind, randomized pharmacological MRI study. Int J Neuropsychopharmacol 2011;14(10):1377–88.

[34] Lane RD, et al. Neural correlates of levels of emotional awareness. Evidence of an interaction between emotion and attention in the anterior cingulate cortex. J Cogn Neurosci 1998;10(4):525–35.

[35] Nicola SM, Surmeier J, Malenka RC. Dopaminergic modulation of neuronal excitability in the striatum and nucleus accumbens. Annu Rev Neurosci 2000;23:185–215.

[36] Zaytseva Y, et al. Cannabis-induced altered states of consciousness are associated with specific dynamic brain connectivity states. J Psychopharmacol 2019;33(7):811–21.

[37] Schmahmann JD, Sherman JC. The cerebellar cognitive affective syndrome. Brain 1998;121(Pt 4):561–79.

[38] Harrington DL, et al. Does the representation of time depend on the cerebellum? Effect of cerebellar stroke. Brain 2004;127(Pt 3):561–74.

[39] Bossong MG, et al. Acute effects of 9-tetrahydrocannabinol (THC) on resting state brain function and their modulation by COMT genotype. Eur Neuropsychopharmacol 2019;29(6):766–76.

[40] Fischer AS, et al. Impaired functional connectivity of brain reward circuitry in patients with schizophrenia and cannabis use disorder: effects of cannabis and THC. Schizophr Res 2014;158(1–3):176–82.

[41] Grimm O, et al. Probing the endocannabinoid system in healthy volunteers: cannabidiol alters fronto-striatal resting-state connectivity. Eur Neuropsychopharmacol 2018;28(7):841–9.

[42] Whitfield-Gabrieli S, et al. Understanding marijuana's effects on functional connectivity of the default mode network in patients with schizophrenia and co-occurring cannabis use disorder: a pilot investigation. Schizophr Res 2018;194:70–7.

[43] Wetherill RR, et al. Cannabis, cigarettes, and their co-occurring use: disentangling differences in default mode network functional connectivity. Drug Alcohol Depend 2015;153:116–23.

[44] Cheng H, et al. Resting state functional magnetic resonance imaging reveals distinct brain activity in heavy cannabis users—a multi-voxel pattern analysis. J Psychopharmacol 2014;28(11):1030–40.

[45] Pujol J, et al. Functional connectivity alterations in brain networks relevant to self-awareness in chronic cannabis users. J Psychiatr Res 2014;51:68–78.

[46] Behan B, et al. Response inhibition and elevated parietal-cerebellar correlations in chronic adolescent cannabis users. Neuropharmacology 2014;84:131–7.

[47] Denier N, et al. Resting state abnormalities in psychosis compared to acute cannabinoids and opioids challenges: a systematic review of functional imaging studies. Curr Pharm Des 2012;18(32):5081–92.

[48] Hammoud MZ, et al. Influence of Delta9-tetrahydrocannabinol on long-term neural correlates of threat extinction memory retention in humans. Neuropsychopharmacology 2019;44(10):1769–77.

[49] Bhattacharyya S, et al. Increased hippocampal engagement during learning as a marker of sensitivity to psychotomimetic effects of delta-9-THC. Psychol Med 2018;48(16):2748–56.

[50] Norris C, et al. The bivalent rewarding and aversive properties of Delta(9)-tetrahydrocannabinol are mediated through dissociable opioid receptor substrates and neuronal modulation mechanisms in distinct striatal sub-regions. Sci Rep 2019;9(1):p9760.

Chapter 7

Epidemiology

"New Health and Human Services data confirm some of our worst fears about marijuana normalization."

Kevin Sabet, president of Smart Approaches to Marijuana, a lobbying organization, in a press release.

"Federal data shows youth marijuana use isn't increasing under legalization."

Kyle Jaeger, Science and Health section of Marijuana Moment.

These opposing quotes are from two press headlines issued on the same day (August 20, 2019) discussing the identical federal data released from the 2018 annual National Survey on Drug Use and Health.

This chapter examines some cannabis-related questions from large-scale, population level studies, and surveys. Who is using cannabis? Why? How are these nationwide trends changing over time? What does epidemiology have to tell us about cannabis and risk for psychosis, cannabis dependence, IQ decrements, lung cancer, violence, and motor vehicle accidents? Who is at risk for these problems? Is cannabis a gateway drug? How are these trends likely to change if the United States decides on nationwide legalization of recreational marijuana?

Cannabis is the world's most commonly used illicit substance. One in seven US adults, in a nationally representative survey of over 16,000, used marijuana in 2017, while almost 9% said that they had used the drug in the past 30 days. 20% reported using marijuana in the prior year if they lived in a state where recreational use was legal, versus 12% in states where neither medical nor recreational cannabis was allowed [1]. Breaking down use by administration method, almost 13% of the population reported smoking marijuana, 6% consumed edibles, almost 5% vaped, and close to 2% said they had used concentrates. Focusing only on those who used, over half (55%) preferred smoking. Extrapolating these numbers to the US teen and adult population, around 37.5 million US individuals reported past-year use and approximately 24 million endorsed use in the past-month.

Meanwhile, a huge social experiment and epidemiologists dream come true is going on over the border in Canada since legalization occurred in October 2018. It's too early to divine anything at this point, but this is a venture akin to the ending of alcohol prohibition, that will be intensely studied in the United States to help guide social policy over here.

Weed Science. http://dx.doi.org/10.1016/B978-0-12-818174-4.00007-0

Epidemiology as a branch of science can be tremendously useful in answering questions on a population level that are not otherwise addressable. For example, if schizophrenia usually occurs in 1 in 100 people in the population, and cannabis smoking in adolescence theoretically doubles this risk, then you will still need to study thousands of adolescents to find a sufficient number who have or have not smoked cannabis, and follow them along to see who ultimately develops this rather uncommon illness. In very basic terms, epidemiologists perform these kinds of investigations in defined populations, (such as all young adults in a particular city) to figure out among other agendas, who develops particular illnesses (e.g., men), when (e.g., in adolescence or old age), and where (e.g., in close proximity to a toxic waste dump). Once the epidemiologists have gathered this basic information, then they can estimate such things as how many people in their population are starting to develop a disorder, how many already have it, and what risk factors for the disorder might be. This information is important because it can identify, for example, who's getting sick and where and thus direct treatment resources appropriately, or stop an epidemic, signal that social policy changes need to be initiated.

One technique employed by epidemiologists is to follow a large population of individuals (termed a cohort) over time and carefully document what's happening with them. For example, a cohort could consist of individuals at increased risk for a particular disorder such as Alzheimer's disease, picked because they had a parent with the disorder, or all of the children born in 1 week in a particular country, followed from birth to see who develops which disorders as an adult. The pleasant harbor city of Dunedin in New Zealand is named after Edinburgh, the capital of Scotland from which many of its early immigrants derived, lured by an 1860s gold rush. As one of the main cities in New Zealand, Dunedin was selected by epidemiologists for a birth cohort study of over a thousand infants back in 1972–73. This multidisciplinary health and development study is still running. Whereas most longitudinal study investigators consider themselves lucky if they can get 70% of their participants to follow up long-term, an extraordinary 96% of the Dunedin cohort's 38-year-olds showed up for their 2011 assessment. At this point only about one-third of the original cohort still live in the city, but for the study the epidemiologists gather them from the four corners of the earth and transport the participants back to Dunedin for each wave of follow-up. These occurred initially every 2 years from ages 3 to 15, then at ages 18, 21, 26, 32, and 38. The Dunedin epidemiology study is important in its intersection with the world of cannabis in two major respects. First, since the children were studied many years before they first began to use cannabis, we know what they were like prior to drug use. Second, their drug use has been tracked fairly carefully so that the epidemiologists can speak with some authority on the relationship between youthful cannabis use and subsequent effects, and risk for developing psychotic illnesses, and on the drug's relationship to IQ. We will return to both of these issues a little later in this chapter.

Because circumstances are changing so rapidly in the United States with regard to cannabis, epidemiology is our most useful tool to track the altered use rates and resulting consequences, for example, dependence, motor vehicle accidents, or emergency room visits. Cannabis is significantly more available because its use has been legalized or decriminalized in multiple states. California legalized medical cannabis in 1996, followed by more than 30 subsequent states. Colorado and Washington legalized recreational marijuana in 2012, followed by another dozen states. Not only is cannabis more available, but also over the last 20 years, it has been viewed as increasingly harmless, and thus more acceptable: consequently use has increased [2].

One might think that it would be straightforward to exploit this natural experiment enabled by comparing contrasting and sometimes neighboring states that do or do not legalize marijuana, and seeing, for example, whether crime or addiction rates change differently in states that legalize. However, as we see later, analyzing these trends is not quite that straightforward.

Deborah Hasin at Columbia University has published much of the important work in this field looking at the number of individuals using cannabis in the United States and the types of problems they experience, based on large-scale epidemiologic surveys. She stresses that "… cannabis use is a health behavior that is important to study, just as alcohol consumption is a widely studied health behavior, even though not all persons who drink alcohol have drinking problems" [3]. Hasin also mentions some caveats in weighing this type of epidemiologic evidence. For example, epidemiologists often compare US state A, where recreational marijuana was legalized in a particular year, versus adjoining state B where the drug is still illegal, on such metrics as the change in driving while intoxicated arrests, crime rates, or emergency room visits. However such laws can pass in a state on a particular date, but actual marijuana sales may not occur for a year or more thereafter, (as happened in Massachusetts). So, the epidemiological clock needs to start when the first dispensaries opened, not when the legislators voted. Another issue is that individuals can easily drive over the border from state A to state B, purchase marijuana and drive back home to consume it, thus diluting any between-state differences. States may choose to alter how they define driving under the influence of drugs while a study is underway, for example, initiating blood test-based per-se limits. When that happens, the driving under the influence of drugs (DUID) statistics before and after the change in legal definition are no longer properly comparable. Individuals recruited into epidemiologic studies may report drug use differently when cannabis is legalized. For example, if they feel that cannabis use is more socially acceptable and is legal, then they may be more inclined to accurately report use. Finally, if recreational marijuana becomes legal and affordable, then use patterns of other recreational drugs such as alcohol may change significantly, and these altered patterns observed need to be taken into account by epidemiologists.

As pointed out by Prince and Conner however, cannabis use is almost always measured by having survey responders report only on use frequency, for example, the number of times or of days that they used in particular time periods, (for example, last 7 days, last 6 months) but not the quantity that they used, nor the potency [4]. All three metrics (how much, how often, and what was in it) may be extremely important if one wants to quantify, for example, whether risk for schizophrenia is related to the total amount of THC somebody has used. Epidemiologists need the ability to quantify substance use by employing quantitative measures of such use. For example, for cigarette consumption they employ "pack years" (the number of packs a day you smoked multiplied by the number of years you smoked) to calculate a quantifiable, straightforward measure of lifetime nicotine exposure. For cannabis a comparable measure is not intuitive. If you buy an ounce of weed from your friendly neighborhood dealer, you're often not exactly sure of the true weight, and certainly not clear about the percentage of THC and CBD in your purchase, not to mention other cannabinoids. And as Prince and Conner showed me in their experiment at the Denver social club in Chapter 8, users are extremely inaccurate in estimating the quantities of cannabis products that they load into a joint or dab rig. With alcohol, for example, measurement is very straightforward and translates readily across different types of booze. Alcoholic beverages come in standard sized bottles. One standard bottle or can of beer, glass of wine, or shot of spirits equates fairly accurately to a standard alcohol unit. However, no equivalent metric exists for cannabis. That's particularly unfortunate, as cannabis potency has tripled since 1995 as a direct result of selective breeding for increased THC content [5]. Another issue is that cannabis consumption often involves sharing a joint or bong with others, making your individual dose hard to estimate. Recently a couple of researchers from the psychology department at Washington State University came up with an improved survey measure of cannabis consumption that assessed frequency, quantity, and age of onset in detail. Some methodologic improvements included adding pictures of different quantities of cannabis (e.g., a gram or half a gram), in various forms (e.g., bud, loose herbal material, and "joint"), probing 7 different administration methods from blunts to vaporizers and edibles, clarifying personal use (as opposed to sharing) and typical THC levels in the cannabis used, and examining use of cannabis concentrates such as oils and wax. The instrument is called the Daily Sessions, Frequency, Age of Onset and Quantity of Cannabis Use Inventory, that conveniently if cheekily, collapses into the acronym DFAQ-CU (try pronouncing that out loud) [6].

Although it's straightforward to test subjects' urine for cannabis metabolites to confirm that people who tell you they are using marijuana are in fact screening positive for the drug, this is rarely done for reasons of practicality, expense, and convenience, so that we have to rely on people's word in most epidemiologic surveys.

Yet another complicating issue is that because marijuana legalization thus far has been left up to states, with no coordinated federal policy, states are not especially motivated to regulate or define in any coordinated way, such issues as cannabis quality control, uniform packaging standards for edibles, sale to minors, driving under the influence of cannabis, public consumption or diagnosed conditions for which medical marijuana is appropriate. This results in a hodgepodge of distinctly different regulations in different states and makes it hard to compare many key metrics such as those regarding DWI or problematic effects resulting from edibles in any consistent manner. Finally, correlation (as measured in epidemiologic studies) is not the same as causation. Sales of TVs and rates of motor vehicle accidents may be perfectly correlated, for example, (as income levels rise people can afford to purchase both more cars and more TVs), but TV sales do not cause car crashes. Keeping in mind these various caveats, let's take a look at some of the more important cannabis-related findings that epidemiologists have unearthed.

How many people are using cannabis regularly, what percent of adults and teens and for what reasons?

Cannabis is now widely used in the United States. In 2016, according to the Center for Behavioral Health Statistics and Quality, (CBHSQ) 22.2 million Americans of 12 years of age or older reported cannabis use in the past 30 days [7]. New data are now available for one further year, encompassing 2017, and the number of Americans using marijuana is now 24 million per the NSDUH survey (see below). As the perception of risk of using cannabis falls, (as it has done steadily), use tends to increase, and currently only 15% of 18–25-year-old view cannabis use as risky. Deborah Hasin and her colleagues examined a very large amount of data from two different phases of the National Epidemiologic Survey on Alcohol and Related Conditions (NESARC) from 2002 and 2012–13 along with additional survey findings in the general US population, such as the National Survey on Drug Use and Health (NSDUH) [3]. They discovered that adult cannabis use had increased over the past decade, and within most socio-demographic subgroups, all over the United States, in both rural and urban areas. Both cannabis use and cannabis use disorders grew significantly over this time span, with the biggest increases in men, young adults, Blacks versus Whites, low-income groups, the never-married, and urban residents. These increases began in 2007. The upward trend in adult and adolescent African-American use is notable, because it represents an alteration in a long-established pattern of higher use in whites. These data are consistent with earlier reports from the same group at Columbia University which showed that back in 2001 the prevalence of marijuana use was 4.1% and had more than doubled to 9.5% over the next decade. Equivalent figures for cannabis use disorder were 1.5% at the earlier date, which again had almost doubled to 2.9% in 2012–13 [8].

Adolescent cannabis use in the surveys stayed flat or tended to decrease, (from almost 16% down to 13%), but there was a recent uptick for adolescents who were older and non-white. Work from other investigators has shown that for high school age adolescent, tobacco and alcohol use has decreased over the same time period. Hasin also reported that cannabis use disorder decreased significantly in 12–17-year-olds from 4.3% in 2002 to 2.3% in 2014. These latter figures are especially important, as there has been much anxiety among public health officials that legalization/decriminalization of cannabis would lead to significantly increased consumption in this biologically vulnerable (as reviewed in Chapter 8) youth population. It's reassuring both that such increases are not occurring, and that medical marijuana legalization does not boost adolescent cannabis use, as occurs in adults [9]. Hasin also examined this trend [10] in over a million US adolescents in the 8th, 10th, and 12th grades across the years 1991–2014. Although overall cannabis use increased in states that legalized medical marijuana, 8th-graders decreased their use and no significant change occurred in 10th or 12th graders. A recent meta-analysis of 55 studies tracking cannabis use trends found that adolescent marijuana use fails to increase following cannabis legalization, and helps to clarify that prior reports of higher rates of marijuana use among teens in those states is better explained by generally higher rates prior to the passage of medical cannabis laws [11].

Although cannabis use disorders are growing overall, the risk of dependence formation among heavy marijuana users appear to have declined since 2002. Stephen Davenport at the RAND Corporation analyzed trends in marijuana dependence among daily/near-daily cannabis users, using data assembled from the National Survey on Drug Use and Health (2002–2016) to try and determine why they had changed. He discovered that dependence among these heavy users fell by 39% over the period studied from 26.5% in 2002–04 to 16.1% in 2014–16. Consistent with the overall result, when he analyzed individual dependence symptoms, most of these also showed significant declines. Examples of such symptoms included reducing important activities because of cannabis use, using the drug despite emotional, mental, or physical problems, failing attempts to cut back, spending lots of time obtaining, using, or recovering from marijuana and failing to follow through on limits set on use. It's hard to explain these data, but possible explanations include de-stigmatization of cannabis use, greater legal access, (presumably people spend less time skulking around tracking down their dealer), improved product quality and consistency, or perhaps use of edibles or vaporizers (leading to reduced smoking and use with tobacco). The more cynical possibility is that individuals were just less likely to detect their own cannabis-associated problems or to report them in the survey [12].

A few updates: at the July 2019 Research Society on Marijuana [13] third annual meeting, Silvia Martins from Columbia University reviewed recent unpublished data derived from analysis of NSDUH surveys between 2008 and 2016 examining the impact of marijuana legalization in Colorado, Washington, Arkansas, and Oregon, starting with Colorado's legalization in 2012. Past month and daily use of cannabis stayed close to flat in 12–17-year-olds, 18–25-year-olds, and those older than 26, as did rates of past year cannabis use disorder, except in the 12–17-year-olds, where there was a small, hard- to-interpret increase from 22.8% to 27.2%. An analysis within high schoolers across all states showed slight decreases in most grades post-legalization, with a very small increase in 12th graders, particularly among those who worked (which may give them access to increased funds). Youth were also more likely to use cannabis if they lived closer to a dispensary, and if they were exposed to cannabis advertisements [14], observations that are useful to guide social policy makers.

Because increasing numbers of people use marijuana to help treat medical conditions, it's important to know who these individuals are, how (edibles, smoking) and how much they are using, and what clinical conditions they are trying to ameliorate. Two Nebraska researchers recently surveyed over 169,000 individuals, half of them women, representing the US adult population [15]. Not unexpectedly, adults who reported medical conditions were more likely both to have used marijuana once or more in the last month, and used more than 20 out of 30 days in that month. Adults aged 18–24 had the highest prevalence of current marijuana use irrespective of medical conditions, more than 10 times higher than those age 65 and older. Among people surveyed who had medical conditions, 11% of young adults reported using marijuana on a daily basis. Overall, adults with medical conditions were almost twice as likely to be using marijuana. Those who did were less likely to report using cannabis for recreational purposes. Medical use varied widely among different US states with Alaska's being the highest. There, 38% of adults aged 18–34 with medical conditions were users. The biggest medical marijuana consumers were those with multiple medical conditions, and those illnesses, as stated by the patients included arthritis (64%), skin cancer (62%), kidney disease (57%), stroke (54%), diabetes (52%), and heart attack or coronary artery disease (each around 50%). Remember that many patients had multiple medical illnesses, and used medical cannabis to treat more than one condition, so that these totals exceed 100% by several-fold. Somewhat puzzlingly asthma (46%) and chronic obstructive pulmonary disease (COPD: 44%) were also commonly given as reasons for use, as was depression (49%). The survey raises many questions which it's not able to answer. Were diabetic users employing medical marijuana to help treat the nerve pain which is commonly found in the disorder? Was the depression improved by marijuana use or made worse (as has been reported in some surveys)? The overwhelming majority of COPD subjects (82% of them), smoked their medical marijuana. Was the drug helping improve their condition, or making it worse by irritating their lungs?

Large-scale studies with the ability to survey hundreds of thousands of individuals who mirror the population at large are incredibly useful in documenting national trends and allowing epidemiologists to draw broad, generalizable conclusions. That's not to say that small-scale studies lack usefulness, but such supplementary survey information must necessarily be weighed very differently. Some such data are derived from commercial cannabis companies surveying their customers; despite their small size, such studies are interesting because they can shed light on emerging trends in close to real-time. For example, Canopy Corporation and Tilray in Canada tracked several hundred seniors.

They, and studies in the western United States, suggest that older people comprise around 25% of the overall number of cannabis consumers. The Canadian data pick up an emerging trend not reflected in the US national survey earlier; senior citizens are overcoming their doubts and embracing medical marijuana use. These older baby boomers used cannabis in their youth and now seem attracted to legal marijuana for its medical benefits; for arthritis and other painful conditions, insomnia and various chronic health issues [16]. Similarly, a survey commissioned by the Colorado Department of Public Health and Environment examined past-year marijuana use in 274 community-dwelling residents aged over 60 years. The mean age of their sample was 72.5, with two thirds being women. Startlingly, 45% reported past-year marijuana use, of whom over half reported using cannabis for both medical and recreational purposes. Reasons for use included arthritis, chronic back pain, anxiety, and depression. The marijuana users reported improved overall health, quality of life, day-to-day functioning, and improvement in pain. Those who had used opioids in the past year were more likely to belong to the cannabis user group, likely related to chronically painful conditions [17]. Another small-scale Canadian survey from Kantar Consulting reported that although only 13% of surveyed Canadians report frequent cannabis use, half of those who do are millennials age 22–39 years [18].

Schizophrenia: epidemiologic studies

As we will explore in Chapter 8, cannabis use by young people seems to raise the odds that psychotic symptoms such as hallucinations and delusions will develop, as well as boosting risk for more chronic psychotic illnesses such as schizophrenia. To put things in perspective, only 1 in 100 people develop schizophrenia, and most young cannabis smokers never develop any psychotic illness. So an important question is, which cannabis users are at particular risk of developing psychosis? Epidemiology can help address questions like this [19]. As clearly argued by investigators at Yale [20], as a schizophrenia risk factor, cannabis meets many of the necessary criteria for a causal agent, just as cigarette smoking does for developing lung cancer. These include timing (cannabis use occurs prior to development of psychosis and adolescent exposure may be more significant), dose-response relationships (higher-THC cannabis increases risk more), biological plausibility (as we reviewed previously, cannabis may alter brain dopamine function and dopamine is involved with psychosis), experimental evidence (challenging people with CB_1 agonists can provoke psychotic symptoms), consistency (multiple studies all suggest increased risk), and coherence (epidemiologic findings are consistent with lab studies). On the other hand other factors muddy the bong waters. Psychotic disorders are complicated illnesses that have multiple identified risk factors; cannabis use is only one among many of these. Thus the specificity of cannabis' risk for developing schizophrenia is not likely to be particularly high (unlike its users) although certainly significant. We will review many of these points further in chapter 8.

Prospective epidemiological studies that follow their subjects through the age of risk of developing a disorder and carefully monitor drug use, consistently link cannabis use with increased subsequent risk of both (rare) more serious, chronic schizophrenia-like psychosis, and (common) fleeting psychotic symptoms. Early onset of use, daily use of high-potency cannabis, and consuming synthetic cannabinoids seem to carry the greatest risk, for example [21,22]. But let's dive into a few of these studies to see how they reached these conclusions, and how well their findings hold up to subsequent replication.

Back in 2002, the New Zealand Dunedin epidemiology team that I alluded to earlier, turned their attention to the risk of schizophrenia in their sample of 1037 individuals. The researchers wanted to resolve the question of whether any the increased risk of schizophrenia in cannabis users might be a consequence of psychotic symptoms that existed before the onset of the disorder rather than being due to cannabis. A longitudinal Swedish study of Army conscripts back in the 1980s had suggested that adolescent cannabis smoking increased the risk of later schizophrenia, but left the chicken and egg question open. In other words, were these young men individuals who had recognized the emergence of early psychotic symptoms and were attempting to self-treat via cannabis use? Having access to a sample that had been studied carefully since age 3 (well before any drug use had begun), and where psychiatric symptoms had been assessed every few years, offered the opportunity to unravel that question in a new population. Louise Arseneault led the Dunedin study analysis. She and her coworkers reported that their participants who had used cannabis by age 15 were 4 times more likely to have developed schizophrenia symptoms by age 26, compared to individuals who had either never used cannabis, or used it only once or twice between ages 15 and 18. The effect was stronger with earlier use. After psychotic symptoms at age 11 were controlled for, the risk for having a diagnosed adult schizophrenia-like disorder remained to some extent, but was no longer statistically significant. The crucial numbers in this study were small; only 25 of the total sample had developed schizophrenia by age 26 (10% of the cannabis users versus 3% of the controls).

These early New Zealand epidemiology data were provocative, but left open many questions. Like good researchers, the team kept picking away at the problem. Many teens use cannabis but few develop schizophrenia. Are some individuals at particular risk, perhaps for identifiable biological reasons? A later, 2005 investigation from the Dunedin researchers led by Avshalom (Avi) Caspi [21], asked the question "are some individuals genetically predisposed to be vulnerable to cannabis effects in a way that makes them more likely to develop schizophrenia?" Caspi suggested that normal variation in the gene coding for the enzyme COMT (that breaks down dopamine in brain neurons) might influence psychosis risk. A little background on genetics is necessary to unpack these findings. For many common normal variations in our DNA, there are two normally occurring variants that we can think of as "flavors" (these are technically known as polymorphisms), in the DNA that codes for a particular gene. These "flavors" represent variations in DNA bases. Changing the DNA's base letter can sometimes result in an amino acid swap when the DNA is translated into a protein molecule. The technical name for the DNA difference underpinning such variation is technically termed a "single nucleotide polymorphism," usually abbreviated to SNP, and pronounced "snip." For each of the many possible thousands of SNPs in all of our genomes, each of us inherits one copy of one of the variants from one biological parent and a second copy from our other parent. The variation determines which of two amino acids is expressed in the resulting protein chain. Sometimes these swaps make absolutely no difference in how the gene functions; in other cases (so-called "functional SNPs"), the resulting protein has different physiological properties. It can perform its molecular role either more efficiently, less efficiently or occasionally not at all. For example, sickle cell disease is due to this type of swap affecting the hemoglobin gene. In the case of the COMT enzyme, at base-pair location 158, the two different amino acids produced by the variant SNPs are valine (usually abbreviated to Val) and methionine (abbreviated to Met). Thus there are three possible "flavors" of COMT: Val-Val, Val-Met, or Met-Met. You may be asking why, out of the 30,000 or so candidate genes in our genome, the Dunedin investigators would focus on COMT. The logic was based on the following. The COMT gene is located on human chromosome 22q11; a micro-deletion on is part of chromosome 22 that contains the DNA for COMT, (plus several other genes) results in an uncommon syndrome abbreviated to VCFS. Individuals with the syndrome tend to exhibit facial, heart, IQ, and ENT abnormalities, but relevantly have about a 40% chance of developing a schizophrenia-like illness, or 40 times the population risk. Second, the COMT gene breaks down dopamine inside of neurons, and there is suggestive evidence that excessive amounts of dopamine contribute to the risk of psychosis. And Val-Val and Met-Met flavors of COMT break down dopamine with significantly different levels of efficiency, so the SNP definitely translates into a functional difference. So off the researchers went, drew blood samples from all 26-year-old participants and genotyped their DNA for the COMT 158 Val-Met variant. As we saw earlier, these individuals had already been assessed for psychosis by Louise Arseneault's team. This time around, the researchers assessed adult cannabis use at ages 21 and 26 to see how often the study participants had used cannabis in the last year.

The results of the analysis were striking. The Val gene variant increased psychosis risk in a striking SNP-variant dose-related manner [21]. Merely having one or other variants of COMT by itself had no bearing on schizophrenia risk. The effect of adolescent cannabis exposure was significant, as seen already in Arseneault's study. But there was a significant interaction between one's COMT profile and cannabis use. The greater your Val "dose," the higher was the risk of schizophrenia that youthful cannabis use posed. Thus, if one had the Met/Met genotype, then the risk of developing a schizophrenia-like illness by age 26 was about 4%, whether or not one had smoked cannabis as a teen. If one possessed the Val-Met genotype, then about 2% of non-smokers and about 5%

of cannabis smokers, not really very different, had developed the disorder. But the risk was tending to distribute a tad more toward the Val side. But for those individuals with the Val/Val variant, about 2% of the former teen non-cannabis users were affected with schizophrenia, but if one had smoked cannabis as a teen, then the figure leapt to around 14%. If they used cannabis as teens, carriers of the COMT valine 158 allele were both much more likely to exhibit psychotic symptoms and more likely to develop a schizophrenia-like disorder, whereas cannabis had no comparable influence on individuals with the Met-Met variant.

These numbers suggested what geneticists call a "gene-environment" interaction, that is, one's particular genetic makeup makes one more (or less) vulnerable to some factor in the environment. For example, ethnically Asian people are more likely to have a particular variant of a gene that breaks down alcohol in a way that produces larger amounts of the mildly toxic chemical acetaldehyde, and causes facial flushing and nausea when they consume alcoholic beverages. In other words, their particular gene variant is lying dormant and undetected until it interacts with something (in this case alcohol) in the environment. When it does so, it manifests as red blotches on the face and upper body, a runaway heart rate and vomiting. In the case of the Dunedin sample, the environmental component of schizophrenia risk was cannabis use, and the genetic risk factor COMT genotype "flavor." However neat and convincing this story appears, there are problems with it. Subsequent attempts at replication in other samples do not reliably reproduce this original finding. When very large-scale genome studies searching for schizophrenia-risk genes turn up candidates, COMT is not among the 108 genes discovered. Similar claims for gene/environmental cannabis risk have been made for SNP polymorphisms in the AKT1 gene and psychosis risk, and these also do not replicate reliably. No single schizophrenia risk SNP conveys a high degree of risk for the disorder. The risk from common gene variants such as COMT Val/Met is almost certainly additive, with hundreds of genes each contributing a tiny amount of vulnerability, so that logically a striking effect from a single gene variant would be unlikely. Thus, variations in COMT may be playing some sort of role in genetic risk in the cannabis-psychosis story, but it seems unlikely to be as dramatic as that initially reported from the Dunedin study. One emerging subtlety is that there may be a three-way interaction of COMT polymorphisms, cannabis use, and having been abused as a child [20–23].

Try the following as a thought experiment. We know that both schizophrenia and substance abuse tend to run in families. This suggests that both have a genetic basis, and indeed risk genes for both conditions have been identified. One theoretical possibility is that the genes conferring risk for the two disorders overlap, so that the reason that cannabis use and schizophrenia appear to be related is that they share a common underlying (genetic) cause. Thus this line of argument would suggest that both conditions were caused by the same underlying risk, and not that one (cannabis use) was leading to the other (schizophrenia). Although there is some disagreement here, most studies trying to unravel this question have clarified that cannabis' risk-increasing effects for psychosis are most likely not explained by a shared genetic predisposition for both schizophrenia and cannabis use. Another relevant puzzle piece is that ratios of THC to CBD of (illegally sold) cannabis in the United States increased from 14:1 to 80:1 between 1995 and 2014 [5]. Recall that as chemovars are bred for higher THC, then CBD percentages necessarily fall. Given the potential role of cannabidiol as an antipsychotic, (see Chapter 10) some have claimed that this altered THC:CBD ratio, in addition to the increasing concentrations of THC by itself, is one factor driving increased cannabis-associated psychosis risk. So the next epidemiologic study, we examine on that particular issue—are the amount of THC in smoked cannabis, and how often that cannabis is smoked, relevant to the risk of developing schizophrenia?

Marta Di Forti is part of a large team of epidemiologists, statisticians, and clinical psychiatrists with expertise in psychosis, based at the Institute of Psychiatry and the Maudsley Hospital, both in London, United Kingdom. The research group includes Robin Murray, (now Sir Robin) a prominent psychiatric researcher who has been hunting down the relationship between cannabis and psychosis for the past 25 years. He was a member of the investigators that had examined the data from the 2002 Dunedin sample. Di Forti [22] examined more than 900 first-episode psychosis patients and over 1200 community controls across 11 sites in Europe and Brazil in an observational study. This is the largest and most informative epidemiologic investigation carried out to date, looking at the relationship between cannabis smoking and the new onset of psychosis. The research is particularly interesting because the sample size was big and very different types of cannabis are available at the various locations in Europe that were studied. For example, in Amsterdam, London, and Paris high-THC cannabis is readily available, whereas these chemovars have not yet reached Italian cities such as Palermo, that participated in the study. Statistical models used by the investigators estimated that daily cannabis use was associated with an increase of just over 3 times the odds of developing a psychotic disorder compared with people who would never use the drug (the odds ratio was 3.2). This study excluded individuals if they met criteria for short-lived psychotic symptoms resulting from acute intoxication, such as some of the case examples we will encounter in Chapters 3 and 8.

A quick statistical diversion into odds, likelihoods, and probabilities will help in understanding Di Forti's findings. Ladbrokes is a century-old betting and gambling company in the United Kingdom that is rumored to allow punters to wager on any event. Their offices will provide odds not only on the outcome of racehorses and sporting events, but also political contests and world events. Epidemiologists too use the term "odds ratio" (abbreviated to OR) to quantify risk for an event happening; let's unpack the term a little bit. The difference between probability, likelihood and odds is as follows. Probability is the percentage that something will occur; if you flip a coin repeatedly, the probability over a million coin tosses that it will come up heads is 50-50, or 0.5. Probabilities are always expressed as some number between zero and one. Likelihood speaks more to the probability of a certain outcome occurring; for example, theoretically in your coin-toss experiment you could come up with heads 10 times in a row, but the likelihood is low. The odds of an event are simply a ratio of the probability that the event will occur divided by the probability that it won't. In the sort of case-controlled designs favored by epidemiologists (population X lives near a toxic waste dump, population Y lives 25 miles away, is population X at increased risk of developing lung cancer?), we can't really calculate the probability of disease accurately in each of the exposure groups. Therefore the relative risks cannot be pinned down precisely. However we can compute the odds of the disease in each of the two groups, and then compare the two values by calculating the odds ratio. Unlike probabilities, odds ratios are open-ended; if I smoke a pack of cigarettes a day for 50 years my OR for lung cancer may be 20, if I kiss someone with an active swine flu infection, my OR for developing the disorder may be 100. I'm merely guessing at this latter number and certainly not volunteering for any study that aims to answer this question. Incidence risk ratio is a related statistic. Incidence rate is the number of events divided by how long someone is at risk for something occurring; incidence risk rate puts a number on the rate of a disorder in people exposed to something divided by the rate in the unexposed individuals; it's a helpful estimate in cases where the risk is relatively low.

For example, the risk of developing schizophrenia is somewhere around 1 new case in every 100 people. The equivalent risk in individuals who smoke high-THC cannabis daily, beginning in their teens, may be two such cases. Theoretically, one of these affected individuals may be somebody who was already destined to develop the disorder (although there is no current way of knowing that for sure). The drug may have served to nudge them in the direction of the illness a little earlier. The second case may be somebody at lower inherent genetic risk of schizophrenia, or at slightly elevated genetic risk but with more environmental protective factors such as being raised in a loving, stable home, who is nevertheless propelled down the road to the disorder as a result of cannabis use. By this logic, the incidence risk ratio for schizophrenia is two for cannabis-exposed individuals.

So, back to Marta Di Forti. Her study determined that the OR for new-onset psychosis increased from 3.2 with daily cannabis smoking to 4.8 for daily use of high-THC cannabis types. In other words, the risk for a psychotic illness increased almost 5 times in individuals who smoked cannabis with a THC content of 10% or greater every day, compared to people who never used. The odds among users of high-potency cannabis overall (i.e., adding together people who used daily and those who used less often) were 1.6 times higher. In a recent radio broadcast on this issue, Dr. Di Forti made the dosing analogy of high-THC content cannabis to the link between alcohol consumption and liver damage. Drinking a bottle of whiskey a day may be more toxic to your liver than 20 bottles of beer [24], although the total amount of alcohol consumed may be equal.

To clarify the status of the patients, these new-onset psychosis individuals in her study consisted of cases whose key symptoms such as hallucinations were non-transient and sufficiently serious to have come to the attention of a psychiatrist, although not necessarily diagnosable as schizophrenia. There was no evidence that these individuals had exhibited any mental illness prior to their cannabis use, nor, (unlike the Swedish study we discuss next), did they have strong family histories of schizophrenia. Their genetic status with regard to polygenic risk scores for schizophrenia is important, but was not measured here however. (A polygenic risk score sums up all of the individual risks carried by each SNP variant associated with schizophrenia across hundreds of thousands of genes to arrive at a composite risk number for a particular individual). Di Forti's study also tried to control for the effects of stimulant medications such as cocaine, as well as possible confounds from socioeconomic status and educational level.

Overall she determined that the adjusted incidence rate for psychotic disorder was related both to use of high potency (i.e., high THC) cannabis r=0.7, and to daily use r=0.8, and these two effects seemed independent of each other. The epidemiologists went on to calculate that if high-THC cannabis were no longer available, then 12.2% of cases of first episode psychosis could be prevented across the 11 sites, rising to 30% of cases in London and 50% in Amsterdam. Interestingly, and in distinction to some prior studies, starting to use cannabis by age 15 only slightly increased a person's OR for psychotic disorder, and this small boost in risk was not independent from how often he or she used or how much THC was in their cannabis. Short of testing people's urine and hair for THC metabolites (which would tell us a little more regarding their recent historical use) or actually testing cannabis samples from each participant supply for CBD and other cannabinoids as potentially protective or risk-modifying factors, this study was carried out in an extremely rigorous fashion. The fact that it was performed in multiple cities across Europe adds to its generalizability. Because of the impossibility for ethical and practical reasons (as discussed in Chapter 2) of conducting a random population challenge experiment, the Di Forti observational study presents important epidemiologic evidence for links between cannabis use and psychosis risk.

Because Marta DiForti and her team were not able to gather detailed information about the cannabis-using individuals who developed psychosis *before* they became ill, they were however unable to address one vital question. That is, were these individuals likely to have developed the disorder anyway, and cannabis merely nudged them along the path to illness. Or, on the other hand, did it truly cause the illness in people who would never have developed chronic psychosis to begin with? This is an important issue to address, and it is not just a theoretically interesting but unresolvable chicken/egg paradox, but one that is actually answerable. And indeed it has been answered. A very recent study sheds important light on this issue. Kenneth Kendler at Virginia Commonwealth University is a scholarly, much-published psychiatrist, schizophrenia researcher, and geneticist. He and his collaborators conducted a detailed study of over 7500 individuals in Sweden, land of socialized medicine, meticulous record-keeping and careful patient follow-up, to address this exact chicken/egg question [25].

First, they identified people who had developed a short psychotic illness in the context of drug use that lasted for a month or more. These are individuals like James in the next chapter, who developed an acute persisting psychotic state that in his case was clearly linked to cannabis consumption. They followed all of these 7606 individuals using a carefully detailed Swedish national health registry and figured out what their psychiatric diagnoses were an average of 7 years later. They also tracked whether these individuals were working, and identified all of their both close and distant relatives. And by searching the health registry, the researchers determined how many of these relatives carried a diagnosis of schizophrenia.

The most important findings are as follows: the drugs that were associated with the initial, relatively short-lived psychotic episodes were mostly alcohol, cannabis, and stimulants such as cocaine and amphetamine. No surprises there; all of these drugs are known to increase risk for psychosis. Overall just over 10% of the individuals with a brief psychotic episode occurring in the context of substance use went on to develop schizophrenia. These latter were people who resembled Janet (in the next chapter) who had a chronic psychotic illness that qualified for a diagnosis of schizophrenia. This 10% figure is significantly higher than the average 1% population risk for schizophrenia, so that experiencing the earlier brief substance-induced psychosis episode had boosted these individuals' risk for schizophrenia by 10 times. Interestingly the patients who had converted to schizophrenia by 7 years after their initial episode were more likely to have used cannabis than other drugs (nearly 20% of such people went on to develop the chronic illness), versus less than 5% of those who had become psychotic after abusing alcohol. The risk for stimulant users lay somewhere in between these percentages. As a group, the 7000+ people who started off with a diagnosis of substance induced psychosis were much more likely than average to have family members who themselves were substance abusers. Again, no surprises; substance use tends to run in families. But more interestingly, they also had higher numbers of family members with a diagnosis of schizophrenia than the average person in the street. But when the 10% who converted to schizophrenia were compared to the 90% who didn't, those who had later developed the more serious illness had the same number of family members with schizophrenia as regular schizophrenia patients who had not abused drugs. In other words, they were at significantly increased genetic risk of developing the illness compared to the population at large because they tended to have many close relatives with schizophrenia. In fact they had many more affected relatives than did the 90% who never progressed to chronic psychosis. Another factor that put individuals at risk of developing schizophrenia after their initial acute psychotic episode, was being sufficiently sidelined by that first illness to have to apply for permanent disability.

So what are we to make of from Kendler's Swedish data? One obvious conclusion is that while substance use (especially cannabis) may be pushing people
in the direction of psychosis, those who develop a permanent, serious illness
following use of the drug seem to have been already at risk for the disorder
based on their family history of schizophrenia. So the drug likely did push them
a bit further and faster along a path to illness that they were presumably already
on due to their genetic risk. But cannabis most likely "does not *cause* schizophrenia, (*my emphasis*) even among individuals who have developed psychosis
as a consequence of such substance abuse" [26]. That's not to say that cannabis is blameless, or that we can ever know for sure based on family history,
who will or will not develop schizophrenia. But we can say that those who
develop chronic psychosis in the context of cannabis risk are not just random
individuals. They are people who started off life at increased genetic risk of
schizophrenia by virtue of having family members with the same illness. There
are two other practical lessons to be drawn from these facts. One is that people
who develop a short-lived psychotic illness related to cannabis use that lasts a
month or more should quit using cannabis, because their risk for developing
schizophrenia is 10 times higher than that of the general population. The other
is that if somebody has a family history of schizophrenia, then they are taking
a calculated risk if they use cannabis, and probably shouldn't be tempting fate.

Does using cannabis lower your IQ?

The trope that "dope makes you dopey" is frequently invoked, but what's the
evidence for this? For example, chronically cannabis-intoxicated daily users
may be portrayed as overly mellow and slow on the uptake, but do these effects
vanish if they abstain from the drug for a few days or a few weeks? University of
Pennsylvania investigators reviewed almost 70 cross-sectional studies of cognitive functioning in a total of over 2000 adolescent and young adult cannabis
users [27]. They concluded that "associations between cannabis use and cognitive functioning… are small and may be of questionable clinical importance
for most individuals. Furthermore abstinence of longer than 72 hours diminishes cognitive effects associated with cannabis use." Although this review
was restricted to cross-sectional studies, because of its scope it does provide
useful information. But what happens when we follow the same set of individuals across time, beginning before they ever used cannabis? Relevant evidence comes again from the Dunedin study. Madeline Meier and her coworkers
looked at cannabis use in participants at ages 18, 21, 26, 32, and 38 years [28].
Neuropsychological testing was carried out at age 13, before people began using
cannabis, and again at age 38. These researchers reported that people who used
cannabis persistently had declines in their neurocognitive test scores over time
and reported more subjective thinking problems, even after controlling for years
of education. Cognitive impairment was most obvious in adolescence-onset
cannabis users, and the more people had used the drug the greater their degree

of cognitive drop. For individuals who had smoked heavily as adolescents and then quit, stopping use did not fully restore their missing IQ points back to baseline pre-drug levels. The investigators tested the participants' cognitive abilities in a great deal of detail, looking at verbal and performance IQ as well as almost 20 tests of psychological domains comprising memory, information processing, processing speed, reaction time, perceptual reasoning, verbal comprehension, and vigilance. The advantage of this longitudinal study design was that each participant could act as his or her own control. Given that everybody started with a different IQ score as a child, it was straightforward to see whether peoples' scores went up or down over time in relationship to their cannabis use. The researchers were also able to examine dose effects, in terms of whether an individual had used cannabis at the time of one, two, three, or more study evaluations. They found that study participants who had never used cannabis had a very slight increase in IQ over time, whereas for users there was a steady decrease in IQ proportional to the number of study occasions that a person had smoked cannabis. For example, 38 people who were using the drug at three or more study waves had a decline of around six IQ points. Since IQ captures general function across a large number of cognitive domains, one question is whether any specific area of ability seemed particularly affected. When Meier and her crew probed this question, carefully adjusting for childhood IQ and the subjects' sex, all domains seemed equally impaired. The bottom line was that persistent cannabis use over 20 years was associated with cognitive decline, that more use was associated with bigger losses, and that the effect was restricted to adolescent-onset cannabis users.

The investigators tried to rule out alternative explanations; they calculated effects of recent use of cannabis (i.e., were their subjects stoned when being tested), persistent tobacco dependence, use of alcohol or other recreational drugs or schizophrenia. Any or all of these possibilities could impair IQ. The statistics get very complicated, but they suggested that cognitive effects stayed significant when adjusted for all of these potential confounding variables. The Dunedin scientists wondered whether the neuropsychological impairment associated with cannabis had any effects on their subjects' everyday functioning. They checked out that possibility by asking each participant in the study to identify individuals who "knew them well." With the permission of the subjects, these informants were then contacted and completed a checklist asking among other things whether the study member seemed to have shown problems with their attention and memory over the last year. Those with persistent cannabis dependence were identified by their compadres as having significantly more problems in those areas. So, case closed, you would think. Smoking dope makes you dopey. But not so fast.

Within a year of the Dunedin report, another paper was published in the same scientific journal challenging Meier's findings and suggesting that socio-economic status might be a factor explaining both changes in IQ and likelihood of using cannabis as an adolescent. [29]. To quote from that later publication

"This model… would predict reduced IQ in so far as heavy, persistent adolescent-onset cannabis use involves a culture and norms that raise the risk of dropping out of school, getting entangled with crime, and other such behaviors…. Because the effect in this case would be a result of culture rather than pharmacology, it would also have different policy implications." In other words, cannabis might be along for the ride, but is not causal IQ-wise. Further criticism of the Dunedin finding came from a different direction. Epidemiologists: meet geneticists. Nicholas Jackson from the University of Southern California, William (Bill) Iacono from the University of Minnesota and collaborators from Sweden and Pennsylvania looked at data from two longitudinal studies of adolescent twins, numbering around 800 and 2300 individuals respectively [30]. The twin samples were studied respectively at the University of Southern California and recruited from LA area school districts, and at the University of Minnesota drawn from two twin cohorts derived from the general population. I will not resist the temptation to allude to the Minnesota twins. Because twins mostly come from the same family backgrounds, (unless separated early in life) and are very similar to each other genetically (in fact identically so in identical twins), they constitute the perfect controls for each other. If one twin smokes cannabis and the other doesn't, for example, then you have a nice tightly controlled mini experiment that tends to rule out confounding factors such as family background and socioeconomic status. Notionally, this is somewhat similar to the thought experiment I've referred to where half of the population is chosen to be deliberately exposed to a drug and half to placebo. With epidemiologic twin studies, a researcher is able to look across his or her entire twin sample and summarize all of the relevant effects. For the Minnesota/California samples, the researchers had measured their twins' intelligence between 9 and 12 years of age, before any involvement with cannabis, and repeated these IQ measures between ages 17 and 20 years. As in the Dunedin study, marijuana users had lower test scores relative to cannabis non-smokers, and showed significant IQ falls over the time of the two measurements. But here's where the similarity in findings ends. Not only was there no relationship between how often people used cannabis and their change in IQ, but the cannabis-using twins showed no more IQ decline than did their cannabis abstinent siblings, whose IQ measures also fell. How could that possibly be the case? Getting high in adolescence, as the investigators point out, "occurs within a broader delinquent context in which alcohol and other drugs are used." And when the investigators included binge drinking and other drug use in the model, the effect of marijuana use on vocabulary decline entirely disappeared. So for a start, IQ decline no longer seems to be specific to marijuana use but rather related to general substance involvement. These findings seem reminiscent of what we found in the freshman college student sample that I will discuss in the next chapter. So why would IQ also decline in the abstinent co-twins of the cannabis-using subjects? The explanation favored by Jackson, Iacono, and their colleagues was as follows. First of all in the baseline assessment before any subject even used marijuana, future cannabis users already had significantly lower IQ subtest scores. Thus cannabis did not lower their IQs; they were low from the get-go. This observation undermines the temporal association we discussed earlier. Next, the investigators became convinced by the data that at least part of the relationship between marijuana use and IQ drops, is that there are common underlying risk factors that account for both phenomena. For example, numerous studies show that being behaviorally disinhibited (e.g., displaying conduct disorder, delinquency, acting out, and excessive risk-taking) are behavioral traits that predict both substance use disorder and lower IQ. Social factors such as coming from families that do not value education and are more lax in supervision of children, may be a factor as well, as is the possibility that delinquent kids are less likely to perform well academically due to attitude and school absence, and are more likely to use drugs. So, went the argument; cannabis had not caused the plummeting IQs. A collection of socioeconomic and family factors had led to both. Faced with two apparently contradictory findings from two well-conducted studies, how can this issue be resolved? One difficulty in comparing the twin study to the Dunedin sample data is that the latter looked at 38-year-old individuals who had used cannabis over a significantly longer time period than had that of the 17–20-year-old twins. Perhaps prolonged exposure and the passage of time are necessary for effects to become more manifest and measurable. Or perhaps longitudinal cohort studies and twin studies are somehow inherently not comparable. But this is where our story gets even more interesting.

For those of you who might still believe that science is not exciting and lacks battles, in 2018 Madeline Meier, Avi Caspi, Louise Arseneault and many of the original Dunedin investigators decided to look for themselves at associations between adolescent cannabis use and cognitive decline in a large sample of almost 2000 twins from the British Environmental Risk Longitudinal Twin Study. In other words, they crossed the methodologic divide and adopted the other side's research design, perhaps in the hope that their scientific rivals' findings might fail to replicate. The British teens that they chose to study were members of a nationally representative cohort of twins born between 1994 and 1995. Here, IQ had been measured at ages 5–12 and 18 with additional testing of executive function at the last time point. Cannabis use and dependence was assessed in the 18-year-olds. So what did they find—an effect more like that from the Dunedin study, or one more closely resembling the Minnesota/California results? Adolescents who had used cannabis (compared to those who didn't) had lower childhood IQs before they ever smoked the drug, and also lower scores at age 18. IQ scores in those with cannabis dependence were nearly 6 points lower at age 12, and 7 points lower at age 18 than those without such use, but didn't decline across that time period. Cannabis-users also had worse executive function overall at age 18, but crucially, as in the Minnesota study, no lower than their non-cannabis using co-twins. The investigators concluded that "Short-term cannabis use in adolescence does not appear to cause IQ decline or impair executive functions, even when cannabis use reaches the level of dependence. Family background factors explain..." (these relationships). This finding is particularly notable since the investigators identified no effect of cannabis on IQ, even when their teen subjects had reached a more serious level of drug use (dependence) than previously studied, where one would expect any negative consequences of the drug to be more apparent.

The most parsimonious explanation that helps reconcile these rather different findings is that adolescent cannabis use may not have a detectable impact on cognition unless it occurs at very high levels and/or over many years. A rather sobering study comes from an integration of data across three large longitudinal studies from Australia and New Zealand [31]. The number of participants (depending on the individual factors analyzed) varied from approximately 2500 to around 3800. The study looked at the maximum frequency of cannabis use before age 17 and how the subjects were faring in the community up to age 30. The outcome was dismal. There were dose-response relationships between how often the individuals had used cannabis as adolescents and unfavorable adult outcomes. These ranged from their being less likely to complete high school or obtain a university degree, and more likely to have adult cannabis dependence, to be using other illicit drugs, and to have attempted suicide. However the conclusions of this study must be tempered by the caveats we discussed earlier, namely Rogeberg's hypotheses about rebellious adolescent culture and the many subsequent adverse outcomes associated with an impulsive, acting-out youth. Again, in other words, the argument would be that antipodean cannabis users' dismal adult outcomes were driven, along with their cannabis use, by their pre-existing culture and temperaments. A final epidemiologic investigation on cannabis use and IQ assessed 15 and 16-year-olds from over 2000 teens participating in a British longitudinal study of children and parents [32]. After statistical adjustment, those who had smoked cannabis more than 50 times did not differ from never-users on either IQ or educational performance. The relationship between cannabis use and both outcomes pretty much evaporated when cigarette smoking was considered, and in fact there were strong relationships between cigarette use and educational outcome, that persisted even when the cannabis users were excluded from the analysis.

The bottom line across all of these studies is that while cannabis use in teenagers may be having detrimental effects on IQ, the results are hardly definitive and are complicated by many potentially confounding background factors. For that reason, many of the investigators in the field are awaiting the outcome of the recently-begun Adolescent Brain and Cognitive Development (ABCD) study that is following 11,000 US 10-year-olds in a very large-scale national epidemiologic sample with serial IQ testing and brain imaging. This cohort study will capture the trajectories of normal brain and IQ development beginning prior to any substance use, and document any longitudinal consequences.

To conclude this chapter, we touch briefly on a number of miscellaneous epidemiologic studies that examine related topics that didn't quite fit with the broader questions we investigated earlier, but nevertheless are relevant and interesting.

Are medical marijuana laws associated with changes in either nonmedical prescription opioid use or prescription opioid use disorder?

A Columbia University epidemiology group led by Luis Segura studied 627,000 participants between 2004 and 2014 from the National Survey on Drug Use and Health. After medical marijuana laws were passed, there were tiny changes in the prevalence of non-medical prescription opioid use (4.32%–4.86%) and in the rates of prescription opioid use disorder (15.4%–14.76 %), none of which were significant. These findings were consistent irrespective of age or racial/ethnic group and contradicted the hypothesis that people would tend to substitute marijuana for prescription opioids [33].

Does opening a marijuana dispensary affect crime rates?

Proponents of marijuana legalization claim that anything that diminishes the black market in cannabis will promote public safety, whereas law enforcement agencies tend to claim that dispensaries significantly boost crime rates. For example, they are concerned that federal restrictions preventing dispensaries from accessing banks legally will attract robbers to these cash-only businesses. An interesting study from the IZA Institute of Labor Economics and the RAND Corporation explored the effects of marijuana dispensary laws on crime in all 58 California counties, since California legalized medical marijuana in 1996. They found a significant relationship between the granting of dispensary allowances and property crime rates; such crimes fell, not rose where dispensaries opened. There was no effect on violent crime. These data are consistent with some recent studies reporting that dispensaries may help reduce crime by occupying previously vacant buildings and employing security staff in these areas. It's also possible that increased legal cannabis availability is associated with less alcohol consumption, thereby reducing alcohol intoxication-related crime. Also feasible is that the legal system can focus less on illicit marijuana sales and thus more on other criminal activity. The authors also found some positive association between new dispensaries opening and increased DWI arrests, but the crime database does not specify which substances were involved in such cases [34].

What are trends in terms of type of use?

The RAND Corporation began following a cohort of 2500 6- and 7-year-olds, who are now aged 24. The CDC's Youth Risky Behavior Surveillance initiative also tracks such children. Renée Johnson from Johns Hopkins University and her colleagues recently presented data from these and other sources examining modes of marijuana consumption among high school students in Colorado, Washington, and Oregon [35]. The trends are that concentrates are increasing as a percentage of total sales, in part driven by heavy users who prefer higher THC concentrations, but also influenced by the fact that many more blends are available in concentrate form that are impossible to find in marijuana flower (such as 1:1 ratios of THC to CBD). Concentrate users are generally seen as "further along the path" of cannabis use. Many teenagers mix-and-match: of those who used cannabis, 87% smoke, 2% use edibles as an "add-on" 5% vape, and 4% use "dabs." So-called "polymodal," that is, mix-and-match use is common.

Who is at risk for cannabis use disorder (CUD)?

Across the various surveys mentioned at the start of this chapter, especially NESARC data, it seems apparent that those at greatest risk for CUD are younger heavy users, those with clinically painful conditions, and people from lower socioeconomic strata. Magdalena Cerda from NYU has analyzed data showing increased CUD rates in states that legalize marijuana, particularly in those aged 26 and older [35].

We have reviewed how survey research methods and other epidemiologic study strategies enable collection of informative data from large populations. These approaches can provide information on diverse topics including cannabis use trends, emerging risks for outcomes such as psychosis, cannabis use disorder, and cognitive deficits, and are able to identify populations at especially high risk for particular problems. I have focused both on examples where trends are relatively clear and unambiguous, and others that clearly represent scientfically feuded territory, where answers are still unclear. Sharper-eyed readers will have noticed that I did not discuss emerging trends in motor vehicle accidents, or driving while under the influence of cannabis in this chapter. The major reason for that, as I will explore in Chapter 8, relates to the difficulty in fishing for clear and informative driving data in the murky waters of many changing definitions and methods of ascertaining driving while under the influence of drugs, and of sparse suitable standards of comparison for accident rates.

References

[1] Steigerwald S., et al. Smoking, vaping, and use of edibles and other forms of marijuana among U.S. adults, 2018. Available from: https://annals.org/aim/article-abstract/2698115/smoking-vaping-use-edibles-other-forms-marijuana-among-u-s.

[2] Compton WM, et al. Marijuana use and use disorders in adults in the USA, 2002-14: analysis of annual cross-sectional surveys. Lancet Psychiatry 2016;3(10):954–64.

[3] Hasin DS, Shmulewitz D, Sarvet AL. Time trends in US cannabis use and cannabis use disorders overall and by sociodemographic subgroups: a narrative review and new findings. Am J Drug Alcohol Abuse 2019;45(6):623–45.

[4] Prince MA, Conner BT. Examining links between cannabis potency and mental and physical health outcomes. Behav Res Ther 2019;115:111–20.

[5] ElSohly MA, et al. Changes in cannabis potency over the last 2 decades (1995-2014): Analysis of Current Data in the United States. Biol Psychiatry 2016;79(7):613–9.

[6] Cuttler C, Spradlin A. Measuring cannabis consumption: psychometric properties of the daily sessions, frequency, age of onset, and quantity of cannabis use inventory (DFAQ-CU). PLoS One 2017;12(5):pe0178194.

[7] Results from the 2017 National Survey on Drug Use and Health: Detailed Tables. Rockville, Maryland; 2017. Center for Behavioral Health Statistics and Quality.

[8] Hasin DS, et al. Prevalence of marijuana use disorders in the United States between 2001-2002 and 2012-2013. JAMA Psychiatry 2015;72(12):1235–42.

[9] Cerda M, et al. Medical marijuana laws and adolescent use of marijuana and other substances: alcohol, cigarettes, prescription drugs, and other illicit drugs. Drug Alcohol Depend 2018;183:62–8.

[10] Hasin DS, et al. Medical marijuana laws and adolescent marijuana use in the USA from 1991 to 2014: results from annual, repeated cross-sectional surveys. Lancet Psychiatry 2015;2(7): 601–8.

[11] Key substance use and mental health indicators in the United States: results from the 2017 National Survey on Drug Use and Health. Rockville, MD: Center for Behavioral Health Statistics and Quality, Substance Abuse and Mental Health Services Administration; 2018.

[12] Davenport S. Falling rates of marijuana dependence among heavy users. Drug Alcohol Depend 2018;191:52–5.

[13] Research Society on Marijuana. Available from: https://researchmj.org/.

[14] Shih R.A. Young adults who live near medical marijuana dispensaries use marijuana more often, have more-positive views, RAND Organization. 2019. Available from: https://www.rand.org/news/press/2019/06/17.html.

[15] Dai H, Richter KP. A national survey of marijuana use among US adults with medical conditions, 2016-2017. JAMA Netw Open 2019;2(9):pe1911936.

[16] Subramaniam V. The new grey market: as older users warm up to cannabis, pot companies want to learn more, Business Financial Post. 2019. Available from: https://business.financial-post.com/cannabis/the-new-grey-market-as-older-users-warm-up-to-cannabis-pot-companies-want-to-learn-more.

[17] Lum HD, et al. Patterns of marijuana use and health impact: a survey among older Coloradans. Gerontol Geriatr Med 2019;5.

[18] Hicks B. Nearly half of Canada's cannabis consumers are millennials, according to new study, Civilized Life. 2019. Available from: https://www.civilized.life/articles/nearly-half-of-canadas-cannabis-consumers-are-millennials-according-to-new-study/.

[19] Gage SH, Hickman M, Zammit S. Association between cannabis and psychosis: epidemiologic evidence. Biol Psychiatry 2016;79(7):549–56.

[20] Radhakrishnan R, Wilkinson ST, D'Souza DC. Gone to pot—a review of the association between cannabis and psychosis. Front Psychiatry 2014;5:54.

[21] Caspi A, et al. Moderation of the effect of adolescent-onset cannabis use on adult psychosis by a functional polymorphism in the catechol-O-methyltransferase gene: longitudinal evidence of a gene X environment interaction. Biol Psychiatry 2005;57(10):1117–27.

[22] Di Forti M, et al. The contribution of cannabis use to variation in the incidence of psychotic disorder across Europe (EU-GEI): a multicentre case-control study. Lancet Psychiatry 2019;6(5):427–36.

[23] Castle D, Murray RM, D'Souza DC, Marijuana and Madness. 2nd ed. Cambridge, UK, Cambridge University Press; 2018.

[24] Krasny M. Understanding the link between cannabis use and mental illness, 2019. Available from: https://www.kqed.org/forum/2010101872329/understanding-the-link-between-cannabis-use-and-mental-illness.

[25] Kendler KS, et al. Prediction of onset of substance-induced psychotic disorder and its progression to schizophrenia in a Swedish national sample. Am J Psychiatry 2019;176(9):711–9.

[26] Tandon R, Shariff SM. Substance-induced psychotic disorders and schizophrenia: pathophysiological insights and clinical implications. Am J Psychiatry 2019;176(9):683–4.

[27] Scott JC, et al. Association of cannabis with cognitive functioning in adolescents and young adults: a systematic review and meta-analysis. JAMA Psychiatry 2018;75(6):585–95.

[28] Meier MH, et al. Persistent cannabis users show neuropsychological decline from childhood to midlife. Proc Natl Acad Sci USA 2012;109(40):E2657–64.

[29] Rogeberg O. Correlations between cannabis use and IQ change in the Dunedin cohort are consistent with confounding from socioeconomic status. Proc Natl Acad Sci USA 2013;110(11):4251–4.

[30] Jackson NJ, et al. Impact of adolescent marijuana use on intelligence: results from two longitudinal twin studies. Proc Natl Acad Sci USA 2016;113(5):E500–8.

[31] Silins E, et al. Young adult sequelae of adolescent cannabis use: an integrative analysis. Lancet Psychiatry 2014;1(4):286–93.

[32] Mokrysz C, et al. Are IQ and educational outcomes in teenagers related to their cannabis use? A prospective cohort study. J Psychopharmacol 2016;30(2):159–68.

[33] Segura LE, et al. Association of US medical marijuana laws with nonmedical prescription opioid use and prescription opioid use disorder. JAMA Netw Open 2019;2(7):pe197216.

[34] Hunt PE, Pacula RL, Weinberger G. High on crime? Exploring the effects of marijuana dispensary laws on crime in California Counties, 2018. Available from: https://www.iza.org/publications/dp/11567/high-on-crime-exploring-the-effects-of-marijuana-dispensary-laws-on-crime-in-california-counties.

[35] Johnson RM, et al. Usual modes of marijuana consumption among high school students in Colorado. J Stud Alcohol Drugs 2016;77(4):580–8.

Chapter 8

Toxicology

"...The stuff I got'll bust your brains out baby,
Ooh, it'll make you lose your mind..."

Robert Johnson, *Stop Breakin' Down Blues*[a]

"The basal ganglia of the brain are likewise stimulated, causing a 'welling up'
of the primitive emotions. Fear, sex, aggressiveness, elation come to the surface,
all colored by the imaginative processes that have been set in train by the psychic
areas. These cause.... sex delirium in acute cases and sex perversion in mild and
chronic cases; aggressive outbursts; grandiose delusions."

By, Johnson D.M. *Indian Hemp: A Social Menace* [1].

It's not about how many times you fall, it's about how many times you get back up.
Patrol officer: Sir, that's not how field sobriety tests work.

This chapter tries to address many questions. How potentially harmful is cannabis as a drug—and how does it rank in this respect with other drugs from alcohol to heroin? What happens to THC once it gets into your body? How long does it stay there, and how is it metabolized and excreted? How do laboratories test for the presence of cannabis in your urine, blood, and saliva? What are the various potential ill-effects of cannabis on physical and mental health? What about possible marijuana contaminants, such as pesticides or fungal spores? In the case of chemovars, how do we measure the content of different cannabis constituents? What else should be measured in the plant to standardize the cannabis that we consume? How is this information passed along to consumers? Does the concentration of THC in cannabis have anything to do with health effects? What are long-term consequences of cannabis use on the brain? While pondering these questions in the summer of 2019 during a visit to Denver to observe cannabis research, I decide to take a break in order to explore what was new in the weird and wonderful world of cannabis products. Who is using which consumables, where, how, and with what consequences? Here are a few impressions of what a legalized cannabis future might look like.

****

Out on Denver's "Green Mile" of cannabis themed businesses, the dispensaries I poke my head into have myriad merchandise on offer. I'm hungry, so things

---

a Permission to reprint lyrics from Robert Johnson's *Stop Breakin' Down Blues*, provided by Hal Leonard LLC, Milwaukee WI.

Weed Science. http://dx.doi.org/10.1016/B978-0-12-818174-4.00008-2

159

that first catch my attention include fast-acting edibles incorporating encapsulated forms of THC that are rapidly absorbed from the G.I. tract, e.g. Pixie Stix, containing fruit-flavored sweetened THC powder. Dutch Girl Stroopwafels are lemon or cinnamon-flavored crunchy cookie sandwiches enveloping a thin layer of chewy THC-enriched caramel containing 9 mg apiece. Other edibles include CannaCubes cherry micro lozenges and Cannavative's 'indica gummiez'. There are seemingly endless varieties of vaping devices for sale everywhere. Some are very high-end, aimed at well-off weed consumers who want to make an impression on fellow cannaseurs. For the traditional flower user on-the-go who dislikes mess and is clueless when it comes to grinding and weighing, one company seems to have taken a hint from Keurig, and markets flat-topped cones of 1-gram "pre-bowls" in different varieties to pop directly into your pipe.

But my wandering and gawking isn't addressing some of the scientific questions that I want to explore. Many cannabis researchers rely on subjects' self-reported measures of weed consumption—basically, how much of what do you use and how often do you use it? People's memories are notoriously fallible for all sorts of things, including substance use. Furthermore, this is a substance that can interfere with short-term memory. Accurate quantity and frequency estimates are important to research in a variety of contexts. For example, if you are trying to assess the relationship between how much cannabis people are using and whether they are feeling less anxious, or having more memory difficulties, or showing increased risk of developing psychosis, then having an accurate metric of the amount of drug used is just as important as measuring the outcome. Sure, it's possible to take blood or hair samples and analyze them as markers of cannabis use reaching back days or months, but the process is somewhat invasive and the lab tests are expensive. Self-report questionnaires are constantly being refined. But the key question of how accurate people are in estimating the amount of drug that use can be answered empirically. This is a basic issue that I first raised in Chapter 2. One approach to addressing this issue is as straightforward as sitting cannabis users down with some of their favorite product and asking them to estimate how much of it is in a standard unit, such as a gram or an ounce. Knowing of my interest in this topic, two collegial cannabis researchers at Colorado State University have invited me to sit in on one of their measurement experiments at Denver's Cultivated Synergy meeting space.

Private clubs in Denver have official designation as private residences, so that cannabis can be legally consumed inside their doors. At some such venues, vendors demo their new products, and can distribute free samples, although they are not allowed to sell them there. The demos are a legal strategy to distribute "testers" and to introduce potential consumers to sample products that they may subsequently decide to purchase at their local dispensary. This arrangement is a nice compromise to deal both with Denver's public consumption law, that prohibits cannabis use in public spaces, and also legal requirements that forbid selling weed product outside of legal dispensaries. I'm about to enter one of these private clubs.

Stepping inside the door of Cultivated Synergy, one encounters a pleasant open-design meeting and display space with interesting art on the walls, that is furnished with easily movable desks, tables, and screens. This is one hub of cannabis culture and identity in Denver where concentrate makers, marijuana users and merchants of various cannabis products rub shoulders, while consumers use their own products or sample the giveaways. This space is a cannabis-related business location during the day and a private club for cannabis users at night. A speakeasy it's not. It's friendly but definitely businesslike. I show my driver's license to Cecile, the event manager, sign myself in and get an official wristband for the evening.

Mingling with the population here I find myself amid the rich dank odor of mixed terpenes emanating from multiple sources, with people using various rigs, pipes, and vaping devices as well as the expected joints. I'm impressed by how normal and casual everything seems. People have abandoned the furtive, conspiratorial mien that was an inseparable part of cannabis use in the past. Weed use here is no different than sipping a beer at a social or a cup of coffee at Starbucks. Although alcohol is available, few people are indulging and nobody seems to be drinking more than a single beer throughout the evening. The clientele may be stoned but they are calmly chatty, chilled, and friendly. Inside a typical bar at this time of the evening, people would be likely garrulous, disinhibited, loud, and a little pushy. The mood here is vastly different and I don't see anything resembling the cannabis equivalent of being "all liquored up." Cecile tells me that in the 3 years that Cultivated Synergy has been open, there have been zero confrontations from guests. Having known several bar and pub owners over the years, you would never hear such a claim from one of them.

Nighttime events here include well-attended budtender appreciation evenings, and "Dab and Dine" events, where cannabis dabs with different terpene profiles are paired with appropriate foods. Cecile tells me it's easy to recruit chefs for the dab and dines (D & D's) because many of the local chefs are enthusiastic cannabis users. She hands me a pile of recent menus. Some of the latest D & D's have featured such dishes as shrimp sautéed with fresh dill and drizzled with a lobster cream sauce atop toasted crostini, paired with the flavors of "Flo Live Sugar" concentrate blend, whose dominant terpene is terpinoline. According to the tasting menu, "The floral, lemongrass and pine flavors of the Flo pairs well with seafood, offering a sweet and crisp experience." Similar events have also paired Terps with ciders and other food varieties. Quoting from another concentrates-meets-delicacies menu, "Harambe Adhesive Honey Bucket, a distinctive cannabis flavor showing rich dark tones of umami and back-of-palate intrigue............is a natural companion to a dark chocolate dessert. The boudine is a perfect complement to the heft of flavors but lightness of aftertaste."

But Cultivated Synergy tonight is not about gourmet tastings. I have been invited to witness a research study by two faculty members from Colorado State University's Psychology department. Brad Conner is an associate professor and director of addiction counseling, and his colleague and collaborator Mark Prince is a junior faculty member. They make an intuitively complementary scientific pairing. Brad is more energized, cerebral, and detail-oriented while Mark at first blush appears to be more of a laid-back, thoughtful observer. He turns out to be a skilled schmoozer, persuader, and dealmaker who helps the science move along. Tonight they've taken over the space, together with a gaggle of their graduate and undergraduate students to continue their research on measurement accuracy. At three separate screened-off tables, individual newly-arrived marijuana users from the club are making estimates of quantities of different forms of cannabis, from dabs to flowers. Just as plate size in a restaurant can skew one's view of portion size, the psychologists are interested in how different cues affect judgments of cannabis amounts.

For example, at one table a student researcher is asking a volunteer to estimate 1 gram of flower and to load it into an unused glass pipe. The pipe is then weighed and the amount of cannabis recorded, out of sight of the volunteer. At another table, the instructions are similar except that the remit is to roll a 1-gram joint. On the table there are visual cues as examples, consisting of life-sized images of various amounts of cannabis, for example flower, rolled cigarettes, etc., along with their corresponding weights, photographed next to familiar-sized objects such as bottle caps. At the final table, the cues are pre-weighed actual chunks of bud and rolled joints in different labeled weights, providing real-world actual 3-D examples. Brad and Mark are determining how accurate these regular cannabis users are at estimating the quantities they typically consume, whether it be dabs or flower, and whether their accuracy is helped at all by the presence of different types of cues. Because of the legalization of recreational marijuana in Colorado, many people in-state are addressing important questions regarding predictors and consequences of drug use in various populations. In past experiments similar to tonight's, Brad and Mark have shown that budtenders, whose daily jobs involve repeatedly weighing quantities of different cannabis products as accurately as possible, and who are almost invariably cannabis consumers, are extremely inaccurate in estimating cannabis amounts under these research conditions. And this isn't (as you might imagine) because they arrived at the event already terminally stoned. The method used here captures volunteers at the start of the evening before they've had much of a chance to catch a buzz. In addition to the weighing experiments, all the volunteers are completing interview surveys about their cannabis use. I am the proverbial fly on the wall observing what's going on, buzzing from table to table and listening in. The volunteers seem happy to participate, and one makes the comment that she has "never been so open about my weed use to anyone before."

I look for the source of cool air and see that a large fan is blowing from the balcony as the atmosphere in our space thickens palpably with exotic plant particles. People who have completed the weighing experiments hang out and begin using their own cannabis. Some in the counter area start using concentrates. Over in their corner, a Bee-Nails electronic box is heating a vaporizing coil mounted atop a compact water pipe. A friendly middle-aged businessman cleans the rig meticulously with a Q-tip, before depositing a caramel-colored chunk of concentrate the size of a sesame seed on the hot metal. To accomplish this, he is using a small, skinny metal surgical-looking spoon that would not look out of place in a nail salon or orthodontic suite. He's happy to tell me all about how this setup works and why he likes to use it. The local manufacturer's blended concentrate contains 66.15% THC and zero cannabidiol (CBD), he tells me. He dials up the device temperature to 582°F. This is sufficient to vaporize all of the volatiles instantaneously but not to fry or char anything in the mixture. Before the vapor can waft away into the room, Mr. Businessman caps the cup-shaped chamber deftly with a metal cover, from whose half-dome top protrudes a non-conducting metal handle that avoids transmitting heat to the user. He says that he usually prefers to set the device at a lower temperature to appreciate the terpenes a bit better, but tonight is more about a quick buzz than a leisurely experience. Watch out for "lung burn" at high temperatures, he warns. A quick mental calculation tells me that the amount of THC Mr. B is currently sucking into his lungs so efficiently from the bong is theoretically sufficient to stop a charging lion in its tracks, yet he not only fails to appear intoxicated, but continues to discourse informatively on dab-related topics. This has to be an example of the body's capacity for THC tolerance in action. "Always check the label to figure out the THCA to THC ratio as well as the overall content," he continues. He runs through descriptions and ratings of different products and manufacturers faster than I can scribble things down. This brand of shatter beats that variety of crumble by a mile but is inferior to someone else's new version of budder that has a better blend of terpenes and is cheaper. Half a gram of dabs provides up to about 20 hits, and at around 75% THC on average, that totals 375 mg of the magic molecule.

The different tools lying around serve different purposes. Some are crafted to slice and dice the product, others to scoop and dump it efficiently. One is designed to deal with syrupy concentrates, another with hard and brittle ones. "Heat this one before using, don't heat that one." Mr. B is exhaling surprisingly large clouds to have originated in such a tiny dab. While we chat, a gaggle of dab hands forms around the rig to sample the demos and schmooze. In turn each user selects a blob of concentrate ranging from roughly the size of a pinhead to that of a plump grain of basmati rice, nicely covering a spectrum from micro-dabbers to gourmands. There's a fair amount of semi-intense coughing going on and most people are drinking lots of cool bottled water, but contrary to my expectation nobody is sweating like a horse, looking pole-axed, or forgetting how to use their limbs normally. Nobody's brain falls out. Mostly, everyone is chatting pleasantly, laughing, and generally behaving, as my mother would have put it, "like a normal person." The conversation ranges over a number of interesting, if random topics. I learn that many consumers still tend to buy from street dealers, both because the price is reasonable and they know the dealer socially. Plenty of the assembled club patrons use small amounts of alcohol, but generally not at the same time as using cannabis. Nobody here is a cigarette smoker, but one person is a nicotine-vaporizer user still trying to quit. A woman in her late 20s describes a cannabis concentrate company in Arizona that features their own terpene bar. Their dispensary allows the average person to mix-and-match cannabis products in the way that they might blend custom pipe tobacco. Down in Arizona, just as an upscale cosmetics and beauty products store will allow you to mix your own perfumes or add your individual choice of essences to shampoos or body oils, this outfit encourages you to mix personal vaporizer cartridges to suit your mood preferences. The consumer bellies up to the custom bar, and crafts his or her own personal blend in terms of relative percentages of THC, CBD, and other cannabinoids, and the presence and amounts of half a dozen terpenes. Want more focus and creativity? Add a dab of limonene. Wish to combat short-term memory loss? Blend in more alpha and beta pinene. Some in the little gathering find the concept appealing, while others diss it as mildly pretentious.

Meanwhile the fan is blowing effectively from the balcony, and I seem to be avoiding a contact high. A number of thoughts strike me about tonight's cannabis users. Here they are quietly enjoying a good time, but is their experience devoid of risks? How many of them will be driving home from the event, and if so to what extent is their driving ability compromised by the beer? Are any of the women here tonight pregnant? Is cannabis exposure safe for their fetuses? Are people in the crowd familiar with one or another of the psychosis studies, and has any of them hallucinated or become paranoid for a few days after using? Is anybody here dependent on cannabis? And just how dangerous is this stuff compared say to alcohol, or cocaine, heroin, or tobacco? The remainder of this chapter takes up all of these questions and focuses on the potentially toxic side effects of cannabis, addressing the issues from a rather different perspective than that of an epidemiologist. These are more the sorts of questions that front-line clinicians tend to ask.

✻✻✻

What are some major concerns regarding health effects of cannabis to the user, and to other people?

Automobile driving

This is a topic with important public health and policy implications. As recreational and medical cannabis are increasingly legalized, inevitably more people will use these substances and, as a consequence, more people will be driving under the influence of marijuana. The nature of the cannabis-related products is changing too, with increasing THC to CBD ratios, and availability of edibles and concentrates that are capable potentially of producing profound and long-lasting intoxication. Edible's effects may also catch drivers by surprise. While common sense tells us that marijuana intoxication likely impacts driving ability adversely, some users deny or minimize this. "I just drive a little slower," or "I'm more relaxed when I drive stoned" are responses that I hear often from my driving research subjects. My usual response is to ask them if they would be willing passengers on an airliner where the pilot was intoxicated on cannabis.

According to the National Highway and Traffic Safety Administration, drugged driving is on the rise. While the number of annual fatal vehicle crashes in the United States has trended down in recent years, as has the number of alcohol-involved deaths (thanks to the police being more serious about enforcing DWI rules), the number of driver fatalities involving positive drug tests has increased. The overall death totals began falling around 2006 from around 36,002 to around 30,000, but the proportion of fatal motor vehicle accidents involving drugged driving has risen from about 5% to about 20%. Numbers are not directly comparable however; the number of individuals tested for drugged driving and the nature and sophistication of the testing used has changed significantly over time. Also, the presence of certain drugs on postmortem testing does not necessarily imply that the individual was intoxicated at the time of death (for reasons that we will discuss later); this relationship varies for different drugs. A 2016 meta-analysis that pooled together data from multiple separate studies to total almost a quarter of a million individuals, concluded that cannabis intoxicated driving constituted a low-to-moderate magnitude risk, with the odds (recall odds from Chapter 7) of being involved in an accident being increased about 1.36 times compared to driving sober, a minor increase. To put things in proportion, the odds of being involved in an accident with a blood alcohol concentration of 0.10 is somewhere around 20 times greater than driving sober, and that of driving very high on amphetamines about 55 times. The summary of the literature shows that compared to the sober state, marijuana-intoxicated drivers vary their driving speed more, have slower reflexes causing them to take a little longer to brake, begin driving a tad later when a light turns green, tend to weave a little bit more, and are somewhat slower in avoiding other vehicles so that they're more likely to collide with them. But even small-scale differences are important if it's your kid who is out on their bicycle. The weight of evidence for many types of studies—epidemiologic, testing chronic marijuana smokers, conducting laboratory studies of consequences of acute dosing, strongly support that marijuana use causes detectable deleterious effects both on driving and cognitive test performance. This brings us into territory of interest to law enforcement authorities, which is how useful roadside neurocognitive tests are in detecting recent marijuana use and consequent driving impairment. When I mention "recent use" and "driving impairment" do not assume that those are identical, however. What people used, when they used it, and the type of driving they are engaged in, all have some bearing on the issue of driving impairment.

When I first became involved seriously in this research field, I was surprised by how many key questions were only partially answered. Some of these unresolved issues include what cognitive and behavioral aspects of simulated motor vehicle driving are impaired acutely by marijuana use? What are the most sensitive cognitive, motor and biosensor tests to detect these impairments, and how feasible is it to get them out of the research laboratory for purposes of roadside administration? How relevant are these kinds of tests to driver performance and safety in the real world? Once someone is impaired by an acute dose of cannabis, how long does it take for them to come back to their sober baseline? What's the relationship of cannabis-caused driving impairment to the levels of THC and its major metabolites in blood or saliva? Are any of these measures related to brain activation patterns measured in a functional MRI or by an EEG device? And what's the best naturalistic test environment to answer these questions? None of the experts seem to know what the time course of THC-related driving impairment is. If you smoke half a gram of 25% THC premium weed at 9 a.m., when are you safe to drive? Impairment on cognitive tests is not the same as actual driving impairment, so it's important to validate any candidate test procedures against actual or simulated driving. The relationship between alcohol-related driving impairment when combined with cannabis impairment is obscure. Consider this; if you're below the legal alcohol limit and smoke a joint, are the combined effects determined by simply adding the separate intoxications together, or does impairment somehow multiply, so that the combination is much worse? Finally, no one seems entirely sure what the specificity of impairment patterns caused by THC is compared to those due to opioids or Valium-like drugs. Cannabis is complex. Not only is there variation from one chemovar to another, with THC not being the only relevant psychoactive compound, but the over 400 compounds in the cannabis plant produce a massive number of possible combinations. Some chemovars are advertised as sedating, others as wake-me-up's. Do these produce equivalent impairment of driving ability? In addition there are multiple cannabis products with different percentages of THC and different absorption characteristics, from edibles to dabs.

It is important to understand why the earlier-mentioned questions remain unanswered, so let's review a few of the complexities. There are not very many well-controlled studies of cannabis-impaired driving. Also, cannabis impairment is significantly more complex to study than alcohol impairment when it comes to driving. Let me explain. With alcohol, there are very straightforward proportionate, relationships between how much someone has had to drink, the levels of detectable alcohol in blood or breath, and driving impairment. For cannabis, there is no clear connection between how much drug someone has used, the resulting concentrations of THC and its metabolites in breath, blood, or saliva, or between these concentrations and actual driving impairment. We will explore the underlying reasons for why these circumstances are so different for alcohol versus cannabis a little later, but part of the explanation lies in the complex pharmacokinetics of THC versus those of alcohol. Because THC and its metabolites tend to hang around in the body long after use, it's unclear how to interpret body fluid sampling after fatal and nonfatal motor vehicle accidents. Also, rates of such accidents are subject to a skewed, agenda-driven motor vehicle administration reporting in relationship to marijuana legalization. The pro legalization side will tend to play down apparent increased accident rates following cannabis legalization, whereas it's in the interests of the anti-legalization side to exaggerate such numbers. Unlike alcohol, where there is a set of well-established, evidence-supported, agreed-on roadside sobriety testing paradigms, not only are these procedures not appropriate for testing cannabis-impaired drivers, but nobody is entirely sure what the best tests for stoned drivers actually should be. Yet another problem is that there is a lack of sober baseline testing on presumptively impaired drivers. Maybe they were awful drivers when sober.

Let's contrast alcohol DWI to the situation with cannabis in a bit more detail. Alcohol's story is straightforward. As we all know, it is extremely soluble in water, and most of our bodies consist of water, so that an alcohol dose fairly quickly and readily diffuses throughout all your tissues pretty much equally. Thus the concentration of alcohol in your brain, your blood, and your breath are very similar to one another and also directly related to how much you drank. If you take a swig of alcohol and I wish to plot your blood or breath alcohol concentration (usually abbreviated to BAC) over time, I need to know only three things. These are how much alcohol you drank, your sex, and your body weight. With those three facts, I can be extremely accurate in predicting how high your peak BAC will be and how slowly it will fall to zero. If I know how much food was in your stomach when you took the drink I can be even more accurate. In addition your BAC is very tightly and directly tied to your motor vehicle crash risk.

None of these things is true for cannabis. THC is fat-soluble not water soluble. So rather than floating around your body in an equally distributed manner, there's a big initial THC peak in your blood a couple of minutes after smoking that rapidly diminishes minute by minute as the chemical is absorbed into body fat. Since there is lots of fat in everybody's brain covering the axons of neurons, the brain is one of those places that enjoys a high concentration of THC. Once that initial spike quickly comes down to very low levels, THC and its metabolites persist in the body for very long periods of time, gradually leaching out of fat stores. If you're a heavy cannabis smoker, and then quit, those chemicals may still be detectable in your blood or urine 2 weeks or more after you last used. That's a major problem for law enforcement. There is no easy way to tell if a low level of THC or its metabolites in your blood or saliva represents the fact that you smoked a joint 20 minutes ago and your driving is currently impaired, or that you last used 10 days ago and you have long been at your behavioral baseline. Unlike alcohol, THC has no smooth decrease in its blood levels; just a quick spike followed by a gradual decrease from a lowish level. To make things even more complicated there is no clear relationship between those blood levels and your driving impairment, because of THC's fat solubility. What's going on in your brain is in no way reflective of levels in your blood or saliva. Essentially those blood measures are uninformative. Not only that, but driving impairment-related behaviors due to cannabis are not straightforwardly related to dose. Remember the bi-phasic relationships for anxiety and THC dose, where low doses were calming but higher doses had the opposite effect? There are similar complex dose relationships for your heart rate, how much you feel the drug's effect subjectively, and perhaps some complex behaviors related to driving. The whole issue of cannabinoids' fat solubility makes blood testing hard to interpret. As people in the research world like to say, "there is no BAC for THC." If alcohol behaved like THC, imagine that you were drinking in a bar 10 days ago and had 6 beers, but haven't had a drop of alcohol since. The police pull you over stone cold sober at a routine roadside stop and your BAC is still positive, so they arrest you today for traces of the beer you consumed a week and a half ago. If that seems bizarre and unfair, it's precisely the situation with cannabis. Based on drug level, the police have no way currently of distinguishing between someone who used the drug recently and is driving-impaired, or days ago, and is completely sober. The same problem applies to postmortem blood samples taken from drivers killed in motor vehicle accidents. A positive sample does not equate to the person being incapacitated by cannabis at the time of the fatal crash. And this is something that is generally ignored by opponents of liberalizing marijuana legislation, when they cite driving-related consequences of legalization based on such blood level data. To help address this technology gap, the company Hound Labs is working on a "cannabis breathalyzer," that helps to measure concentrations of a THC metabolite that disappears from breath a couple of hours after somebody smoked cannabis. Theoretically that instrument might be a useful guide for live drivers and a way of distinguishing between relevant recent and irrelevant distant use. There are considerable practical problems in devising such a breathalyzer. Soon after swallowing a couple of alcoholic drinks, your breath is teeming with millions of alcohol molecules that are pretty easy to detect on a relatively straightforward screening roadside device. In comparison, the number of relevant THC metabolite molecules in breath is tiny, requiring a fairly sophisticated and very sensitive detector.

THC, CBD, and their metabolites such as 11-nor-9-carboxy-THC (THCCOOH), make their way to hair follicles and thereby into your hair where they can be detected albeit somewhat expensively, by laboratory testing. You can cut a hair sample into segments, and since people's hair grows at a relatively uniform rate, a specialist laboratory can analyze chemical concentrations up from the root toward the tip to detect who smoked what and when. Hair analysis is a reasonable way to confirm long-term abstinence, or perhaps in Rastafarians (who do not cut their dreadlocks), to examine very long-term cannabis use, but obviously is useless in people who shave their heads [2].

How do current procedures for driving under the influence of drugs (DUID) play out? In general, a law enforcement officer pulls over your vehicle either for "probable cause" if, for example, you are weaving when driving, or perhaps randomly at a police checkpoint. The police ask you to exit your vehicle and then administer roadside tests for impairment that are based on established procedures for alcohol intoxication. Unfortunately, these tests are in all likelihood not relevant in detecting cannabis intoxication. If you perform abnormally on roadside screening, for example, not being able to walk a straight line, you are generally asked to blow into a breathalyzer in order to test your BAC. If that's normal then the police may decide to have your blood drawn for drug testing at a nearby facility. The average time between law enforcement making that decision and the needle entering your vein for the sample is 90 minutes. If THC shows up, laws in 13 states prohibit driving with any amount of the substance detectable in plasma. But we've already seen that THC can hang around in your blood for a couple of weeks after you last smoked, so the situation is exactly parallel to our example earlier of police charging you with a DUID 10 days after your last drink. A handful of states specify a legal THC cutoff level above which driving is illegal. In Colorado, Montana, and Washington, this value is 5 ng/mL of blood; in Nevada and Ohio, the value is 2 ng. There is no scientific basis for these measurements. The remainder of US states prohibits driving while "incapacitated by" or "under the influence of" marijuana; these are slightly different legal standards, but both boil down to a somewhat vague prohibition on driving while high. Thus there is considerable and confusing variability among state DUID laws for cannabis.

Some jurisdictions employ Drug Recognition Experts (DRE's). These are police officers trained in the US Department of Transportation's, National Highway Traffic Safety Administration's (NHTSA'S), and Drug Evaluation and Classification Program (DECP) standard protocols. Such individuals are certified to conduct examinations on drug-impaired drivers, usually at precincts or jails. The process involves a multi-step standardized procedure that combines data from several sources: medical (e.g., pulse rate), psychophysical (e.g, subject sways when standing on one leg), and observational (what the DRE sees when interviewing the subject) [3]. In my experience and that of many other marijuana researchers, the problem with DRE's is that there is significant variability from one of these evaluators to another in performance when it comes to accurately identifying cannabis-intoxicated drivers. Since it's not yet clear to anybody what are the most reliable tests for identifying cannabis-impaired motor vehicle drivers, under properly controlled conditions, the DRE's perform about as well as the rest of us, which is only moderately. On the other hand, they do consistently well in ID-ing drunk drivers.

If the current relationship between cannabis intoxication and impaired motor vehicle driving is obscure, (which it is), then the obvious thing to do is to test-drive individuals who are stoned. It's neither legal nor ethical for scientists to get someone intoxicated on marijuana and then have them drive in traffic on a major highway. But if you want to measure validly how impaired somebody's driving is, you really do want to test them under realistic circumstances. This is where simulated driving enters the picture. Driving simulation is the gold standard for identifying and quantifying relevant impairment. It provides a controlled, safe environment that translates ideally into real-world driving. You can easily mimic scenarios that are unethical or impractical to test in real life, such as a puppy suddenly running into the road. With a skilled programmer you can easily manipulate the environment, including the behavior of other vehicles, whether it's day or night or numerous different weather conditions. To avoid your research subjects learning a particular scenario after they've run through it a couple of times, you merely program a series of variations into the driving sequence. This raises the critical issue of validation. Let's say that an investigator has designed some supposedly state-of-the-art simulated driving hardware and software. How will he or she know whether actual drivers on a real road drive in a manner that's comparable to the way that they perform on this experimental set up? Fortunately there is a way to answer that question.

Virginia Tech's Smart Road is a 2.5-mile long limited-access highway located in Blacksburg VA, that mimics an interstate and is used for advanced vehicle testing. Part of it is connected to a public road. The Smart Road is used by Virginia Tech's Transportation Institute for testing autonomous vehicles, experimental road surfaces, and boasts a movie-set like array of portable features, including reconfigurable buildings, alleyways, intersections, and removable line markings. Hovering above the whole enterprise is something that looks like an airport control tower from which everything can be observed and reconfigured. For example, some of the permanent "light poles" contain water sources that can make it realistically "rain" or even "snow." Buried beneath the tarmac are hundreds of sensors that communicate with computer equipment stored in the trunks of numerous experimental test vehicles. The vehicles themselves are outfitted with dual-controls devices like those in learner driver cars, as well as cameras to monitor eye movements. The dual-control allows one of the experienced Blacksburg driving scientists to take over the wheel or pedals at short notice, if the driver's performance starts to become worrisome.

More than 25 years ago I organized a scientific expedition down to Blacksburg and had a series of volunteers consume alcohol (sufficient to push them over the legal BAC limit) and placebo at different times, and then drive both an actual Tech Institute vehicle on the Smart Road, and also separately navigate a computer-based realistic mockup of the Smart Road on a driving simulator. We were able to show that the inebriated volunteers made the same sorts of errors (exceeding the speed limit, speeding up around curves, weaving) on the simulated road, as they did on the "real" highway, thus validating the driving simulation paradigm [4].

If Virginia Tech has the coolest real vehicle test environment, then the best driving simulation equipment is the National Advanced Driving Simulator (NADS) located in Coralville Iowa. Until recently it was the largest ground vehicle driving simulator in the world. NHTSA put millions of dollars into developing the NADS (the developers obviously didn't give much thought to the acronym), which is truly impressive to behold. The simulator itself is a large dome that looks like a scaled-up white NASA lunar probe or an extra-terrestrial vehicle that is about to land in Area 51. It is sufficiently large to have an entire vehicle lowered into it, and when I use the term "vehicle," this can be anything from a standard sized sedan to an Abrams tank. The dome is perched on stubby legs that are attached to a motorized platform that in turn rides on rails that allow the dome to rotate nearly 360 degrees and to accelerate for 60 feet.

Inside the dome one can project a realistic virtual reality highway, or indeed any desired terrain. This combination of features makes anyone who's "driving" a car inside the dome, (and I can attest to this) feel exactly as if they are driving a real car on a real road with all of the sensory cues that are missing from most driving simulators. When I drove a real car on an actual road soon after being inside the NADS dome, the genuine experience possessed a slight aura of unreality, as if the real thing wasn't quite sufficient.

Marilyn Huestis, a doyenne among cannabis toxicology researchers, ran a series of cannabis-intoxicated driving experiments, testing smallish numbers of cannabis or placebo dosed subjects with and without alcohol on the NADS. She was able to document that stoned drivers showed detectable weaving, even at fairly modest doses of the drug, and that volunteers using both alcohol and cannabis made a greater number of errors. [5–7]. What's needed next for these sorts of experiments is to show more definitively how these types of errors relate to on-road driving performance.

One of my lab's contributions to the field of driving under the influence of marijuana is to try to understand what's going on at both a brain and a behavioral level in cannabis-intoxicated driving. Our experiments involve cannabis-using volunteers (both men and women to examine sex differences) who get high either regularly (i.e., almost daily) or occasionally (about once a month) consuming either high or low THC cannabis or placebo [provided by the National Institute on Drug Abuse (NIDA), and grown at their official supply site in Mississippi]. Subjects visit the laboratory on 3 separate days a week apart to receive 1 of their 3 doses. Parenthetically, NIDA likes to emphasize that their name is the National Institute *on* Drug Abuse, not the "National Institute *of* Drug Abuse," which latter conjures up a hopelessly inebriated group of government scientists avidly consuming a large quantities of illicit substances. Since NIDA generously funds much of my research, I will definitely venture no further in poking gentle fun at them.

Back to our volunteers. Once they've consumed whatever dose they are receiving on a particular day (and everything is double-blind and randomized, so neither we nor they have any idea what that is) via a desktop vaporizer, they rotate through a series of periods of simulated drives inside the MRI scanner so we can examine their brains, then driving on a regular driving simulator outside the scanner, taking different varieties of cognitive tests, and providing us with regular samples of blood and saliva for laboratory testing for THC and its metabolites. All of these procedures are repeated multiple times throughout the 7-hour study day. Thus we can tell when the drug is exerting its maximal effects, as well as when these wear off. Driving in the MRI scanner involves a series of fairly realistic simulations measuring different traffic patterns of varying complexity. These driver challenges may be very simple, (can you drive in a straight line while simulated wind gusts occasionally buffet your vehicle?) or quite complex, (can you safely pass a stalled vehicle in your lane when this involves waiting for a gap in busy oncoming traffic?). The steering wheel, gas and brake pedals are all realistic and conceptually familiar to anyone who's ever played a high-end video driving game, except that nothing contains any ferromagnetic material, so that it is safe to use in the MRI scanner. One immediate question is when one of our participants is involved in a virtual "accident" with another vehicle, what were they looking at, at the time of the crash? Were they actually peering straight at the other vehicle, but just not reacting in time to avoid the collision? Or were they instead distractedly staring at the virtual sky (such pretty virtual clouds) and not even noticing that they were about to hit another car? To parse these different events we use an infra-red eye tracker so that we can see precisely what our subjects are looking at, at any given time.

An important related question is, what's needed for roadside testing to be effective in identifying cannabis-intoxicated drivers? This information could be provided by a so-called "field sobriety test" administered at the roadside by the police, or a suitably specific and accurate assay of THC or one of its metabolites in a body fluid. What forensic toxicologists dream about is a really accurate and sensitive blood, urine, or saliva test that indicates specifically only recent cannabis use. That would be one that identifies accurately someone who got high within a couple of hours, and doesn't yield false positive results that may incorrectly imply very recent use but are instead due to indulgence many days prior. The toxicologists would like their assay to indicate such recent cannabis use, whether or not the person generally smoked 5 times a day or only 1 day a month. They would add to their wish list that the test was also positive for a few hours after use of edibles, and could distinguish between somebody who'd actually used cannabis personally rather than inhaled secondhand smoke (e.g., if their friends were toking up in a car while the testee was driving but he or she hadn't indulged). Finally, their dream assay should be capable of being processed rapidly; say within 24 hours, and not be wildly expensive. Slower, more expensive drug tests might be fine for screening Olympic athletes (where half of all positive drug tests are for cannabis) but are not feasible for routine roadside testing.

In their search for informative cannabis tests, toxicologists can perform interesting experiments such as having volunteers smoke joints, inhale from vaporizers, or consume edibles and then harvest their subjects' plasma and urine samples over the ensuing minutes, hours, and days, for example [8–10]. These sorts of scientific efforts have yielded valuable information, but nothing quite yet that meets criteria for their dream assay. This is what Hound Labs wants their breathalyzer to provide.

Here are some of the conclusions the field in general has reached. THC glucuronide is a chemical produced when THC has first been metabolized in the body and is then transformed chemically so that it becomes water-soluble. Once that happens, this chemical is rapidly eliminated in your urine and it is no longer detectable. So that if THC glucuronide is still floating around your body and shows up on a blood test, it is one biological indicator of recent use in both chronic and occasional users. But not everybody who has used cannabis recently will test positive for THC glucuronide, so that there are false negative test results. In other words, the person did use weed but the test result is negative. In addition to THC-COOH, THC-OH (see further) in urine may provide information on recent consumption. A number of minor cannabinoids including THCV, CBD, and CBG are detectable in saliva (or as toxicologists refer to it "oral fluid") after acute use, but are not particularly helpful in the context of roadside testing because they can hang around for a day or more. The THC metabolites 11-hydroxy-THC (11-OH-THC) and 11-nor-9-carboxy-THC (THC-COOH) are useful, but carboxy THC is present in only one thousandth of the concentration of THC in oral fluid—it provides a good measure of actual smoking versus contact high since the latter will produce some THC in oral fluid but no carboxy. The latter indicates that the subject has actually been smoking and cannot be due to exposure to secondhand smoke. If subjects are taking oral dronabinol (synthetic, legally prescribable, FDA-approved THC), then the THC itself never appears in the blood, but you can still detect its carboxy metabolite. Bottom line—while there are a number of useful blood and saliva tests, none yet really has demonstrated the characteristics needed for quick identification of cannabis-impaired drivers.

If performance on the standard roadside sobriety tests for alcohol, such as walking a straight line heel-to-toe while reciting the months of the year backwards, or accurately touching your finger to your nose with your eyes closed are not especially sensitive to cannabis effects (and they are not), then are there equivalent areas of impairment that would be appropriate for detecting acutely cannabis-impaired drivers? Many research groups, including my own, are working on this problem. Some test candidates still being assessed in my laboratory include using the accelerometer that's buried inside of your iPhone (the technology that detects when you flip the device from horizontal to vertical and ensures that the display remains upright). Since the device reads out several hundred times per second, it could theoretically pick up a characteristic signature body sway due to acute cannabis intoxication. Another technology is a simple portable EEG device that transmits information about your brain to a handheld laptop via Bluetooth. The EEG cap takes less than a minute to put into place and begin recording, so it's feasible to use at the roadside. Other technologies being explored include measuring the eye's response to flashes of light, a person's ability to interpret rapidly whirling or flashing patterns, or to perform accurately on PC-based games and puzzles. All of these approaches have potential drawbacks of one sort or another. For example; if you are trying to assess cognitive abilities, does the test work at the side of a noisy highway at night with many potential background distractions? Does the EEG device work on somebody with hair extensions or dreadlocks? Does your eye respond to flashes of light differently if you have had cataract surgery? So right now, the field of developing straightforward roadside sobriety tests for stoned drivers has attracted widespread interest, but has yet to come up with the killer app.

Another wrinkle in the DUID area is the fact that a fairly solid proportion of individuals who use cannabis do so in the context of alcohol consumption. We saw this use pattern among our 2000-person college student study, (discussed later in this chapter) and previously noted that some of the Colorado social club attendees consumed small amounts of beer with their cannabis. Many surveys bear out the same tendency. I mentioned a little earlier that how the intoxication effects of the two drugs add up when they are consumed together is not well understood. These types of drug-drug interactions can be complicated. For example, the metabolism of cocaine users who concomitantly drink alcohol, brews up a combined compound called 'cocaethylene' that both prolongs and amplifies cocaine's intoxicating effects several-fold. Evidence to date suggests that alcohol may multiply cannabis' relatively mild deleterious effects on automobile driving [5] although other studies suggest few if no interactions between the substances [11,12]. The critical experiments, parsing these effects by exploring a wide range of doses of both drugs alone and in combination, have yet to be carried out, so that like many other aspects of cannabis-impaired driving there are more questions than answers.

Psychosis and delirium

Let's switch gears. With the recent legalization in the many US states of medicinal and recreational cannabis, concerns have emerged about cannabis as a risk factor for psychosis. The numbers are not trivial; in 2017, 567 people were treated at Vancouver-area hospitals for cannabis overdoses or related mental health issues [13]. Even if the percentage of people affected is low, which is probably the case, as more and more people find cannabis products easily available due to legalization, then the number of cases is bound to rise. But to put things in perspective, around 5 people in British Columbia actually die every day from direct or indirect alcohol effects [14]. We looked at psychosis from an epidemiological risk point of view in the prior chapter. Let's dive into the clinical aspects of the different psychosis and related entities here.

As reviewed by Wilkinson [15], and others, four distinct psychosis-related syndromes can be provoked by acute and/or chronic cannabis use. This fact alone complicates the broad statements that "cannabis causes schizophrenia." I will describe examples of each of these syndromes further, so that readers can have an idea of what typical cases look like. Both an acute (i.e., hours-long) psychosis with prominent positive symptoms (e.g., hallucinations or delusional beliefs) and an acute delirium with disorientation, waxing, and waning levels of alertness (sleepy 1 minute, alert the next) accompanied by short-lived cognitive problems can be provoked by cannabis intoxication, including recreational use of synthetic cannabinoids such as "K2." Furthermore, an acute psychotic state lasting days-to-weeks beyond the acute intoxication period, (subjects whom Kenneth Kendler ID'd in the first wave of his Swedish sample) and lastly a chronic psychosis resembling schizophrenia that is not linked to recent cannabis exposure (such as those Kendler studied in his second wave) are also well-described. Recall from the prior chapter that most new cases of schizophrenia are not linked to cannabis consumption, and that cannabis exposure alone is neither necessary nor sufficient to cause a persistent psychotic disorder, although it may be a significant environmental risk factor for some individuals.

Delirium

Let's begin with delirium. The New York Times editorial writer Maureen Dowd wrote of her experience with too high a dose of cannabis edibles in that newspaper [16]. Her experience with nibbling on an edible, figuring out the absence of an effect meant that she needed to eat more and then doing so, led to a hallucinatory state that lasted 8 hours, where she was "panting and paranoid," unsure of where she was, and finally convinced that she had died.

Ah, delirium. You've no idea what time it is, or how you got here, wherever here actually is, which seems distinctly open to question. It's very difficult to think straight. Time extrudes like cookie dough, and the stitching on your coverlet has transformed itself into a swarm of angry spiders. Subjectively delirium is confusing and frightening. As one cannabis deleriant related, "I didn't know my name or who I was or where I was or what it even means to be human and have body and a brain. I didn't understand time or space, life or death. It was very metaphysical" [13].

Objectively, delirium is brain state identified by a combination of symptoms that can occur acutely whenever our brain is compromised from a huge variety of factors originating from outside the brain itself. Candidates include drug intoxication, (e.g., from cannabis or synthetic cannabinoids) drug withdrawal (e.g., from alcohol or Valium-like drugs, such as the "DTs," short for delirium tremens), infections, liver failure, malnutrition, or sensory deprivation (like being in an ICU). Individuals with immature brains (e.g., children with fever) or aging brains (elderly individuals with dementia) are particularly vulnerable to delirium, sometimes known as "sundowning" in older people with Alzheimer's disease. Objectively, delirious individuals become inattentive, drowsy, cognitively impaired, disoriented, hallucinated (often visually), experience delusions (often paranoid), have disturbed sleep-wake cycles, and disorganized language. A psychiatrist colleague of mine loves to relate the story of how he was called to consult on an elderly patient on a medical ward who was referred by his internal medicine specialist in the following manner. "You have to see my patient immediately. He's a man in his 70s who has acute heart failure and he was rational when I admitted him two hours ago. Now he's totally out of his tree. He thinks he is at work in the fire department and is trying to operate some kind of pumping equipment that actually isn't there. He keeps trying to get out of bed. His speech makes no sense; it sounds like he's quoting from James Joyce's *Finnegans Wake*. He can't think straight or talk right and he won't listen to reason. What's going on?"

Delirious patients are located at any given time somewhere on a continuum between being wide-awake, straightened up and flying right on one extreme and being in a coma on the other. Their clinical course frequently waxes and wanes, wandering along this continuum of consciousness over the course of hours like a drunken sailor. On an EEG, their brain waves are grossly disturbed, usually extremely slowed down and scrambled. Their short-term memory and attention is disrupted and they are commonly drowsy. The syndrome generally rolls in like a storm front fairly rapidly, typically over hours, but occasionally over the course of a few days. Focusing in specifically on delirium due to cannabis, what happened to me on my 20th birthday as described in the Introduction to this book, and to Maureen Dowd in her hellish Denver hotel room are pretty much textbook-typical examples of this diagnosis.

How does cannabis delirium come about? Let's look at a few specifics of my own case as described in the Introduction. The THC in hash mostly exists in the acidic form, (THC-A) and must be de-carboxylated by heating it to exert any psychoactive effects. If one were to eat a bunch of raw buds, no high would ensue. So my gently warming the hashish with a quarter-bar of butter in a cheap slightly battered saucepan accomplished that. Because THC is extremely fat-soluble, dissolving it in butter efficiently prepared it to be absorbed from my gut. So, what went wrong for me, Maureen Dowd and large numbers of edible consumers every year? Since my edibles episode devolved into a seeming interstellar misadventure, a suitable inquiry might be likened to the scientific commission convened to uncover the cause of the Space Shuttle *Challenger* disaster. To misquote slightly Richard Feynman's conclusion in that commission's report, "reality must take precedence over folklore, for nature cannot be fooled." What was underlying reality here? Some back-of-the-envelope calculations reveal that if we assume that the THC content in the 2.5 g (or 2500 mg) of my high-quality Lebanese hash was around 15% (it was high-quality), then there were 375 total mg of THC in my mini-birthday cake. A typical oral dose in edibles such as a chewy cannabis gummy bear or one square of a chocolate edible bar is 5 to 15 mg of THC. For a newbie, that's usually more than a sufficient quantity for a nice mellow buzz. In an entire large gourmet cannabis chocolate bar you might find 20 of those 5 mg squares, yielding a total of 100 milligrams of THC. So conservatively I had consumed 25 times the appropriate dose, or nearly 4 entire chocolate bars worth. A rough equivalent would be aiming to drink a single beer and instead consuming an entire case of 24 bottles in one sitting. Or meaning to order one espresso at Starbucks, but instead chugging down an entire full-to-the-brim Venti cup of pure espresso, followed by another two double shots. In short, too much of a good thing.

Another mistake of mine was to forget that eating cannabis is a completely different beast than smoking it. For a start, timing of the drug's effect differs completely. Smoking marijuana gives you an immediate buzz, (lung to blood is an extremely, fast and efficient absorption method), that has pretty much dissipated a few hours later. With edibles nothing much happens for around 40 minutes, and once the experience starts it can last 4–6 hours or more, ideal for long movies and for concerts. Not knowing the appropriate dose or the typical onset time for edibles is a common rookie mistake, and "overdose-by-impatience" is exactly the resulting trap that both Maureen Dowd and I fell into. Never assume that because nothing has happened half an hour after sampling an edible, that you under-dosed and therefore need to ingest a lot more. What's the science behind the need to hurry up and wait? When THC is swallowed it's absorbed from the gut (which takes a while) and then goes to the liver, an organ designed by evolution to break down any chemicals and toxins that get into the body into simple harmless compounds that can be efficiently excreted. The first thing that the liver does to metabolize delta-9 THC (and again this takes more time) is to convert it into a chemical called 11-hydroxy delta-9 THC. The latter does not exist in the cannabis plant. Once manufactured in the liver, the 11-hydroxy circulates back into the blood, where it enters into the brain even more efficiently than regular delta-9, so that it's a highly businesslike and effective chemical with which to get wasted. Some cells in the gut are also capable of metabolizing THC to 11-hydroxy, and send it on to the liver. (By contrast, the blood levels of 11-hydroxy after smoking are pretty minimal, so it's mainly the delta-9 that's getting us buzzed). This absorption and metabolism takes about an hour, which is why there is no immediate high from using edibles. All of this continuing cycle of absorption, recirculation of cannabinoid compounds, their re-metabolism in the liver and further recirculation accounts for the some of the ebb and flow of THC and its metabolic offspring in the blood, and the waves of intoxication that are typically pronounced following edible ingestion.

Back in 1970, the year of my resinous birthday cake, Louis Lemberger and his collaborators, the famed NIH scientists Julie Axelrod and Irv Kopin, took a batch of delta-9 THC[17], labeled it with a small, safe amount of radioactive tracer and injected it into human volunteers. They showed that the drug hung around in the blood for more than 3 days. Its metabolites began to appear in plasma within 10 minutes after the injection, where they took up residence, swirling through the bloodstream along with the parent compound. Delta(9)-THC was ultimately completely metabolized to 11-hydroxy and other compounds, with the radioactive metabolites continuing to be excreted in urine and poop for more than a week. A couple of years later, Lemberger had the bright idea of administering 11-hydroxy directly to human volunteers intravenously, and in a separate experiment, comparing it directly to the same dose of intravenous THC [18,19]. In these two later experiments, volunteers, blinded to what substance they were receiving, reported that the 11-hydroxy compound got them both to their peak high more quickly (in about 4 minutes) than did THC (about 15 minutes) and that they were significantly more buzzed when they got there. So the bottom line of all of this examination of THC metabolism is that edibles take longer to get you stoned, because the THC has to be absorbed and metabolized to have its effect via that route. But once the intoxication starts, because in part due to the conversion to 11-hydroxy, gram-for-gram the average person is going to get more intoxicated and for a much longer time, than if they smoked flower or hash.

My third, birthday-related mistake was what social scientists like to call "expectation bias." Given my prior experience with cannabis, I was prepared for the magic birthday cake to float me away gently into a peaceful little bower, where colors shimmered softly, musical waves danced beautifully and happy clouds floated peaceably by. Unicorns frolicking sedately might've been a nice extra. Being blasted at the speed of light into a seeming perpetual abyss of delirium was definitely not part of the game plan. It never occurred to me that such a possibility might crop up, so that when it arrived in the blink of an eye, there was no psychological plan B.

Enmeshed with these events is yet more science. Smoking or vaping cannabis is a faster way to get high than using edibles, because the heat quickly decarboxylates the THC and other compounds and sucking the smoke or vapor into your lungs results in immediate absorption from lung capillaries and quick circulation in your blood straight to your brain. Smoking is not terribly efficient, because up to half of what you're inhaling from a joint, for example, may just float away into the surrounding air and never make it into your lungs. But you can straightforwardly titrate your dose in real time because of the rapid onset of drug effects. If your high starts to feel a little excessive, you can just quit puffing. But once the edible enters your gullet, the horse has figuratively left the barn, even if the effects don't show up for another hour. If it turns out that you misjudged your dose, there are no mulligans. Adding more complexity, there's also the tricky concept of "bioavailability." If you choose to inject THC intravenously as Lemberger did with his subjects back in the 1970s and Deepak Cyril D'Souza does at Yale, you can be sure that 100% of the dose gets into the person's bloodstream. That's much more efficient than smoking cannabis as we just discussed. With edibles, somewhere between 5% and 25% of the drug may make it into your bloodstream from your gut. But that amount varies significantly from person to person, depending on your individual metabolism, what else you have eaten recently and other factors. If you really wanted to both increase bioavailability from the gut and reduce person-to-person variation, taking THC by suppository would be far more efficient. But it's not hard to predict that this is never going to be a popular trend, and that the company that makes Preparation H is not destined to be a major player in the cannabis market anytime soon. Yankee ingenuity being what it is, weed entrepreneurs have harnessed the process of nano-emulsion to solve both the variable absorption problem and the frustratingly long wait time for edibles to do their thing. Thus, quick-acting edibles may be in our future. Next, let's move on from delirium to psychosis.

Short-lived, acute psychosis

What happened to Arjun during Holi is a typical example of a very short-lived psychotic state. In very simple terms, the term "psychosis" describes individuals who, while wide-awake, (unlike delirium) have abnormal perceptions and/or beliefs. They may perceive things that seem completely realistic to them, but originate purely from inside their brain and not from any stimulus in the outside world (e.g., they experience visual or auditory hallucinations). Other people, either in addition to hallucinations or occurring by themselves, have bizarre, irrational, beliefs that are both inconsistent with their own cultural group and impossible to argue the person out of because they are so utterly convincing. Such beliefs are termed delusions. In Arjun's case, his brief psychosis was clearly precipitated by using a larger-than-usual dose of cannabis in the context of his personal worries and upsets (mental set). His symptoms disappeared within an hour or two with no treatment other than reassurance in the hospitality tent. In other instances, a different syndrome of more persistent psychotic symptoms following cannabis can last for days to weeks, as described next. These are the type of cases I mentioned earlier that are admitted to the psychiatric hospital at Ranchi in the month following Holi, as well as the individuals that were followed by Marta DiForti in her European cannabis/psychosis study, and by Kenneth Kendler in Sweden.

Acute, persisting psychotic state

An example of this syndrome is illustrated by a (composite) patient of mine. 20-year-old James smoked weed occasionally and generally liked the feeling it produced. It made him chatty and happy, somehow simultaneously relaxed and energized, and listening to music when stoned was always awesome. So when friends offered him a fatty of "really dank weed" before a concert, James didn't think twice before indulging. Or in taking a couple of extra bong rips before the group hit the road. It was about half-hour into the performance before things began to feel a little off. The atmosphere felt charged with meaning, as if something super-important were just about to happen. The feeling wasn't exciting so much as menacing. The order of songs being played was definitely not random, and that together with the way that they were announced made it very clear to James that the band was referring to him specifically and the nasty breakup he and Meaghan had the month before, in a mean, snarky mocking way. Not only was it obvious to him what was going on, but the entire crowd had very clearly caught on to the message. People around him were smirking, nudging each other, pointing at him, and signaling each other about him in a code of eye blinks and shoulder shrugs. The music from the stage was doing something very unpleasant to his brain by making it vibrate in a way that interfered with his thoughts. James had to get out of there in a hurry, and as he did so he experienced some kind of panic attack. His heart was racing, he felt shaky all over, his chest hurt, he was dizzy and the whole concert hall felt unreal in a way that made him think he was going crazy. He made it back to his apartment and immediately took a few Xanax to help him relax and to sleep off whatever this was. When he woke up at 3 am, he now felt convinced that the neighbors were monitoring his every move and had somehow bugged his apartment. When he listened very carefully, he could hear some kind of subsonic whispering or chatter that he knew intuitively was coming from communication devices. James knew both that whatever was going on had to be illegal and that there was a police station two blocks south of where he was. So he snuck rapidly out of his apartment and made it safely to the precinct, despite the blue car that seemed to be following him. The police heard his story, recognized what was likely happening and persuaded James to accompany them to the local hospital's emergency department. He felt safe there. The psychiatrist on call interviewed him the next morning, when it was clear that the symptoms of paranoia from the night before seemed to have increased in intensity. James was alert—in fact hyper-alert, visually checking out everything in the environment. He was absolutely clear that nothing like these experiences had ever happened to him previously. Although he spoke in a whisper to avoid being overheard, his speech was coherent, memory unimpaired, and he was perfectly oriented in terms of knowing where he was, the day and date, so that this was clearly not a case of delirium. His neurological exam was squeaky clean. Blood and urine tests were positive for THC but no other drug. The psychiatrist acknowledged James' anxiety and distress; she offered him a low dose of an antipsychotic medication, which he accepted, but several hours later he seemed increasingly upset and agitated at the emergency room staff "monitoring my thinking" and "messing with my mind." He now also refused to eat or drink because of suspicions that people wanted to "drug and poison me" and refused further medication because "the problem is not with me, and I'm not the crazy one." It was now over 24 hours since he had smoked the cannabis. James continued to insist that the drug had nothing to do with his symptoms, and that "all of this is really going on right now." The psychiatrist was able to persuade James to admit himself voluntarily to the psychiatry unit on the grounds that he would be safe there. His admission diagnosis was officially one of "substance-induced psychotic disorder (cannabis) with delusions, with onset during intoxication." After several days of treatment with antipsychotic medications on the unit, all of his symptoms began gradually to fade away, and James was able to accept that the cannabis had likely precipitated his symptoms. At times, thereafter he seemed a little unsure whether the experiences at the concert and subsequently had really occurred, but he was clear that they had now ceased and he was definitely back to baseline a few weeks later. Several months later, follow-up showed no return of any psychiatric symptoms. James was clear that what he'd undergone was extremely frightening, and was equally clear that he did not want to take the risk of using cannabis in the future.

Our fourth and last, category of cannabis-associated psychiatric illness is the most serious, and the entity that most concerns legislators and public health officials. Let's begin with a case vignette.

Chronic psychosis resembling schizophrenia

Janet (again a composite patient), was brought for psychiatric consultation and diagnosis by her parents, who'd been alerted by her roommates. Concerned about what they heard Mom and Dad flew in from California. Her friends had been concerned by several weeks of Janet's increasingly withdrawn behavior, suspiciousness, failure to show up for weekly meetings in her graduate program, and a worrisome neglect of her personal hygiene, grooming, and eating. Recounting meeting with her in the apartment 2 days before, mother wept in describing how Janet, who had always been rather prim and fastidious squatted down in front of her parents and nonchalantly changed her tampon. It was this startling and deeply upsetting change in her customary behavior more than her odd statements and preoccupations that pushed them to seek help for her.

What emerged from Janet's evaluation was that she had no family history of psychiatric illness, and no known risk factors for psychosis. Her birth and development were unremarkable except for walking and speaking earlier than either of her sisters. She'd always been above average in IQ and mathematically talented. Janet skipped grades in school, took advanced placement classes in high school and gained easy admission first to an undergraduate STEM program and most recently to a university graduate program in condensed matter physics in the Midwest. Placed with older kids in high school, Janet had begun smoking marijuana, most likely based on a number of factors, that included seeking peer acceptance, feeling relaxed and less socially anxious when using, because it helped her "get in the groove" and focus on programming, music felt more meaningful and her first serious high school boyfriend had been a regular user. Her subsequent marijuana use had been steady over several years and heavy on weekends but Janet was insistent that it had never adversely affected her energy or motivation or provoked any concerning symptoms. In the last few years she had begun vaping cannabis concentrates, but had tapered off and then discontinued use several months previously for financial and other reasons.

Careful questioning elicited the fact that her psychotic symptoms had probably started about a year previously. On her morning jogs, she had the strong sense that there were coded messages somehow embedded in the refreshing breeze that blew from the lake. As she ran and the air currents swirled around her face and in her hair, there was an inescapable feeling that somebody somewhere was trying to convey important information to her. Initially she had no idea either what was attempting to be transmitted or its origin. From these relatively mild initial beginnings, the urgency, detail and context of beliefs had multiplied to the extent that they preoccupied her almost entirely. Janet's delusions had now evolved to a state of extreme complexity, with bizarre ideas and beliefs nested systematically inside yet other delusions in a convoluted labyrinth of madness. She'd been savvy enough not to betray any of these frankly odd ideas to anybody else, even to close friends, but was surprisingly willing to talk about them when asked directly. The basic story was this. When Janet was 7, an alien spacecraft had landed in a field near her parent's house. She had communicated with these beings, who had selected her from all earthlings based on her "talents and intuition" but had then "gone silent" until the last few months, when they re-contacted her both telepathically and through coded messages via songs on the radio and TV advertisements. They could beam messages directly into her head and make their thoughts manifest, to the extent of making her think and do what they willed. (She had never mentioned the belief about the events at age 7 to anybody then or later, and this was likely a false memory or backdated delusional belief rather than anything that had occurred to her until recently). Janet had invested slowly increasing amounts of time and energy over many months in a quest to interconnect these alien messages, a series of meaningful coincidences in her life and equipment data that she was analyzing as part of her graduate thesis on superconducting materials. She had concluded that as a test of her abilities and perseverance she'd been placed in a form of programmed, simulated reality, where it was her task to puzzle her way out, make key logical connections, solve an "extra-planetary algorithm" and ultimately to save humanity from a competing race of malign aliens. As part of this process, everyone around her including her roommates and parents had been replaced by extremely realistic-appearing fake doubles. However what had begun as what seemed to her at the time a challenging and perhaps even exciting mission, had now devolved into a terrifying insoluble and inescapable web. Every car that passed by, every Facebook post, every plane that flew overhead only supplied additional confirmation that she was trapped. Because of her intelligence, Janet had managed to conceal her illness from friends and parents for many months.

Once she was referred for psychiatric help, any logical attempt by the psychiatrist to question her delusional system was met with a seemingly logical counter-argument. The truth became apparent of the old saw that it's impossible to argue someone logically out of a belief that they were never argued into rationally to begin with. The alien-algorithm delusions were one major part of Janet's disorder, but alongside of their development had been a gradual erosion of her sly sense of humor, vitality, get up and go, enthusiasm for classical music, passion for sci-fi, interest in social relationships, appreciation of nature, and ultimately her daily self-care. Her parents said that she had lost her essential "Janet-ness" and that she seemed emotionally flat and somehow deadened. To them this erosion of her joie de vivre was far more distressing than any ideas about beings on other planets.

A thorough workup revealed that Janet didn't have a brain tumor or any obvious physical explanation for her psychiatric symptoms. There were no longer traces of THC or any other drug in her body. She was not delirious, and had no mood disturbance to suggest severe depression or bipolar disorder. Janet met criteria for schizophrenia as a diagnosis of exclusion, since there was nothing else to explain her symptoms and they had existed for a significant amount of time. Given her lack of a family history of psychosis or any other known risk factor for her disorder, the clinicians caring for her concluded that her significant use of cannabis since her mid-teens likely played a role in raising the odds that she would develop the disorder.

Janet responded reasonably well to several months of treatment with antipsychotic medications. Her web of delusions gradually dissolved, and she was able to return to her graduate program with a somewhat reduced work burden. But the medications failed to offer significant help for her shrunken liveliness and animation, diminished spontaneity and blunted emotions. However, a peer support group, exercise program and therapy with trusted counselor helped manage them to some extent. She has not resumed cannabis use. Two years later she has not returned to baseline.

Acute human laboratory challenge studies

One obvious question is whether it is possible to reproduce psychosis symptoms and laboratory by giving volunteers sufficient doses of cannabis or cannabinoids. The answer to this is a clear yes. A recent review [20] emphasizes the advantages of well-controlled laboratory research studies in allowing fine-tuning and oversight of dose, delivery and proper characterization of administered pharmaceuticals compared to a placebo. This information should not be too unexpected, given the uncertainties and inaccuracies associated with tallying contents of marijuana purchased at dispensaries (and certainly on the illicit market), users' imprecise and subjective estimates of how much cannabis they consumed in a session and distortions of memory related to recalling retrospectively how a particular dose of drug made them feel. Laboratory administration is also safer in that subjects can be observed and reassured when necessary by trained professionals. What happens to volunteers when we give them large doses of THC or other CB_1 receptor agonists under controlled laboratory conditions? It is clear that CB_1 agonists can mimic symptoms of psychosis. Randomized, double-blind, placebo-controlled crossover laboratory studies showed that cannabinoid agonists including plant-derived and synthetics, when administered acutely can mimic the subjective experience of psychosis in a dose-related manner. These manifestations can be evoked in otherwise healthy individuals. Such transiently produced symptoms include numerous negative, positive, and cognitive phenomena including hallucinations, conceptual disorganization (confused thinking), perceptual distortions, delusions, emotional blunting and withdrawal, and slowed thinking and movement. These symptoms have a distinct time course, are related to dose and are not explained merely by how sedated the subjects are. Deepak Cyril D'Souza vividly describes experiences of healthy individuals who were administered intravenous THC in his laboratory at Yale. One of his experimental subjects believed that the clocks in the laboratory were being deliberately slowed down to confuse him, as part of the experiment. This seemed to be a delusional misinterpretation of subjective time slowing induced by the drug. Another subject believed that even after the drug infusion was long over, that THC was still being administered surreptitiously through the blood pressure monitor. When that was taken away to allay his worry, he now insisted that the drug was being given to him through the bed sheets.

A similar study was carried out by Marco Colizzi and his colleagues at the Institute of Psychiatry in London [21]. Sixteen healthy participants were given acute intravenous injections of just over 1 mg of THC and placebo on separate, randomized, placebo-controlled dose days. Twenty minutes following the injections virtually every one of them had at least mild symptoms of psychosis, but fewer than one in five had moderate-to-severe changes. Similar to volunteers in the Yale studies, typical drug-induced symptoms included conceptual disorganization, which affected almost everybody. For example, asked what an apple and a ball have in common, subjects gave illogical responses such as: "You can eat the apple, but not the ball," or "You can put the apple in the ball." Other frequent symptoms that occurred in roughly half to two thirds of subjects, included were hallucinations, suspiciousness, and paranoia. Examples include "The ventilator's noise is louder… the noise is actually rain, it's raining inside the room, I can see it and feel it, there is a black sky with seven blue drops…." One volunteer stated "What have you done to me? I understand, you want to make me paranoid with brainwashing questions…." Another said: " I feel I am all over the place and can't stop laughing, thinking you will expose me, I will say something stupid or strange." A lower percentage of Colizzi's volunteers, roughly 20%–25% also reported symptoms of grandiosity, hostility, and delusions. Quoted examples include: "Is this real? Is this a fake interview made by a fake doctor, like a Truman show?," or "I can understand things better and look for details, I am superior to others," "The injection changed me into someone with increased abilities," and "I thought you were going to attack me, people are entering the room to check on me." Compared to the placebo injections, these volunteers significantly increased their scores on standard symptom checklists used to gauge illness severity in patients with psychotic disorders. They also both reported and were rated by the researchers as having significant increases in negative symptoms of psychosis, including reduced rapport, lack of spontaneity, emotional withdrawal, and both concrete and stereotyped thinking. Here, subjects made statements such as "I can't follow my thoughts, I'm not able to think," or "I'm not interested and I am not willing to talk, I don't care…," and "My mind went blank, empty, with no thoughts."

The University of Colorado at Boulder (motto: "Be Boulder") will not allow cannabis to be consumed on its premises for research purposes, despite the drug being legal within the state for recreational and medicinal use. UC Boulder marijuana researchers therefore designed an ingenious work-around. Dr. Kent Hutchison and colleagues in the UCB Psychology department use a research vehicle known as the "Canavan." This fully equipped motor vehicle, that is essentially a behavioral pharmacology laboratory on wheels, can be parked outside of the research subject's dwelling. Participants can then smoke their own supply of cannabis in their homes, and immediately walk outside and into the van, where they can be studied by assessment of their blood levels of THC and CBD and behavioral tests.

THC certainly has psychoactive properties and can mimic both positive and negative symptoms of schizophrenia and psychosis. It's worth emphasizing that nobody outside of research laboratories chooses to take THC intravenously, but it would be instructive to look at blood levels of THC and its major metabolites compared to individuals who inhale concentrates through dabbing, as these conceivably could be similar. Similar experiments show transient symptom worsening in individuals with schizophrenia, even in those taking antipsychotic medications [22]. So, what are the underlying biological mechanisms by which the drug is producing these effects?

How might THC cause psychosis?

Researchers are beginning to document how cannabis use impacts the endocannabinoid system, but how cannabis increases risk of psychosis is not yet well understood. One obvious hypothesis is that acute THC challenge results in dopamine (DA) release, and the DA hypothesis of schizophrenia argues that so-called "positive" symptoms of psychosis such as hallucinations and delusions are due to excessive brain dopamine in particular neural circuits. The dopamine hypothesis is based on several observations. Drugs that cause dopamine to be released from neurons in large quantities or to build up inside neural synapses, such as amphetamines and cocaine, can precipitate psychosis if taken in sufficiently large quantities. Second, all known antipsychotic medications block DA receptors. Third, most never-medicated schizophrenia patients when given intravenous amphetamine, release excessive amounts of pre-synaptic DA, as evidenced on PET scans. Last, at least for first-generation antipsychotic drugs, the more powerful the medication at blocking DA D_2 receptors, the lower the average dose of the drug needed to be effective in ameliorating positive psychotic symptoms. Or in simple terms the more powerful the drug in blocking DA, the less of it is needed for an effective dose. It's important to understand that use of the word "positive" in this context implies not good and desirable, but the addition of some phenomenon not normally occurring to one's mental life. In other words it's shorthand for "extra, unusual mental events." In the same way "negative" psychosis symptoms refer to the subtraction of something normal from mental life, typically emotional range, enthusiasm or whatever it is that helps us get out of bed in the morning. The presence of these negative symptoms of schizophrenia is what was so upsetting to Janet's parents.

DA-containing neurons are located in different parts of the brain. We encountered them previously in reward circuits in Chapter 5, and midbrain movement and sleep centers in the Psychology section. We saw there that these DA neurons interact with the ECS in multiple ways. Many DA cells communicate with (or in the case of some reward regions, are members of) the striatum, so-named because parts of it are striped. The striatum consists of a number of deep brain nuclei that are the input to the remainder of the basal ganglia, many of whose cells are modulated by DA. If the DA hypothesis of cannabis psychosis is correct, then we would expect THC to release large amounts of DA in this region. In other words the hypothesis would be that cannabis or constituent chemicals such as THC would act analogously to amphetamine, as mentioned earlier. That simple, straightforward conjecture turns out not to be the case. Human PET studies from Sagnik Bhattacharyya, [23], and Robin Murray at the Institute of Psychiatry in London UK, show that acute administration of THC releases only unimpressive amounts of striatal DA, and also that striatal DA is low in chronic cannabis users [23]. Similarly, studies disagree as to whether drugs that block DA receptors (e.g., DA D_2 receptor antagonists such as commonly prescribed antipsychotic medications), reverse acute psychosis-provoking effects of THC in healthy subjects. Certainly, effective long-term treatment with antipsychotic drugs does not prevent an acute THC challenge from causing a flare-up of positive symptoms in patients with chronic schizophrenia [22]. Interactions between the cannabinoid and glutamate neurotransmitter systems have also been proposed as a mechanism whereby cannabis triggers psychosis [20]. Evidence for this conjecture is mainly from animal experiments, but also supporting the idea are some human spectroscopic data, where accurate chemical measurements have been made using an adaptation of MRI scanners that can tune into and measure amounts of brain molecules [24]. But, no clear answers.

Focusing more closely on negative symptoms of altered reward and social functioning, marijuana use patterns in adolescence influence the functional organization of different brain regions (i.e., how their activity is coordinated on fMRI scans) in early adulthood, measured while participants perform tasks that entail rewards. Also, lower functional connections between medial prefrontal cortex "brake" regions and the nucleus accumbens ("reward and salience" region) at age 20 were related to prior teen cannabis use, and also associated with poorer social functioning at age 22 [25].

Amotivation: As the old joke states, "People say nothing is impossible, but I do nothing every day." Separate from the issue of chronic psychosis, some studies suggest that chronic cannabis use is associated with persistent psychosocial problems such as impaired school and occupational performance that overlap with negative symptoms. It's often suggested that these difficulties may be due to chronic drug-related effects on either thinking and/or motivation. The stereotype of the 1960's hippie reclining on an eastern-themed cushion listening to acid rock or a contemporary stoner with terminal couch lock, living in his parent's basement and driven to do little more than play video games and eat junk food is a trope familiar to all of us. In fact some precursor of this cliché dates back at least to the 18th century, where tales of torpid, impoverished, unemployed Egyptians poleaxed by hashish entered the popular imagination. The Indian Hemp Drugs Commission also reported that heavy cannabis use was associated with reduced motivation. The term "amotivational syndrome" is frequently applied to this state, that is supposedly characterized by detachment, loss of drives and ambition, lethargy, flat emotions ("like, whatever man") and impaired working memory, attention, and judgment. Is there any truth behind this stereotype? And if people with the supposed syndrome quit cannabis use, does the presumed amotivation disappear, or does it persist?

We immediately run into another chicken and egg problem. If you start off life with low socioeconomic status or chronic poverty and find it a struggle to get even a low-paying job, then chronic intoxication may seem like a reasonable life choice. One way to try and dissect this conundrum is to bring volunteers into the laboratory and to dose them for prolonged periods of time with cannabis and see what happens. Setting aside the legal, ethical and logistical problems involved with this, we could conduct such a hypothetical experiment and try and assess whether the overall motivation of our subjects begins to dwindle. If it does, then is it dose-related? Another, more feasable approach might be to find a population of chronic cannabis users and to withdraw them from the drug under controlled circumstances. Do they gradually morph into Energizer bunnies? And if not, does that reflect their baseline pre-drug state, or presumed irreversible brain damage due to chronic cannabis toxicity? Animal studies shed some light on these questions, as they provide a situation where previously drug-free subjects can be dosed chronically over a long period of time compared to placebo. Such investigations also help dissect motivation from cognitive issues. In other words, does an animal given cannabis chronically not perform well on a task because of impaired learning or because it just can't be bothered to engage. Mason Silviera and his colleagues at the University of British Columbia in Vancouver investigated this phenomenon of cognitive effort by treating male rats with acute doses of THC, CBD, and both drugs together. The rats had been trained to perform both a more difficult, challenging task to earn a large sugary reward, and an easy low-payoff one. THC administration resulted in the rats becoming what the investigators described as "cognitively lazy." They were still perfectly able to complete the cognitively demanding task, (their accuracy, attention, and decision-making were fully intact) but essentially they couldn't be bothered to try. In other words, on THC they didn't give a rat's ass. Interestingly, their altered performance correlated with CB_1 receptor density in their little ratty prefrontal cortexes. Since these drug exposures were acute, they are not particularly informative about chronic effects. Experiments from the early 1990s involving chronic exposure of rhesus monkeys to marijuana smoke showed some effects on reducing motivation, but these were not definitive. Overall, the evidence for an enduring human cannabis-related amotivational syndrome seems to be equivocal or not very strong [26–29].

Let's return to cannabis psychosis and its underlying biology. Another route to understanding how cannabis might be related to psychosis comes through an examination of the brain's electrical activity. Whenever any of us encounters something in the environment that's either different or seems significant, our brain fires off a characteristic signal one third of a second later. This brain wave, known as the P 300, occurs in all mammals, and is basically a "surprise" or salience message saying "Hey there- pay attention! Something important may be going on here." If you are connected to an EEG machine, an experiment as simple as listening to a series of regularly spaced beeps through headphones can elicit a P 300 wave for each beep that's a little different from the rest. When I say that the P 300 is found universally across the mammalian kingdom, one of our former junior faculty tested this hypothesis by going to the local aquarium and gently attaching an EEG electrode the size of a toilet plunger to the head of a cooperative killer whale. The whale was provided with a series of identical underwater click sounds with occasional, slightly different-toned "oddball" clicks. The whale responded to the latter by generating a classic P 300 response. Reductions in the size of the P 300 peak are often seen in people with psychotic illnesses, although this finding is rather nonspecific and also associated with several other disorders. Drops in P 300 magnitude are provoked by cannabinoid agonists such as THC. Interestingly, the size of those falls for different cannabinoid compounds correlates closely with their psychosis-mimicking effects. Some researchers have theorized that particular cannabis-mediated disruptions in EEG frequency synchronization patterns between distant brain areas might also be related to positive psychotic symptoms. At Yale, Jose Cortes-Briones [30] hypothesized that positive psychosis symptom-like effects of acute THC doses are associated with increased cortical "noise" (defined as EEG randomness levels) in healthy human volunteers. Because EEG data are much cheaper and more straightforward to collect than approaches using MRI or PET scanners, using them as cannabinoid research probes an area of research that we will likely see more of in the future.

Dual diagnosis patients

For reasons that are not well understood, people who already have schizophrenia are at increased risk of subsequently developing substance abuse disorders of all sorts, at a rate (at a rate almost 5 times that of the general population). This includes cigarette smoking and cannabis use. Part of what accounts for these numbers is that people with schizophrenia in general have a tendency for making disadvantageously risky decisions, and an unfavorably altered balance in their ability to process delayed rewards and punishments. These factors tend to push them in the direction of substance use more generally. But let's focus down just on cannabis. One simple hypothesis for increased cannabis use in schizophrenia seems to be that such patients are seeking short-term relief from symptoms, and a boost in mood. They don't think too much about the later consequences of an increased likelihood of a relapse, worsening of their hallucinations, and delusions, and more frequent illness recurrences that also accompany their cannabis use [31,32].

Schizophrenia treatment trials with cannabidiol (CBD)

So, what are we to make of the observations on THC and psychosis that we have just reviewed? Let's integrate them with some additional important information. First, the DA hypothesis of schizophrenia fails to explain some significant features of the illness. About one in three patients diagnosed with schizophrenia does not produce excess amounts of pre-synaptic DA when challenged with amphetamines prior to any physician-ordered treatment. Anti-psychotic drugs that block DA receptors can interfere with these measurements, so that it's important to conduct such DA studies on never-medicated patients. Perhaps as expected, one in three patients with schizophrenia pretty much fails to respond to DA-receptor blocking antipsychotic medications. The obvious implication is that nothing is wrong with the DA system in these individuals. Thus, something else is likely causing their illness. Second, there may be something awry in the ECS of people with psychosis, whether or not they have ever used cannabis [33,34]. Such changes detected to date include altered levels of the endocannabinoid ligand anandamide in their cerebrospinal fluid [35]. The status of such ECS abnormalities is currently unknown. Are they an integral part of the illness, or some secondary phenomenon? Do they affect some patients with schizophrenia or all of them? An obvious question is whether somehow intervening to redress such a brain endocannabinoid imbalance might help treat schizophrenia symptoms. This idea occurred to several schizophrenia clinicians and researchers pretty much simultaneously. Since THC and other CB_1 partial agonists can provoke psychosis, and CBD has opposite effects in the human brain to THC under a number of circumstances [36], then CBD suggests itself as an obvious strategy for treating patients with psychotic illnesses, including schizophrenia.

Furthermore, the argument is often made that 1960s-style cannabis was "safer" from the point of view of psychosis risk because it had a significantly lower percentage of THC and correspondingly higher percentage of CBD. Because CBD is a relatively safe compound with almost no side effects, treatment trials in patients with schizophrenia seem appealing and low-risk. Early results in small, preliminary pilot studies seemed to augur well. Encouraged by these reports suggesting that CBD had antipsychotic effects, [37] a recent study from the UK by Philip McGuire [38] demonstrated that 88 individuals with schizophrenia who were maintained on their baseline antipsychotic medication while adding 1000-mg of oral CBD capsules daily for 6 weeks, experienced significant improvements on both measures of positive psychotic symptoms (such as hallucinations and delusions), and on their clinicians' impressions of the patients' improvement and illness severity compared to placebo capsules. The patients who received CBD also had a tendency toward improvement in cognitive tests and in their overall functioning. However, a carefully-executed attempted replication by Douglas Boggs [39] at a somewhat lower CBD daily dose (600 mg/day) failed to demonstrate any therapeutic effect at all in 36 schizophrenia patients compared to placebo.

Recently, the Maudsley group including Robin Murray, Sagnik Bhattacharyya, and Philip McGuire, looked at the effect of CBD administration in 33 individuals who were starting to show symptoms of psychosis, but not yet enough of them to qualify for a diagnosis of schizophrenia [40]. Nineteen healthy control subjects also participated. All participants received a one-time single oral dose of either CBD or placebo. People in the early psychosis group were felt to be too sick or too paranoid to go into the MRI scanner on two different occasions. Thus the study design entailed subjects not being their own controls. In other words, each subject, whatever their diagnosis, received only one pill (either CBD or placebo) once. After receiving the medicine (or placebo) subjects went into the scanner and engaged in a game-like task for which they received a reward of money if they could respond to stimuli fast enough. The main finding of the study was that a single dose of CBD changed brain activity in the high-risk psychosis patients in several regions that included the striatum (mentioned above), medial temporal cortex, and midbrain. In each of these regions, the brain activation levels for the psychosis risk group on CBD was somewhere in between the response of the healthy control individuals who received no drug and the remaining psychosis risk patients who received placebo. Although less than ideal in design, this experiment implies that CBD may be tending to normalize dysfunction in selected brain regions, all of which we have noted are implicated in psychosis. This in turn may suggest at a brain level how CBD is exerting its therapeutic effects.

Currently a giant question mark hangs over this whole area of potential treatment. Since CBD is relatively inexpensive and has few side effects, it could potentially benefit thousands of patients with psychotic illnesses. However, until there is convincing evidence that it actually works as hoped, this remains a theoretical possibility.

Are there biological differences in individuals with psychosis who have histories of significant adolescent cannabis use?

Some related questions, such as how many individuals would we have to keep away from cannabis to prevent one case of schizophrenia, and what population risk assessments tell us about the major cannabis-related risks for psychosis, were discussed in Chapter 7. As a followup to those issues though, let's look at some studies that address the underlying biological differences between individuals with psychotic illnesses with and without prior heavy cannabis abuse histories. One research strategy has been to examine the "fifty shades of gray matter" question. For example, one study [41] investigated over 100 individuals with various psychotic disorders, and looked at their histories of cannabis use. Using structural MRI's, scientists contrasted the density of gray matter in the brains of people with schizophrenia and psychotic bipolar disorder with and without a history of adolescent cannabis use. As many previous MRI studies in psychotic individuals have shown, the patients as a whole had lower total brain and regional gray matter density compared to healthy controls. Interestingly though, patients who had used cannabis in adolescence had significantly lesser gray matter reductions, an effect that was particularly prominent in schizophrenia. Similarly, Hanna [42], compared scores of cognitive abilities in approximately 100 psychotic individuals from the same psychosis study, to healthy controls. Cognitive test abilities were significantly higher in patients with histories of adolescent cannabis use compared to those without prior cannabis use. Again, this effect was significant only in schizophrenia. Together, these results are consistent with a hypothesis that early cannabis use may precipitate psychosis in those who had less prominent risk factors in the areas of compromised cognition and brain anatomy before they became sick. It is also possible that adolescent cannabis use could be defining a distinct psychosis subgroup, who show typical psychosis symptoms, but that are associated with less cognitive and brain structural damage.

Outside of psychosis risk, another issue raised by Janet's story is how chronic cannabis use in adolescence affects cognitive and academic performance. Before attempting to answer these questions, let's take a look at some relevant background terms.

How do we define "abuse," "dependence," and "substance use disorder?"

Substance abuse (e.g., as in the American Psychiatric Association's Diagnostic and Statistical Manual version IV-TR) [43], is generally defined as drug or alcohol use that is maladaptive, in the sense of leading to clinically significant impairment or distress as manifested by school problems, physical hazards (such as DWI), legal problems (e.g., arrest for disorderly conduct), and continued use despite recurrent social problems (e.g., physical fights). Substance dependence includes all of the above, plus the development of tolerance (needing to use more and more of the substance to get the same effect), withdrawal effects when the substance is stopped (such as feeling significantly anxious or having sleep disturbances, or the shakes), as well as possible compulsive use. In addition, dependence can include features such as using more of the substance than intended, trying to quit but failing to do so, especially in the face of knowing that the substance is causing problems (such as depression), and when using the substance takes up more and more of the person's time and attention so that other activities such as schoolwork and intimate relationships suffer. Binge drinking is defined on one hand by intent, that is using alcohol with the primary intention of getting drunk, and on the other hand, by physical quantities, that is, ingesting (somewhat arbitrarily defined) large quantities of alcohol (5 alcohol units for men or 4 for women) in a short period of time (3 hours). The broader term "substance use disorder" has multiple definitions, including abuse and dependence, but for alcohol sometimes also encompassing binging.

Adolescence is a high-risk period for beginning alcohol and substance use and for substance-related problems. The biological mechanisms that might explain why some people can use substances in moderate amounts without apparent problems while others transition to problem use is only partly understood. Part of substance abuse is genetic, (about 50% is the accepted figure), and the rest environmental. For example, if you live in a "dry" US state where alcohol is hard to obtain, that environmental factor has a profound effect on alcoholism rates. It's precisely this sort of statistic that worries opponents of recreational cannabis legalization, in terms of their concerns about increased access leading to more use and subsequent diversion of the drug to vulnerable teenagers. Teenagers are felt to be particularly vulnerable as a class because their brains are still developing actively into their 20s. Brain connections are being sculpted in all sorts of ways leading to greater neural efficiency, with greater predominance of more "mature" future-directed frontal lobe function over more "immature" emotion-driven, impulsive limbic circuitry. Given the known role of the ECS in brain development, including neural pruning, that we reviewed in Chapter 5, the fear is that cannabis and synthetic cannabinoids will interfere with these adolescent developmental processes to the long-term detriment of brain development. The hypothesis is that younger brains are more vulnerable, and that substance use actually alters their developmental trajectories, something that's true for cannabis in animal models [44,45]. Many, (but not all) studies in high schoolers strongly suggest that individuals who begin marijuana use before they reach 18 have problems in a number of cognitive areas including attention, IQ and executive functioning, and those who keep using across all the years of high school have lower standardized test scores. Again students who smoke cannabis during college end up with lower GPAs [46–50].

The genetic half of substance abuse risk may be mediated through different kinds of inherited physiological response to the drug (e.g., through variation in enzymes that metabolize a substance or the brain receptors that it binds to), or also through DNA that conveys a particular personality style of being more impulsive, disinhibited, sensation seeking, and reward-driven. Certainly, this type of impulsivity is associated with more frequent substance use, in greater quantities and with more negative consequences.

Substance use patterns, particularly for alcohol, peak during the late teens and early 20s when individuals are more open to exploration of new activities and less worried by long-term consequences. Using substances for the first time peaks at age 18 and is mostly over by age 20. Freshman year is a time of exploration, and heavy substance use is a part of this process for a significant proportion of college students. At least for alcohol, the highest prevalence of dependency is seen in 18–24-year-olds, after which young adults seem to "mature out" and many stop using entirely or severely moderate their substance use. College is an interesting environment for substance use, in part because it's a safe ecosystem that can reduce the consequences of binge drinking or weed smoking. Certainly while in college, students experience rapid increases in substance use. Since two thirds of students report consuming alcohol in the past 30 days, heavy alcohol use could be considered normative, (although illegal). After alcohol, marijuana is by far the most frequent substance of choice among college students. National surveys of 18–25-year-olds report over 50% lifetime cannabis use, with one third of individuals in this group using in the last year and one in five in the last month. Furthermore 60% of alcohol-drinking adolescents report using alcohol and cannabis simultaneously. Heavy recreational substance use impacts college achievement and likely cognition. Students who are intoxicated or hung over are more likely to miss classes, fall behind in school work, and get poorer grades. They may be less motivated to study, have fewer opportunities to study (because they are out carousing) or have more difficulties when they do attempt to study because they are hung over, or perhaps exhibit longer-lasting cognitive impairment. These relationship patterns are well-established for alcohol use but more obscure and much less studied for cannabis. Hints on what might be going to come from a variety of sources, but studying college students is a natural place to go to seek this information. Lets look at some examples.

A Dutch team investigated how legal cannabis access affected student performance under a set of unique circumstances. The city of Maastricht in the Netherlands declared a ban on allowing University students to purchase cannabis at its licensed marijuana shops. The prohibition started on a known date, so that it was straightforward for the researchers to explore academic grades of local students before and during its implementation. Interestingly, the academic grades of students who could no longer legally purchase cannabis increased significantly. Given what was presented earlier on freshman substance use, it's unsurprising that these academic improvements were driven by younger students but perhaps less intuitive that the effects were more marked for women and individuals with lower grades. The newly abstinent student body showed the biggest grade effects for courses that demanded more numerical and mathematical abilities. The research team dissected the students' course evaluations, and concluded that their improved performance was due to understanding the course material better as opposed to bucking up their study habits [51].

Ten years ago, funded by the National Institute on Alcohol Abuse and Alcoholism (NIAAA), I organized the Brain and Alcohol Research in College Students (BARCS) study that examined over 2000 freshmen students admitted to Trinity College and Central Connecticut State University over several successive years. Such a sample size is decently powered to answer some major research questions, and including students from two demographically different colleges with varied populations made the results more generalizable. The study was designed with two goals in mind. The first was to search for predictors, to identify which of these 18-year-olds was at increased risk for substance abuse and dependence before this occurred. The second was to examine consequences of substance use once it did occur. For example, what effects (if any) did 2 years of binge drinking or heavy marijuana use have on mood, college grades, IQ tests, brain structure and function, and EEG patterns. Because alcohol and marijuana are the two most abused substances in US colleges we were interested in capturing not only their individual effects, but also their combined influences, and not only at one moment in time but also over the span of 2 years. We managed to recruit 99% of each incoming class as study participants. At the first study visit, the students had cognitive testing (16 different tasks), and assessments of 10 measures of impulsivity, and told us about their current alcohol and substance use. They underwent psychiatric diagnostic interviews, provided information about their family histories of alcohol and drug use, gave us their own prior academic and substance use histories and donated a DNA sample. They also told us specific facts about themselves that we used subsequently to confirm their identity in later online questionnaire sessions. What was the name of your favorite pet when you were aged 5? How many brothers and sisters do you have? etc. Soon thereafter at their second study visit, 450 randomly-chosen students underwent structural and functional MRI scans, computerized impulsivity testing, an EEG, and an even more detailed structured substance abuse interview. Each month thereafter for 2 years, all 2000 research volunteers went online to a highly secure study website using individualized passwords to fill out detailed information about their drinking and drug use that explored quantity and frequency of use (e.g., " In the last month I got high twice a week, used a gram of cannabis flower each time"), consequences of intoxication (e.g., " I binge drank and got into a fight twice, was stoned and had unprotected sex once"). The students also told us about their mood and anxiety levels and sleep patterns. We knew that it was the subjects themselves and not, for example, their roommates who were completing each monthly survey, through their incorrect answers to a few of the many personal identifier questions that we had asked them to answer at the first interview. If they provided a different name for their pet goldfish at age 10, we noted that. We followed these students' college grades each semester. For those who had MRI scans, we first acquired a detailed structural MRI, then measured their resting state fMRI, had them participate in a virtual maze task that measured their spatial abilities, a visuo-spatial measure of recalling abstract shapes, then a task to assess impulsivity, and finally assessed the brain's response to alcohol and soft drink cues. Our interest in spatial and visuo-spatial measures was prompted by the fact that the hippocampus plays an important role in mediating these skills, and that this brain region may be especially vulnerable in teenagers to effects of both alcohol and cannabis.

At the end of 2 years, everything our subjects had completed at the study's beginning (except for the DNA) was repeated. For example, if he/she had an MRI session at the start of the study, then that was repeated at 24 months. What did we discover? When we looked at substance use patterns, three clusters of students emerged. The first was a group of students who rarely drank and seldom if ever used marijuana. Let's call them near-teetotalers or substance virgins. The second was a group of individuals who drank moderate to high amounts of alcohol but used little or no cannabis. Third was a cluster of students who used medium-to-high amounts of both substances. To my surprise, the expected group of pure stoners who got high on cannabis on a regular basis but disdained alcohol barely existed at these two colleges. While I knew many such people back in the 60's, they were an endangered species in our college sample. Although the three groups started off with highly comparable pre-college standardized test scores (SAT's etc.), in the first semester, the second group of freshmen (moderate/high alcohol, no/low cannabis) had lower academic grades than the first group (near-teetotalers). The differences between these two groups dwindled away over time, so that their grades became comparable. In contrast, the third student group (those who used medium/high amounts of both alcohol and cannabis) not only scored lower in the first semester, but also had consistently lower grades across the entire 2 years that we followed them. As our Dutch colleagues in Maastricht had found for cannabis, our students who curtailed their substance use over time had significantly higher academic GPAs than their colleagues who maintained their high levels of alcohol and cannabis use.

What could we say about the effects of heavy cannabis use alone on academic grades? In this case, absolutely nothing, because the stereotype I had conjured up, a group of red eyed students clustered around a bong, saying things such as "I never drink, alcohol tastes bad, it makes people aggressive, it's an addictive poison, it killed my granddad, etc." was as rare as hen's teeth and therefore couldn't yield meaningful data [52]. But, that's Science for you. You go out and carefully recruit a large diverse sample, study them in great detail and find out that one of the more interesting questions that you wanted to address is not answerable. This certainly doesn't mean that the study is fatally compromised, just that most data sources, including this one, are necessarily incomplete in some respect, and that no research is perfect. We accepted that it was the best data source we had for now, and strategized how to improve our design the next time around. Incidentally, our missing cannabis-purist/alcohol-disdaining college students certainly exist currently in other contexts. I have met them in Colorado and in California. For whatever reason, such individuals didn't make it into the freshman classes that we studied in Connecticut.

In general, exactly as in our study, teens and young adults who use both alcohol and cannabis have worse performance than near-abstainers on tests of complex attention, memory, processing speed, and visuo-spatial functioning [53]. Groups don't tell the whole story however. Many of us have come across individual heavy substance users who are seemingly immune to the expected deleterious effects, perhaps due to genetic luck that has equipped them with resilient constitutions and healthy adaptive physiologies. People like Janet who can consume large amounts of cannabis and maintain respectable GPAs certainly exist, because heavy cannabis use is a population or group risk for college academic grade decline, not an absolute determinant for an individual. Cannabis smokers won't like the analogy, but significant numbers of people smoke a couple of packs of cigarettes a day for years until late life and never develop lung cancer. Certainly, rats can develop behavioral and brain tolerance for heavy marijuana use, so that its cognitive and other effects diminish over time [54]. All of these observations undermine a simplistic "cannabis rots your brain" message, but do not entirely negate it.

Disordered use, dependence, and addiction

A sub-population of mainly young individuals misuses cannabis, finds it hard to quit, and develops addiction/dependence. As with all substances for which there is problematic use, it is not only the THC that constitutes the problem, but both the relationship the individual has with the substance, as well as idiosyncratic reactions that some individual users experience as a result of using the substance. Let's look at another case to understand what goes into the mix here.

John is a 22-year-old man employed in the "gig" economy in car repair. He works for a local garage/order repair shop and for friends on a cash-only basis. John is brighter than average, and as a young teen wanted to be an engineer working at the local aircraft plant, but dropped out of school in the 10th grade due to a combination of factors. He began using cannabis at age 14 with his school friends, as a result of which he found that he was less inclined to study for school, less able to absorb lessons, and attended school less often. He also noticed that over time, he needed to use more cannabis to reach the same level of "high." His grades slipped and he received several academic warnings but managed to hang on. Things were tough at home. His mom drank, his dad worked hard, hung out with his buddies a lot, and was emotionally unavailable. When his parents were together at home, they often fought, and smoking weed made him comfortably numb when they'd scream at each other. Nobody really noticed that things were going worse for John in school or were concerned that he might have a substance use problem. John was arrested a few times for possession of small amounts of cannabis, occurrences made several times more likely by his ethnic minority status. He will readily tell you that his current favorite leisure time activity is smoking marijuana, using the "ODB" (old dirty bong) in his living room. John's definition of "leisure time" is pretty elastic and includes bong hits first thing in the morning after getting out of bed, multiple times during the day, and before going to sleep at night. Getting high might not be at the exact center of his life, but it's pretty close, and certainly up there on his priority hierarchy. A few months ago he was really motivated to quit cannabis when his longtime girlfriend gave him an ultimatum: cut back on your use if you want to stay in a relationship with me. He tried several times, but on each occasion events unfolded in pretty much the same way. He was fine for maybe a day or two, then he would find himself getting angry and irritable, he would lose his appetite and develop a throbbing headache. Getting to sleep at night was a huge struggle, and once he could sleep he'd experience odd and disturbing dreams. His mood would go down a couple of notches, and his mind would feel anxious and restless. He described the state as "sort of being like caffeine withdrawal but scaled up a whole lot," and "not quite as bad as trying to quit cigarette smoking." Nevertheless each of the times that he tried to quit cannabis, after a few days of discomfort and sleep deprivation it just felt a whole lot easier to go back to smoking the ODB. Once you gravitate to old habits you tend to stay there. John's relationship with his girlfriend did fall apart, and he doesn't like to dwell on where his life actually is compared to the engineering job he once wanted. Still, stopping cannabis use just seems out of reach at this point. He's also concerned, at least sometimes, that all those years of cannabis smoking may have impaired his memory and motivation, so that engineering training may be out of the picture forever.

Several things about John's cannabis addiction are typical. He has a probable family history of alcohol abuse in his mother and several of her relatives, and many of the kids in his school and neighborhood were cannabis users, so that both genetic and environmental factors may have pushed him in that direction. His arrests for marijuana possession might have reinforced a feeling of being victimized, (minority-group arrests for cannabis are significantly disproportionate despite similar cannabis use rates in Caucasian and non-Caucasian populations). But John's brushes with the law occurred before age of 18 years and were not accompanied by any referral to teen substance abuse programs, and thus in no way deterred his further cannabis use [55]. Multi-times per day use of cannabinoid agonists such as THC is believed to re-train brain areas related to habits (such as parts of the basal ganglia) and to down-regulate brain CB_1 receptors as measured by PET. It's thus unsurprising that quitting is hard, when your brain is pulling you toward continued use. Stopping heavy cannabis use won't kill you, as can happen with alcohol (delirium tremens (DTs) and alcohol withdrawal seizures have significant mortality), nor is it even non-lethal-but-massively-unpleasant as can occur with going "cold turkey" on opioids. Nevertheless, cannabis withdrawal can be sufficiently dysphoric to deter quitting in a significant number of people, who try [56,57]. It is clinically significant "because it is associated with functional impairment to normal daily activities, as well as relapse to cannabis use" [58]. Interestingly, cannabis withdrawal was first taken seriously only when the CB_1 antagonist drug rimonabant was administered to regular cannabis users; it rapidly precipitated significant withdrawal symptoms in a number of them. A well-written first-person account of marijuana addiction is Neal Pollack's *I'm just a house dad addicted to pot* [59].

How prevalent is cannabis dependence? Somewhere around 9% of users become dependent, and that number doubles for those who start heavy use before the age of 18. To put that in perspective, these numbers are certainly lower than comparable figures for alcohol addiction, but still very concerning. Even if the majority of people who use marijuana don't go on to use other harder drugs like opioids or cocaine, (and most don't) the worry is that if recreational cannabis is legalized nationally, then the drug will be diverted into the hands of teenagers, and that 16%–18% figure will apply to a much larger total number of adolescents. As we saw in Chapter 7, actual rates of adolescent cannabis use have stayed steady, even in states that have legalized medical and recreational cannabis. But the existing percentages are much larger than anyone would want, even though cannabis is much lower down on the list of harmful substances than alcohol, prescription opioids or tobacco. An Australian study suggests that dependent cannabis users were more likely to be aged 18–24 years old, unemployed and to have higher levels of depression than non-dependent users [60]. As mentioned in Chapter 5, some pharmaceutical companies are trying to test cannabinoid receptor drugs to ease the discomfort of cannabis withdrawal. It would be helpful if these efforts were successful, because currently no pharmaceutical or behavioral approach tends to be terribly effective for this purpose [61]. John's concerns regarding long-term effects of cannabis on his motivation and cognitive abilities are not without foundation. A study of US postal workers found lower level of attainment among people who tested positive for cannabis. Outside of acute effects on cognition, long-term use may affect cognitive skills, making it harder to learn and retain information.

Individuals like John who consume cannabis multiple times throughout the day are not rare and a proportion of them continue use into their 30s. John Hughes and his colleagues in Vermont [62] recruited 142 daily marijuana users through advertising and followed them daily for 3 months using an interactive voice response phone system. Briefly, the mean age of the subjects was 33, just under half were employed, 60% were women, and 1/3rd had less than a high school education. The participants used marijuana just over 3 times daily on average. Almost 75% of them used tobacco, and almost all used some alcohol (4–5 drinks a week, with about 1/3rd binge drinking once a month, which was pretty typical for the national average for adults in the same age range). The take-home message of this study was that nearly all the subjects were subjectively chronically intoxicated. They apparently had not developed substantial tolerance to being stoned and were affected by cannabis when awake pretty much whenever the researchers checked in on them.

An important question regarding cannabis abuse and dependence, is what determines who is at risk and who is protected. Brain-related factors hypothesized to be crucial in underpinning abuse liability may include an imbalance between the neurotransmitters DA and glutamate in the brain's reward system, that lead to a "reward deficit syndrome" (RDS). Having a positive family history of any type of substance abuse may be a nonspecific bias toward increased likelihood for marijuana (and other substance use) disorder via this biochemical vulnerability. The conjecture posits that for people with RDS, it takes more reward input to float the affected person's boat, because their reward signals are blunted. Therefore such individuals seek extreme stimulation, just to reach the satisfaction level that most of us feel at baseline. Thus reward deficiency syndrome may be expressed as amped-up thrill-seeking, or sensation seeking. Substance use may be one form of behavior through which this tendency manifests, but high-stakes gambling, hang gliding or extreme rock climbing may be behavioral equivalents. My research group has looked carefully at impulsivity as a risk factor for alcoholism, including in our BARCS freshman student sample. We determined that a combination of impulsivity plus compulsivity traits seems to be such a risk factor. We hypothesize that the first tendency (impulsivity) gets you started with substance use and the second (compulsivity) keeps you coming back for more. This raises the general issue that whatever constitutional, personality, or environmental factors, first encourage you to try a drug may be very different than those that maintain your use. Some of the latter determinants include exposure to childhood trauma and abuse and availability of the drug in your environment. If you are an astronaut on Mars and there is no weed available, then obviously you're not going be able to use any, (unless of course an enterprising interplanetary dispensary got there ahead of you). Additional factors pushing a person toward substance use include being anxious or depressed, beginning use early in life (early habits tend to persist) and sex. In general, men tend to score higher on sensation-seeking scales and seem overall more vulnerable to substance abuse of all types. On the other hand, women, once they begin abusing marijuana and other substances, tend to experience what's called "telescoping." This term refers to the fact that substance abuse in women tends to proceed faster than in men, so that the addictive process from use to dependence unfolds on a faster time scale.

Just as there are risk factors pulling people in the direction of cannabis use disorder, there are protective factors holding them back. A few of the relevant insulating factors, (that were confirmed in part from our BARCS study) include being risk-averse, awareness of family history of addiction (where people were cognizant of their increased risk and therefore took steps to avoid exposure), and involvement in occupational or athletic situations where substance use might impair performance. Other protective factors include initial negative experiences with the drug (especially feeling paranoid or anxious), fear of arrest, awareness of urine screening as part of their job requirements, religious beliefs, a strongly cohesive family background, and a general "meh" factor for cannabis. For the latter, students would tell us "I tried the drug once, and it really didn't do it for me, so I never tried it again."

According to Yifrah Kaminer, a substance abuse professor at the University of Connecticut Health Center, substance abuse of any sort in adolescence raises the odds that the individual will also have other psychiatric disorders including conduct disorder (the young person's equivalent of impulsivity and acting out), major depression or anxiety. Conversely the odds for alcohol and cannabis abuse are 2–3 times greater in those with diagnosed major depression. These syndromes co-occur for several reasons, including substance abuse as self-medication for psychiatric symptoms, as well as use of the drugs themselves causing the emergence of anxiety and depression symptoms. For example, cannabis in adolescence seems to cause short-lived mood increases, followed by more prolonged stretches of low mood [63]. But the point to be emphasized Kaminer says, is that since substance abuse and psychiatric symptoms in teens co-occur and both are multi-dimensional, then they require a multidimensional treatment plan.

More on Structural and functional brain imaging consequences of cannabis use

As mentioned previously, because the human brain is actively developing through adolescence until age of 25, and components of the ECS including CB_1, guide these events, consumption of cannabinoids risks disrupting those processes. Currently, almost 35% of American 10th graders have reported using cannabis. Although rates of cannabis use among high schoolers remain constant for now, one worry is that they may increase if cannabis is legalized at a federal level.

So, what's the actual evidence that cannabis consumption is associated with structural brain changes? This is a more relevant question to ask, since functional brain alterations tend to be much more dynamic, perhaps fleeting and only reflecting recent use. To summarize a much-debated topic, the structural evidence is mixed, but not impressive when it comes to demonstrating cannabis-related brain alterations. Recall that in our 2-year freshman college student BARCS study, we performed imaging on 450 of our students at baseline, and followed up as many of them as we could 24 months later for repeat imaging and IQ testing. While we documented hard-to-miss alcohol effects, (heavier quantities and frequencies of alcohol consumption in our group were associated with significant shrinkage over time of the hippocampus on MRI and poorer 2-year memory function), we were not able to demonstrate any similar effects of cannabis on either brain or memory [52]. Barbara Weiland, at the University of Colorado Boulder, used very high-resolution MRI scans to compare several different measures of brain structure in 29 adult daily marijuana users (average age 27), and 50 adolescent daily users (average age just under 17), with each of these two groups compared to equal numbers of same-aged non-cannabis-using controls. No significant structural brain differences were found between the marijuana users and controls for any of the key regions examined. Not only that, but the degree of difference between the two samples was minimal, so that the size of the difference between the two groups hovered right around zero. As part of their study, the Colorado group also examined a dozen prior published papers that had undertaken similar measures, some of which had reported brain differences in the marijuana smokers. They pointed out that the findings in these earlier papers were inconsistent. No two of them reported quite the same brain areas as affected; also, some reported structural increases in volume, others decreases. In addition, the findings of many of the previous studies were complicated by the fact that the subjects were also alcohol users. As we (see earlier) and many others have shown, alcohol consumption has indisputable effects in shrinking brain structures. The Colorado study, unlike many prior efforts, matched subjects and controls very carefully on alcohol use measures and also controlled for other possible complicating factors that could influence brain volume, including tobacco use, age, sex, impulsivity, and depression [64]. The same group of investigators then performed similar measurements in a group of older adults whose mean age was almost 70, who had smoked cannabis at least weekly for the past year, and on average had used it for close to 24 years. Cannabis users and non-users showed no differences in terms of total gray matter or white matter volumes when age and depression symptoms were taken into account. However the users had larger volumes in a handful of brain regions in parts of the frontal lobe, an area of visual cortex, and a region deep inside the brain concerned with motor movement and learning. The two groups also scored the same on a short computerized cognitive test. The overall conclusion was that once you control for age, cannabis use had no widespread effects on overall brain gray matter volumes [65].

So far then, the studies we've examined have not demonstrated any worrisome brain changes associated with even long-term cannabis use. The exception comes from the IMAGEN study of adolescents that follows a very large European population sample of 2400 adolescents collected across many research sites. The design of the study is such that participants are followed multiple times as they mature, with brain imaging and IQ testing, starting at an age well before alcohol and drug use begins. The investigators can therefore tell whether substance use alters the normal trajectory of structural or functional brain development during the teen years, without being confused by "chicken and egg" questions of whether the brain changes preceded substance use, or vice versa. In this particular IMAGEN study, 46 boys and girls of 14-years of age who had used cannabis only once or twice, were matched with others from the study who had never been exposed to THC. Completely unexpectedly, very large brain regions in both temporal lobes (that included the hippocampus and amygdala), the cerebellum, parts of visual cortex and the posterior cingulate cortex on both sides of the brain, all had significantly greater gray matter volume in the cannabis users. When the IMAGEN investigators went back to look at the previous MRI scans of these individuals, they could not explain these brain differences as a feature that had been there prior to cannabis smoking [66]. How did the investigators try and explain such an unlikely finding? One piece of the puzzle was that the areas where gray matter appeared to have increased after using cannabis corresponded closely to both maps of CB_1 cannabinoid receptors (that other investigators had created using PET scans), and to places in the brain where the CNR1 gene that codes for those receptors were switched on the most. The investigators looked very closely to see whether other factors such as personality differences or psychiatric symptoms might be responsible for the brain differences, but found nothing. In addition, if cannabis kills brain cells, why did the adolescents who'd been exposed to it have more gray matter rather than less, in their brains? One possible explanation is that cannabis has effects on slowing down normal pruning of neurons in the adolescent brain.

Long-term effects of cannabis use

This is a hotly contested topic, as I'll explain. If you study brain function tests, recruit a population of chronic cannabis users into a study, and test them on a series of cognitive and neurological assessments you will undoubtedly find impairments. Many studies have shown this. The big questions are, if you can persuade your subjects to stop using cannabis, will all their test scores return to normal? And how long would you need to keep them away from the drug to make sure that they had returned to baseline—a week, a month, or six months? This issue is a particular problem with cannabis research, since the substance tends to persist in the body (and critically, in the brain), for lengthy periods of time. Not only that, since many cannabis users also use other legal and illegal substances, including alcohol, that can affect thinking and memory, that would have to be taken into account, particularly if the person continued to use them while abstinent from cannabis. Furthermore most individuals (as we've seen) are not especially accurate in reporting their cannabis use, and the amounts and ratios of cannabinoids in the drug not only vary from sample to sample but also have changed across time. Thus it's hard to calculate an accurate relationship between the amount of drug someone has used over the long haul, and any deficits that you might detect. Not only that, different researchers tend to use their own favorite scales and approaches to measure a particular cognitive domain such as memory, rather than all of them converging on a single best assessment method. Many of the existing studies were conducted predominantly in men. Also, it's not entirely clear what the underlying brain mechanisms might be that are responsible for any persisting cognitive deficits.

Given all of this imprecision, one creditable attempt to synthesize the large existing literature on this topic came from Samantha Broyd and her Australian colleagues, that included a couple of researchers that we've encountered elsewhere in this book (Hendrika van Hell and Nadia Solowij) [67]. As we have reviewed previously, they found acute and chronic impairments across verbal learning and memory, and to some extent working memory, large impairments in attention and attentional bias in cannabis users. There were acute deficits in psychomotor functioning and variable impairments in executive functioning such as inhibition, planning, and reasoning interference control and problem-solving. When they examined the persistence of these problems following abstinence, many of the chronic deficits were very much reduced. Those that diminished to a "plus-minus" (i.e. maybe present) level, included problems previously seen in verbal learning memory, and attention. Difficulties in executive functioning such as planning and problem solving didn't shift much over time, but were not markedly impaired in chronic users even prior to abstinence. The researchers concluded that there were likely persistent effects on attention and psychomotor function and possible persistent effects on verbal learning and memory after abstinence. However, they admitted that this whole question of recovery of function after prolonged abstinence is murky, particularly since the vast majority of studies didn't follow the subjects beyond about a month. They also stated that more recent studies suggest that verbal learning and memory impairment may recover with prolonged abstinence. One very large prospective study tracked young adult research volunteers across three periods of data collection spaced 4 years apart. Those researchers found, on tests of immediate verbal memory that former heavy users who had been abstinent from cannabis for a year were more improved cognitively, compared to individuals who continued heavy use. Furthermore, these long-term abstinent subjects did not differ from comparable individuals not using cannabis, on any cognitive measure [68]. However a second study detected impaired attention even after 2 years of abstinence [69]. The very fact that review papers in this section are dancing around a "maybe yes, maybe no" and "yes, but" set of conclusions suggests to me that while cannabis-related cognitive problems may exist, they are subtle and likely to fade over time with continued abstinence.

Whether the proportions of CBD, minor cannabinoids and terpenes in cannabis have any long-term cognitive effects is completely unexplored. Although there are many suspicions that an early age of starting cannabis use, and heavy use in adolescence, may be most damaging to cognition, these are also understudied topics. One advance that will shed some light on this is the newly-funded Adolescent Brain and Cognitive Development (ABCD) study initiated by the National Institutes of Health. This effort will follow 10,000 youngsters recruited from across the United States beginning at age of 10, prior to when most individuals begin substance use, and track them through their teens at least annually with detailed cognitive tests, MRI scans, and investigations of their families, life events, and substance use. Large-scale studies such as these have the potential of addressing many key questions in the field (that we alluded to earlier in discussing the IMAGEN results). These include the what-caused-what problems, of whether brain changes associated with substance use preceded or followed engaging with the drug, and whether normal brain developmental patterns are thrown off course by substance use of different sorts.

How about general health risks associated with cannabis use? A useful review from Germany attempted to synthesize information on this topic [70]. Chronic cannabis use can cause bronchitis and chronic cough, but unlike tobacco smoking there is no apparent link to either lung cancer or throat cancer. Risks for these disorders are probably increased in people who smoke cannabis and tobacco together (spliffs, blunts, etc.) but there seem to be few hard statistics available on this latter question. When I discussed the ECS, I mentioned that cannabinoids may have adverse effects on liver healing; there is some evidence that they can worsen the fatty liver and liver scarring that is part of hepatitis C. Some chronic heavy cannabis users experience repeated episodes of severe nausea and vomiting, the rather rare, so-called "cannabis hyper-emesis syndrome," that is relieved temporarily by very hot showers. Why this intervention works isn't known. Anyone who's ever used cannabis is aware of the fact that it speeds up heart rate; there are rare cases of heart rhythm problems being provoked by use of the drug, as well as very uncommon cases of heart attacks, almost all in individuals who had pre-existing heart disease. To put all of this in perspective, however, compared to the health harms attributable to alcohol or tobacco, those due to cannabis appear to be much less. Harmfulness is an issue we will plunge into in some detail later in this chapter. Over the next couple of decades, we will undoubtedly learn more about possible chronic health effects associated with cannabis consumption, particularly as both THC concentrations and the number of marijuana users rise.

Long-term effects of youthful cannabis use in older adults

Following a research cohort for 2 years to look at longer-term effects of cannabis use (as we did in BARCS) is one thing, but how about measuring the current consequences of the cannabis somebody smoked 50 years ago? Susan Bookheimer, a UCLA neuroscientist, and her collaborators, recently performed a study whose implications might strike fear into the hearts of many of the people who lived through the 60's and are still walking the earth [71]. They say that if you can remember the 60's you weren't there. But let's suppose that you *were* there in the 1960's and 70's (when cannabis use doubled), you *did* inhale, and are now perhaps contemplating retirement. Like many in your age-group, you are concerned about the possibility of Alzheimer's disease. Given your 60's weed use, should you be extra-worried concerning your brain? In other words, you might not remember the 60's but can you remember where you left your keys 5 minutes ago?

To address these questions, Bookheimer recruited her 57–75 year old subjects from the local community and screened out anybody who'd used recreational drugs other than cannabis ever, more than once. She focused in on 24 individuals who began using cannabis before age of 20, smoked it at least 20 days per month in at least 1 calendar year, and had pretty much quit using by age 35. On average her subjects had begun using marijuana at age 17, used for 11 years, and been abstinent for 30 years. None were heavy drinkers. She also recruited an equal number of individuals who said they had never used marijuana or any other illicit substance, ever. All subjects had a physical exam to ensure that they were healthy and a urine screen to rule out anyone currently using recreational drugs. Everyone completed a battery of cognitive tests (focused especially on memory), followed by a research-quality MRI scan, where the research subjects had super-high resolution structural images taken of their hippocampuses. Compared to the group that had never inhaled, the former cannabis smokers had around 10%–15% smaller gray matter volumes, specifically in several hippocampal sub-regions, reductions that were not seen in other nearby brain regions also measured during the scan. Although there were no significant cognitive test differences between the two groups, the former cannabis tokers tended to score lower in every test.

What are we to make of these findings? Significant cannabis use in early life seems to have persisting effects on brain anatomy, at least in the hippocampus, several decades after these individuals had quit using the drug. Inevitably, there are some problems with the study design. The number of subjects was small. Participants were being asked to recall their drug use 40–50 years in the past, which is challenging particularly for a drug that affects memory. Finally, there is a chicken and egg problem, in that we don't know the volume of these subjects' hippocampuses (strictly termed hippocampi), before their first cannabis use, (or non-use in the case of controls). And although unlikely, perhaps people with small hippocampi are more prone to smoke cannabis. But despite these caveats, the study certainly raises concerns. For one thing, the likelihood is that these brain changes were caused by the THC in cannabis. Because current chemovars have a much higher percentage of that substance, they thus may be potentially more harmful than the relatively weak "grass" of the 60s with its 4% THC, full of stems and seeds. Second, the earliest brain changes of Alzheimer's disease attack these same hippocampal areas and their surrounding gray matter very specifically, so those at risk for the disorder may have diminished resilience to its effects, if they're unlucky enough to develop it. These speculations don't have answers at this point, and I certainly don't want to be a scaremonger, because the evidence doesn't warrant that. But this is certainly one of those provocative scientific studies that deserve being re-run in a much larger sample, to see whether the results hold up. If they do, that would be concerning. Although on the other hand, recent small-scale studies in Israel hint that cannabis is a useful treatment for agitation and "sundowning" in elderly demented patients. So if youthful cannabis use hastened your late-life dementia, we may know how to take the edge off your symptoms.

Pregnancy effects

We've already noted that perception of cannabis' potential harms is on the decrease. Perception of harm deters use. Pregnant women are now more likely to use cannabis as a natural substitute for medications prescribed for anxiety and depression as well as morning sickness [72]. Letting any kind of substance that isn't supposed to be there into the developing fetal brain and body is a potential risk. We know that recreational substances such as alcohol and nicotine, as well as some physician-prescribed medications such as thalidomide can reach the fetus via the placenta and the umbilical cord, to interfere with fetal development. Risks associated with various substances include premature births, birth defects, low birth weight, fetal alcohol spectrum disorders, or miscarriages. We've already learned that cannabinoids, being very fat-soluble concentrate in the nervous system, and that anandamide and CB1 receptors guide critical aspects of human brain development. So, how do we determine risk information related to maternal cannabis consumption during pregnancy, and should we be concerned? Well, marijuana is the most common illicit drug used during pregnancy. It may have different effects across the three trimesters of pregnancy, and it is more likely to be used by pregnant teenagers. A survey of over 14,000 pregnant women [73] found past month cannabis use of about 4% during pregnancy (highest at over 6% in the first trimester). Use was doubled (at 14%) in 12–17-year-olds, and also greater in non-Hispanic black women. Another recent study from Northern California surveyed over a quarter of a million women enrolled in the Kaiser Permanente health system. Between the beginning of 2009 and the end of 2017, self-reports of daily, weekly, and monthly cannabis use among women in the year before pregnancy almost doubled. Rates of use during pregnancy itself increased from almost 2% to 3.4% over the course of the study [72].

Another worry is that because of its fat solubility, THC is concentrated in breast milk and passed along to newborns and infants, if their mothers use marijuana [74]. Finally, in a now-infamous investigation, Betsy Dickson and coworkers [75], designed a study where the investigators called 400 dispensaries in Colorado, claiming to be 8 weeks pregnant and experiencing morning sickness. Over two-thirds of the locations called recommended treatment of morning sickness with cannabis products. Medical dispensaries (at 83%) were particularly liable to provide this advice, with the majority of staff basing their recommendation on personal opinion. Overall, 36% of dispensaries stated that cannabis use is safe during pregnancy. While most advised discussion with a healthcare provider regarding cannabis use, fewer than one-third of them made this recommendation without prompting.

Concern is appropriate, but must be balanced with knowledge of actual risks. For example, developmental consequences of some recreational drugs on fetuses have been overblown. Recall the dire warnings in the 1980s regarding the coming epidemic of vast numbers of mentally challenged behaviorally abnormal, "doomed from the womb," "crack babies" that never truly materialized, for example. Similar scaremongering is already appearing for cannabis, for example, the November 23, 2018 British Daily Mail headline, "Could medical cannabis be the new THALIDOMIDE? Fears of a crisis as doctors consider doling marijuana-based medicines out to pregnant mothers despite evidence the drug can damage foetuses" [76]. While cannabis-related experiments on pregnant laboratory rodents are relatively straightforward to conduct, rats are not humans (no matter what you might think regarding your surly coworker or local politician), so that results from animal studies are not directly transferable into the human realm. For example, neurogenesis in the fetal prefrontal cortex takes a full 5 months to complete in humans, versus 1 week in mice. Other precautionary issues are that cannabis may be consumed along with tobacco in ("spliffs" and "blunts") and contaminants such as pesticides or heavy metals can be found in cannabis, that may themselves be independently toxic to fetuses.

A useful comprehensive review of consequences of human prenatal marijuana exposure in relation to later development of neuropsychiatric problems [77] pulls together information from laboratory animal and human cohort studies. They conclude that THC in rats reorganizes the DA system in brain regions concerned with mood and conscious behaviors. CB_1 receptors in our fetuses have distinctly different distribution than those found in the adult human brain. For example, in mid-pregnancy there is an extremely high CB_1 receptor concentration in the amygdala and hippocampus. Women in the Maternal Health Practices and Child Development Project who smoked cannabis during pregnancy gave birth to infants who by 10 years of age had significantly higher depression rating scores than those of otherwise comparable non-exposed women [78]. In animals, THC exposure interferes with the development of other brain neurotransmitter systems including DA and opioid circuits. Although increased risk for schizophrenia in adolescence might be a theoretical concern in later life for cannabis-exposed fetuses, the sort of large-scale epidemiologic studies to show such a relationship have yet to be performed. Bottom line: there may well be risks associated with cannabis use in pregnancy, but their nature, magnitude, and frequency are yet to be determined [79]. This hasn't stopped some scaremongers in the medical world from making dire predictions. Stuart Reece, in Australia, for example [80], uses particularly alarmist language "..massive scale chromosomal shattering or pulverization can occur..... aggressive onset of addiction-related carcinogenesis... downstream genotoxic events including oncogene induction and tumor suppressor silencing.... (THC) in this respect is similar to the serious major human mutagen thalidomide." Paul Merlob in Israel reviewed the evidence for cannabis use during pregnancy affecting the fetus and newborn with regard to issues of fetal growth, stillbirth rates, preterm births congenital malformations, and neurobehavioral alterations in newborns [81]. Once potential confounders such as use of tobacco, alcohol and other recreational drugs had been accounted for, existing studies were either inconclusive or found no marked effects. "Most published studies report inconsistent results... Well-designed studies are needed which should take into account all confounding factors for different neonatal outcomes while accurately measuring quantity and timing of exposure to marijuana." My personal opinion—don't tempt fate. Until we have a better handle on risks, why subject your unborn baby or newborn to possible health problems?

How dangerous is cannabis overall as a recreational substance? And how do its risks compare to those of other recreational substances?

How do we summarize the adverse health effects of recreational cannabis use and quantify the size of the substance's overall risks? More generally, is there any way to rationally compare the harms and risks of using a particular substance compared to other substances? Or, more specifically, how harmful is cannabis use compared to that of alcohol or coffee drinking or tobacco cigarette smoking? These are questions debated by public policy analysts among others, but it's appropriate to pose them here. Part of the problem in addressing these issues is that they are located at an uncomfortable interface between science and politics [82]. Politicians are often elected on promises to reduce drug use and crime. Nixon was a prime example. Admitting that any illicit recreational drug is less harmful than generally supposed will not help their reelection chances. As an example of the political hot potato status of drug harm-related issues, one of the more articulate voices in this debate, David Nutt, a British Professor of Psychopharmacology at Imperial College London, who presented evidence that alcohol was more dangerous than many illegal drugs, including cannabis, was consequently fired from his role as long-time chairman of the UK government's Advisory Council on the Misuse of Drugs in 2009 [83]. But as Nutt says, "Being willing to change our minds in the light of new evidence is essential to rational policy-making" [84]. Certainly having the relative risks of recreational drugs quantified by scientists and drug experts rather than by politicians seems rational [82,84].

How to think about drug-related harm? At first blush if I decide to take a few Valiums to help me get through the day or a shot or two of vodka as an eye-opener before I go to work that could strictly be my personal choice. But if I'm employed as a school bus driver, missile silo officer, or a neurosurgeon, then showing up for work mildly intoxicated can impact other individuals adversely. Also, if I snort cocaine on the weekend in a way that doesn't measurably affect my work performance, am I nevertheless helping a local criminal drug syndicate or ultimately the Sinaloa cartel in Mexico? If one substance such as alcohol rated as generally more harmful than another such as cannabis, but one is legal, how is that weighed into the harm equation? Such questions were first raised almost 50 years ago, by Samuel Irwin, a professor at the University of Oregon. Then, beginning in 2007, Nutt and his colleagues came up with a rational starting point for ranking recreational substances (both legal and illegal) relative to each other, as laid out in several papers [83,85,86], and summarized in his book [84]. They considered the large and diverse means by which drugs can lead both directly and indirectly to harmful consequences, and ultimately how to weight each of these classifications to allow comparison across substances. These investigators proposed that two overall major categories had to be considered: harms to the users themselves (e.g., toxicity or loss of relationships), and harm to others (e.g., crime and economic costs) both to individuals and at a population level. Harms to the users themselves comprised nine different sub-categories, and harms to others a further seven. I will try and summarize this somewhat complicated classification here, because it is thorough and comprehensive. I also acknowledge the thoughtfulness and hard work that resulted in its creation.

The first major item within harms to the user category was *drug-specific mortality*, defined as instances of direct harm to individuals using the drug. An obvious example is deaths from poisoning. This can be gauged by comparing the amount needed to produce psychoactive effects with the fatal dose, to give a safety ratio. For alcohol, this number is 10. If 2 units (i.e., 2 beers, 2 glasses of wine) cause intoxication, then 20 units will kill you. Although the uptick in alcohol-related deaths has been partially obscured by the opioid epidemic, the number of deaths attributable to alcohol in the United States increased 35% between 2007 and 2017, according to a recent University of Washington study. In particular, deaths among women rose 67% [87].

So, when it comes to drug-specific mortality, cannabis is clearly safer than both alcohol and opioids because although huge doses may push you into an extremely unpleasant delirious state, there are virtually no fatalities. Perhaps none at all. This, as we've seen is true in part because CB_1 receptors are not located in parts of the brainstem that control breathing and heartbeat. *Drug-related mortality* on the other hand, counts deaths from chronic illnesses, for example, suicide, alcoholic cirrhosis, tobacco-caused lung cancer, or fatal road traffic accidents. As we saw earlier, cannabis is much less harmful than alcohol when it comes to intoxicated driving, but probably not without risk, given the slightly elevated odds of an accident that we reviewed in the Swedish meta-analysis study. The category of *drug specific damage* includes physical harm short of death due to a drug. The sort of items included here would be holes in your nasal septum due to cocaine use, chronic bronchitis, heart attack or COPD from tobacco, and chronic liver disease from alcohol. Again, cannabis scores relatively low in this category, unless we count individuals who smoke cannabis mixed with tobacco, as is common in both the Netherlands and Jamaica. But that damage is from the tobacco and not the cannabis.

Let's pause for a second to look at some of the alcohol-related statistics of the sort we encountered in the realm of marijuana in the Chapter 7. George Koob, the director of the NIAAA, quotes some statistics for alcohol damage in the United States that illustrate the magnitude of alcohol damage in these categories. In the United States annually, 179 million people, or 66% of the adult population use alcohol. 5.3% of the population, or 14.5 million people are diagnosed with an alcohol use disorder. 88,000 people die from alcohol-related causes of all sorts, with nearly 50,000 of them from acute events including overdoses or injuries, and another 39,000 people from chronic causes such as cirrhosis and cancers. 50% of all liver disease deaths are due to alcohol. There have been significant increases in binge drinking over the past 10 years that are driving higher rates of emergency room and hospital visits, and 5.7% of the population or 14.1 million people aged 18 and over meet criteria for an alcohol use disorder. Extreme binge drinking, where users consume huge amounts of alcohol in a short span of time is also on the increase [88–90]. In comparison, 4.2% of the population misuses opioids and 0.8% of the population or just over 2 million people have an opioid use disorder. There are 47,600 total deaths by opioid overdose annually; 15% of these involved alcohol in 2017. Stacked up against these numbers, the statistics for cannabis barely make a dent in this category. The group of *drug-related damage* captures general harm from drug-related activities and behaviors short of death, for example, acquiring HIV from needle use of IV-administered drugs, nonfatal DWI's, workplace accidents, for example, falling off a ladder when drunk or stoned. Cannabis scores modestly on this item.

Dependence and addiction rates are relatively high for tobacco and for alcohol. As we saw earlier the overall dependence rate for cannabis use is about 9%, so significant. *Drug-specific impairment of mental functioning* includes examples such as drug-impaired judgment leading to unprotected sex or deciding to drive drunk. This category is hard to quantify, but alcohol and cocaine are clearly more harmful than cannabis. Likely due to psychosis risk, delirium and amotivation, cannabis scored detectably in both this and the next category. *Drug-related impairment of mental functioning* subsumes features such as late-life alcohol related dementia, mood disorders secondary to the drug or its associated lifestyle, loss of pleasure and interest in everyday activities, and nonspecific depressed mood, and memory loss. *Loss of tangibles* is a slippery category because in part it depends on legal penalties for a particular drug; currently recreational marijuana is still illegal in many US states, but alcohol in none, for example. Being incarcerated because of possession of a small amount of marijuana is something that can be remedied by changing laws on drug possession, but deciding to spend large amounts of money on cannabis rather than saving for a house would also fall into the category, as would losing your house due to drug use, or if it were seized by the authorities as part of a legal penalty. The final category of harms to users is *loss of relationships:* for example, when John's significant other left him due to his persistent cannabis use, or when partners exit a relationship related to the user's violent behavior when intoxicated. This category is much more problematic for alcohol than it is for cannabis.

The second major overall group of issues defined by Nutt and his collaborators was that of harm to others, considered at the individual and population level. The first item under this heading was *injury* that includes both events that were accidental in nature, such as being involved in a DWI where you are not the driver, or injury that was intentionally inflicted, such as being shot when inadvertently approaching a concealed illegal cannabis growery in a national park. Other examples include being a domestic violence victim or fetus damaged by drugs. Alcohol is far more dangerous than cannabis in this regard [91]. *Crime* is illustrated by consequences of theft to support one's drug habit, growth of drug gangs, or vandalism when intoxicated. Drug laws obviously have a strong impact on this category. For example, during alcohol prohibition in the 1920s, booze-smuggling and speakeasies were tied into the criminal activity of manufacturing and distributing an illegal substance. With the repeal of prohibition this problem disappeared. *Economic cost* subsumes such instances as lost workdays due to intoxication or withdrawal, or costs of drug-related health impairment. Here, cannabis was rated above cocaine, crack, and amphetamines, although dwarfed by scores of alcohol and tobacco. *Impact on family life* includes such examples as neglect of a significant other and family, drug-related violence, or absence due to incarceration on drug-related charges. Cannabis scored modestly here. *International damage* captures such events as fallout from the war on drugs, for example, being threatened or harmed by a drug cartel. This is another legal or political category. *Environmental damage* includes issues such as illegally diverting streams to supply illicit cannabis cultivation, use of harmful fertilizers or pesticides on illegal drug crops, or generating toxic waste from a methamphetamine "cooking" operation. Nutt uses the examples of discarded needles and broken bottles, and aggressive behavior outside of bars. Community decline includes such things as drive-by shootings related to turf wars between drug dealers, crack houses or drug shooting galleries.

Nutt and colleagues developed an algorithm to decide how all of these 16 different criteria should be differentially weighted, that they termed "multi-criteria decision analysis" [86]. He convened an expert panel to score 20 commonly-used recreational drugs classified into 4 very broad groups: stimulants, depressants (that somewhat awkwardly included cannabis alongside alcohol, benzodiazepines, ketamine and GHB), opioids, and empathogens/psychedelics (ecstasy, LSD, and mushrooms). The panel rated each of the 20 drugs from 0 (indicating no harm), to 100 (designating most harm) in each category from across all of the earlier-mentioned 16 criteria. The panel's final summary scores for some of the drugs, ranked in order of harmfulness, were 72 for alcohol, 55 for heroin, 54 for crack cocaine, 35 for methamphetamine, 27 for powder cocaine, 26 for tobacco, 20 for cannabis, 15 for benzodiazepines such as Valium and Xanax, 9 for ecstasy, 7 for LSD, and 6 for mushrooms. For alcohol, harm to others was scored at about double the amount than harm to the user; for cannabis this relationship appeared to be reversed. When you examine the raw scores for harm to others and harm to users respectively, alcohol scores a whopping 85 and 55, respectively whereas cannabis scores a 15 and 25. Some of the panel's considerations require explanation or seem counter-intuitive initially. Tobacco may be very harmful in the long term, but not in the short-term, and outside of evading tax laws on cigarettes, for example, related criminal activity is low. On Nutt's algorithm-based score, cannabis gained many of its points from rankings of both drug-specific and drug-related impairment of mental functioning, "mostly because of the harms associated with smoking, and the drug's links with depression and psychotic symptoms", as well as the development of dependence in around 10% of users [84]. The panel usefully made explicit in their paper [86] not only the separate contribution to the overall scores of harms to users and to others, but each separate ranking for each substance of the 16 subcategories, so it's possible to parse their decisions fairly transparently. Subsequently, the European Union adopted the same approach to create a very similar ranking [92]. The weightings appear slightly different from the earlier British paper, in that the relative proportion of harm to others for cannabis seems to be significantly diminished.

Ranking harms using this sort of approach is rational in terms of considerations that must be weighed by individuals forging policy in the realm of law enforcement, social care strategies, and healthcare. Some of the criticisms that have been leveled against this perspective were raised by the authors themselves, including the fact that they scored only harms and not benefits (e.g., utility of medical marijuana), or that they did not include tax revenues and employment generated by drug sales. For example, if recreational cannabis were legalized, and these criteria were included, the balance would shift significantly in a positive direction. At some point, the law prohibiting a substance may be doing more harm than the drug, a decision that in some ways influenced the repeal of alcohol prohibition in the United States. More generally, the authors did not fully weight the role of a drug's availability and legal status. Ultimately, it's also impossible to avoid subjective value judgments with regard to the weights placed on different types of harm [82]. Others have criticized Nutt and colleagues for not properly taking prevalence of use into account. The panel tried to consider how many people actually used a drug, as a way of assessing its overall impact. For example, if many people use drug A that's moderately harmful, that has to be weighed against the much smaller number of people who use drug B, that's extremely harmful. This quantification turns out to be difficult to implement in practice, as has been pointed out [93]. Other problems with this approach are more subtle. For example, if marijuana concentrates such as dabs were banned because they are assessed to be more harmful than cannabis flower, that might be impractical, since people could always extract concentrates from flower illicitly. Again, if individuals who drive under the influence of marijuana also regularly consume alcohol, then harm scores examining driving should consider co-use rather than ranking each substance individually, since their harms interact.

This is been a long chapter, necessarily to, because cannabis toxicology is an important and complex topic. As one of the first substances used recreationally, precisely because of its psychoactive effects, it's hardly surprising that psychological problems such as psychosis and delirium tag along in its wake. But as a reminder of the complexity of the plant's chemical makeup, it's not entirely unexpected that it also contains a potential treatment for psychosis. Against the old conventional wisdom, it's now clear that cannabis dependency is a real phenomenon, and not just a "psychological" issue. Adolescent cannabis use seems to be associated with reversible cognitive problems, as discussed also in Chapter 7. Also, the drug may possibly cause structural brain changes, but the evidence for this overall seems rather weak. Effects on the developing fetus and infant when mothers use cannabis during pregnancy or while breast-feeding are under-explored, but these practices seem risky and ill-advised. I concluded by trying to gauge the dangerousness of cannabis compared to other recreational substances such as alcohol, tobacco, and opioids in terms of the harms that it causes. It is not a benign drug, primarily because of its tendency to increase risk for psychosis, and secondarily through its effects on motor vehicle driving. Since our neighbor to the North has already taken the plunge in terms of legalizing recreational cannabis, the significant social experiment currently underway there will be closely watched, and inevitably provide much useful information on cannabis-related toxicity over the next few years.

References

[1] Johnson DM. Indian Hemp: A Social Menace. Christopher Johnson, London; 1952.

[2] Zinka B, et al. Can a threshold for 11-nor-9-carboxy-Delta(9)-tetrahydrocannabinol in hair be derived when its respective concentration in blood serum indicates regular use? Drug Test Anal 2019;11(2):325–30.

[3] Hartman RL, et al. Drug recognition expert (DRE) examination characteristics of cannabis impairment. Accid Anal Prev 2016;92:219–29.

[4] McGinty VB, et al. Assessment of intoxicated driving with a simulator: a validation study with on-road driving. In: Proceedings of Human Centered Transportation Simulation Conference. Iowa City, IA; 2001.

[5] Hartman RL, et al. Cannabis effects on driving longitudinal control with and without alcohol. J Appl Toxicol 2016;36(11):1418–29.

[6] Hartman RL, et al. Cannabis effects on driving lateral control with and without alcohol. Drug Alcohol Depend 2015;154:25–37.

[7] Hartman RL, Huestis MA. Cannabis effects on driving skills. Clin Chem 2013;59(3): 478–92.

[8] Brenneisen R, et al. Plasma and urine profiles of delta9-tetrahydrocannabinol and its metabolites 11-hydroxy-delta9-tetrahydrocannabinol and 11-nor-9-carboxy-delta9-tetrahydrocannabinol after cannabis smoking by male volunteers to estimate recent consumption by athletes. Anal Bioanal Chem 2010;396(7):2493–502.

[9] Newmeyer MN, et al. Free and glucuronide whole blood cannabinoids' pharmacokinetics after controlled smoked, vaporized, and oral cannabis administration in frequent and occasional cannabis users: identification of recent cannabis intake. Clin Chem 2016;62(12):1579–92.

[10] Swortwood MJ, et al. Cannabinoid disposition in oral fluid after controlled smoked, vaporized, and oral cannabis administration. Drug Test Anal 2017;9(6):905–15.

[11] Lenne MG, et al. The effects of cannabis and alcohol on simulated arterial driving: influences of driving experience and task demand. Accid Anal Prev 2010;42(3):859–66.

[12] Sewell RA, Poling J, Sofuoglu M. The effect of cannabis compared with alcohol on driving. Am J Addict 2009;18(3):185–93.

[13] Foden BJ. Think cannabis is harmless? So did I. But I know better now, National Post. 2019. Available from: https://nationalpost.com/opinion/opinion-think-cannabis-is-harmless-i-used-to-too-i-know-better-now.

[14] Alcohol-Related Deaths in British Columbia, 2013. Available from: https://www.uvic.ca/research/centres/cisur/assets/docs/report-alcohol-related-deaths.pdf.

[15] Wilkinson ST, Radhakrishnan R, D'Souza DC. Impact of cannabis use on the development of psychotic disorders. Curr Addict Rep 2014;1(2):115–28.

[16] Dowd M. Don't harsh our mellow, dude, 2014. New York Times. Available from: https://www.nytimes.com/2014/06/04/opinion/dowd-dont-harsh-our-mellow-dude.html.

[17] Lemberger L, et al. Marihuana: studies on the disposition and metabolism of delta-9-tetrahydrocannabinol in man. Science 1970;170(3964):1320–2.

[18] Lemberger L, Crabtree RE, Rowe HM. 11-hydroxy-9-tetrahydrocannabinol: pharmacology, disposition, and metabolism of a major metabolite of marihuana in man. Science 1972; 177(4043):62–4.

[19] Lemberger L, et al. Comparative pharmacology of delta9-tetrahydrocannabinol and its metabolite, 11-OH-delta9-tetrahydrocannabinol. J Clin Invest 1973;52(10):2411–7.

[20] Sherif M, et al. Human laboratory studies on cannabinoids and psychosis. Biol Psychiatry 2016;79(7):526–38.

[21] Colizzi M, et al. Descriptive psychopathology of the acute effects of intravenous delta-9-tetrahydrocannabinol administration in humans. Brain Sci 2019;9(4):93.

[22] D'Souza DC, et al. Effects of haloperidol on the behavioral, subjective, cognitive, motor, and neuroendocrine effects of delta-9-tetrahydrocannabinol in humans. Psychopharmacology (Berl) 2008;198(4):587–603.

[23] Bossong MG, et al. Further human evidence for striatal dopamine release induced by administration of 9-tetrahydrocannabinol (THC): selectivity to limbic striatum. Psychopharmacology (Berl) 2015;232(15):2723–9.

[24] Prescot AP, Renshaw PF, Yurgelun-Todd S D.A. Gamma-amino butyric acid and glutamate abnormalities in adolescent chronic marijuana smokers. Drug Alcohol Depend 2013;129(3):232–9.

[25] Lichenstein SD, et al. Nucleus accumbens functional connectivity at age 20 is associated with trajectory of adolescent cannabis use and predicts psychosocial functioning in young adulthood. Addiction 2017;112(11):1961–70.

[26] Volkow ND, et al. Effects of cannabis use on human behavior, including cognition, motivation, and psychosis: a review. JAMA Psychiatry 2016;73(3):292–7.

[27] Pacheco-Colon I, Limia JM, Gonzalez R. Nonacute effects of cannabis use on motivation and reward sensitivity in humans: a systematic review. Psychol Addict Behav 2018;32(5):497–507.

[28] Silveira MM, et al. Delta(9)-tetrahydrocannabinol decreases willingness to exert cognitive effort in male rats. J Psychiatry Neurosci 2017;42(2):131–8.

[29] The health and psychological consequences of cannabis use, 1994. Available from: https://www.health.gov.au/health-topics/drugs?utm_source=health.gov.au&utm_medium=redirect&utm_campaign=digital_transformation&utm_content=drugs.

[30] Cortes-Briones JA, et al. The psychosis-like effects of delta(9)-tetrahydrocannabinol are associated with increased cortical noise in healthy humans. Biol Psychiatry 2015;78(11):805–13.

[31] Krystal JH, et al. The vulnerability to alcohol and substance abuse in individuals diagnosed with schizophrenia. Neurotox Res 2006;10(3–4):235–52.

[32] Castle D, Murray SRM, D'Souza DC. Marijuana and Madness. 2nd ed. Cambridge University Press, United Kingdom; 2018.

[33] Appiah-Kusi E, et al. Abnormalities in neuroendocrine stress response in psychosis: the role of endocannabinoids. Psychol Med 2016;46(1):27–45.

[34] Ranganathan M, et al. Reduced brain cannabinoid receptor availability in schizophrenia. Biol Psychiatry 2016;79(12):997–1005.

[35] Leweke FM, et al. Elevated endogenous cannabinoids in schizophrenia. Neuroreport 1999;10(8):1665–9.

[36] Bhattacharyya S, et al. Opposite effects of delta-9-tetrahydrocannabinol and cannabidiol on human brain function and psychopathology. Neuropsychopharmacology 2010;35(3):764–74.

[37] Leweke FM, et al. Therapeutic potential of cannabinoids in psychosis. Biol Psychiatry 2016;79(7):604–12.

[38] McGuire P, et al. Cannabidiol (CBD) as an adjunctive therapy in schizophrenia: a multicenter randomized controlled trial. Am J Psychiatry 2018;175(3):225–31.

[39] Boggs DL, et al. The effects of cannabidiol (CBD) on cognition and symptoms in outpatients with chronic schizophrenia a randomized placebo controlled trial. Psychopharmacology (Berl) 2018;235(7):1923–32.

[40] Bhattacharyya S, et al. Effect of cannabidiol on medial temporal, midbrain, and striatal dysfunction in people at clinical high risk of psychosis: a randomized clinical trial. JAMA Psychiatry 2018;75(11):1107–17.

[41] Abush H, et al. Associations between adolescent cannabis use and brain structure in psychosis. Psychiatry Res Neuroimaging 2018;276:53–64.

[42] Hanna RC, et al. Cognitive function in individuals with psychosis: moderation by adolescent cannabis use. Schizophr Bull 2016;42(6):1496–503.

[43] Association AP. Diagnostic and Statistical Manual of Mental Disorders. NY: American Psychiatric Association Press; 2000.

[44] Rubino T, et al. Chronic delta 9-tetrahydrocannabinol during adolescence provokes sex-dependent changes in the emotional profile in adult rats: behavioral and biochemical correlates. Neuropsychopharmacology 2008;33(11):2760–71.

[45] Schneider M, Koch M. Chronic pubertal, but not adult chronic cannabinoid treatment impairs sensorimotor gating, recognition memory, and the performance in a progressive ratio task in adult rats. Neuropsychopharmacology 2003;28(10):1760–9.

[46] Pope HG Jr, et al. Early-onset cannabis use and cognitive deficits: what is the nature of the association? Drug Alcohol Depend 2003;69(3):303–10.

[47] Schweinsburg AD, Brown SA, Tapert SF. The influence of marijuana use on neurocognitive functioning in adolescents. Curr Drug Abuse Rev 2008;1(1):99–111.

[48] Solowij N, et al. Verbal learning and memory in adolescent cannabis users, alcohol users and non-users. Psychopharmacology (Berl) 2011;216(1):131–44.

[49] Fontes MA, et al. Cannabis use before age 15 and subsequent executive functioning. Br J Psychiatry 2011;198(6):442–7.

[50] Suerken CK, et al. Marijuana use trajectories and academic outcomes among college students. Drug Alcohol Depend 2016;162:137–45.

[51] Marie O, Zolitz U. 'High' Achievers? Cannabis access and academic performance. Rev. Econ. Stud 84(3) Germany; 2017.

[52] Meda SA, et al. Longitudinal influence of alcohol and marijuana use on academic performance in college students. PLoS One 2017;12(3):pe0172213.

[53] Jacobus J, et al. Neuropsychological performance in adolescent marijuana users with co-occurring alcohol use: a three-year longitudinal study. Neuropsychology 2015;29(6):829–43.

[54] Oviedo A, Glowa J, Herkenham M. Chronic cannabinoid administration alters cannabinoid receptor binding in rat brain: a quantitative autoradiographic study. Brain Res 1993;616 (1–2):293–302.

[55] American Civil Liberties Union. The war on marijuana in black and white. New York, NY: ACLU Foundation; 2013.

[56] Hill KP. Marijuana: The Unbiased Truth About the World's Most Popular Weed. Center City, Minnesota: Hazelden; 2015.

[57] Vandrey RG, et al. A within-subject comparison of withdrawal symptoms during abstinence from cannabis, tobacco, and both substances. Drug Alcohol Depend 2008;92(1–3):48–54.

[58] Allsop DJ, et al. Quantifying the clinical significance of cannabis withdrawal. PLoS One 2012;7(9):pe44864.

[59] Pollack N. I'm just a middle-aged house dad addicted to pot, NY Times. 2018. Available from: https://www.nytimes.com/2018/10/06/opinion/sunday/marijuana-addiction.html.

[60] Swift W, Hall W, Teesson M. Cannabis use and dependence among Australian adults: results from the National Survey of Mental Health and Wellbeing. Addiction 2001;96(5):737–48.

[61] Tirado-Munoz J, et al. Comprehensive interventions for reducing cannabis use. Curr Opin Psychiatry 2018;31(4):315–23.

[62] Hughes JR, et al. Marijuana use and intoxication among daily users: an intensive longitudinal study. Addict Behav 2014;39(10):1464–70.

[63] Cuttler C, Spradlin A, McLaughlin RJ. A naturalistic examination of the perceived effects of cannabis on negative affect. J Affect Disord 2018;235:198–205.

[64] Weiland BJ, et al. Daily marijuana use is not associated with brain morphometric measures in adolescents or adults. J Neurosci 2015;35(4):1505–12.

[65] Thayer RE, et al. Preliminary results from a pilot study examining brain structure in older adult cannabis users and nonusers. Psychiatry Res Neuroimaging 2019;285:58–63.

[66] Orr C, et al. Grey matter volume differences associated with extremely low levels of cannabis use in adolescence. J Neurosci 2019;39(10):1817–27.

[67] Broyd SJ, et al. Acute and chronic effects of cannabinoids on human cognition-a systematic review. Biol Psychiatry 2016;79(7):557–67.

[68] Tait RJ, Mackinnon A, Christensen H. Cannabis use and cognitive function: 8-year trajectory in a young adult cohort. Addiction 2011;106(12):2195–203.

[69] Solowij N. Do cognitive impairments recover following cessation of cannabis use? Life Sci 1995;56(23–24):2119–26.

[70] Hoch E, et al. Risks associated with the non-medicinal use of cannabis. Dtsch Arztebl Int 2015;112(16):271–8.

[71] Burggren AC, et al. Subregional hippocampal thickness abnormalities in older adults with a history of heavy cannabis use. Cannabis Cannabinoid Res 2018;3(1):242–51.

[72] Young-Wolff KC, et al. Trends in self-reported and biochemically tested marijuana use among pregnant females in California from 2009-2016. JAMA 2017;318(24):2490–1.

[73] Volkow ND, et al. Marijuana use during stages of pregnancy in the United States. Ann Intern Med 2017;166(10):763–4.

[74] Bertrand KA, et al. Marijuana use by breastfeeding mothers and cannabinoid concentrations in breast milk. Pediatrics 2018;142(3):e20181076.

[75] Dickson B, et al. Recommendations from cannabis dispensaries about first-trimester cannabis use. Obstet Gynecol 2018;131(6):1031–8.

[76] Adams G. Could medical cannabis be the new THALIDOMIDE? Fears of a crisis as doctors consider doling marijuana-based medicines out to pregnant mothers despite evidence the drug can damage foetuses, 2018. Daily Mail. Available from: https://www.dailymail.co.uk/news/article-6423269/Could-medical-cannabis-new-thalidomide.html.

[77] Alpar A, Di Marzo V, Harkany T. At the tip of an iceberg: prenatal marijuana and its possible relation to neuropsychiatric outcome in the offspring. Biol Psychiatry 2016;79(7):p.33–45.

[78] Gray KA, et al. Prenatal marijuana exposure: effect on child depressive symptoms at ten years of age. Neurotoxicol Teratol 2005;27(3):439–48.

[79] Jaques SC, et al. Cannabis, the pregnant woman and her child: weeding out the myths. J Perinatol 2014;34(6):417–24.

[80] Reece AS, Hulse GK. Chromothripsis and epigenomics complete causality criteria for cannabis- and addiction-connected carcinogenicity, congenital toxicity and heritable genotoxicity. Mutat Res 2016;789:15–25.

[81] Merlob P, Stahl B, Klinger G. For debate: does cannabis use by the pregnant mother affect the fetus and newborn? Pediatr Endocrinol Rev 2017;15(1):4–7.

[82] Kalant H. Drug classification: science, politics, both or neither? Addiction 2010;105(7):1146–9.

[83] Nutt D. Estimating drug harms: a risky business? 2009. Available from: https://www.crime-andjustice.org.uk/publications/estimating-drug-harms-risky-business.

[84] Nutt D. Drugs Without the Hot Air. Minimizing the Harms of Legal and Illegal Drugs. Cambridge University Press, London; 2012.

[85] Nutt D, et al. Development of a rational scale to assess the harm of drugs of potential misuse. Lancet 2007;369(9566):1047–53.

[86] Nutt DJ, et al. Drug harms in the UK: a multicriteria decision analysis. Lancet 2010;376(9752):1558–65.

[87] O'Donnell J. Alcohol is killing more people, and younger. The biggest increases are among women, USA Today. 2018. Available from: https://www.usatoday.com/story/news/health/2018/11/16/alcohol-deaths-emergency-room-increase-middle-aged-women-addiction-opioids/1593347002/.

[88] SAMHSA. National Survey on Drug Use and Health, 2016. Available from: https://www.samhsa.gov/data/data-we-collect/nsduh-national-survey-drug-use-and-health.

[89] CDC. Drug overdose deaths, 2017. Available from: https://www.cdc.gov/drugoverdose/data/statedeaths.html.

[90] CDC. Alcohol-attributable deaths and years of potential life lost—11 states, 2006-2010, 2014. Available from: https://www.cdc.gov/mmwr/preview/mmwrhtml/mm6310a2.htm.

[91] Available from: www.niaaa.gov/alcohol-health/overview.

[92] van Amsterdam J, et al. European rating of drug harms. J Psychopharmacol 2015;29(6):655–60.

[93] Caulkins JP, Reuter P, Coulson C. Basing drug scheduling decisions on scientific ranking of harmfulness: false promise from false premises. Addiction 2011;106(11):1886–90.

Chapter 9

Chemistry, chemical analysis, and extraction. Terpenes to tinctures

---

"Israel is a crossroads for smugglers... who get Lebanese hashish from Jordan through the Negev & Sinai deserts to Egypt. Hence the police vaults are full of material waiting for a chemist."

M. Raphael Mechoulam, Marihuana Chemistry, 1970, Science [1].

What are chemovars and terpenes? What are cannabis concentrates such as BHO, dabs and shatter and how are they made? Why and how do people use them? What are the risks associated with them?

Chemovars

In Chapter 4, we discussed how cannabis breeders develop different varieties to intensify the plant's key chemical characteristics such as concentrations of THC, terpenes, or CBD. So logically an important approach is to bypass botanical descriptions and go straight to chemistry as the bottom line. The essential question is what relevant chemicals and in what amounts and proportions are found in a given cannabis plant? After all, that's what we care about from a medical or recreational point of view. It is the equivalent of listing key ingredients in foods-you really want to know what you are about to consume in terms of calories, salt, preservatives, and carbohydrates. Similarly, for cannabis, quantifying ingredients will provide useful information beyond the fact that the bud you are contemplating is some sort of indica variety that looks prettily plump, resinous, and smells like lemon and strawberry. This bottom-line, list-the-contents objective approach leads directly to the concept of "chemovars" as contractions of "chemical varieties." A little earlier, we discussed using chemical varieties as a direct way to distinguish one type of cannabis plant from another. This approach likely originates with Robert Connell Clark's 1993 book *"Marijuana Botany'"* [2] and was brought to full fruition by Mark Lewis and Ethan Russo [3]. It involves analyzing and displaying the contents of any cannabis variety in an easily-understood graphical format. This checklist includes the species (e.g., sativa dominant), class, type (e.g., flower, resin), principal phytocannabinoids (e.g., THC, THCA, CBGA,

Weed Science. http://dx.doi.org/10.1016/B978-0-12-818174-4.00009-4

CBD, etc.), terpenoids (e.g., myrcene, limonene, pinene, etc. compounds that we will have much more to say about later), the predominant scents and tastes (e.g., fruity, citrusy, floral, earthy), and subjective therapeutic/entourage effects (e.g., relaxation, focus, comfort, energy, calm). Lewis and his co-authors are in the process of patenting their graphical display system under the name PhytoFacts [4]. They plan for this intuitively clear approach to be used both by researchers and as a standard display method in dispensaries. The foundational investigations on which this method is based [5], posed a series of practical questions. Some examples: if you want to develop a "phytoprint," a unique fingerprint or profile to characterize cannabis specimens, what sort of equipment do you need and how do you standardize the assay procedures? What kind of internal standards (stock chemical reference specimens) should you use as the gold standard to compare the same assays across different sites? Once you develop the assays, how repeatable are they? If you repeat the same measurement 5 times in a day, or daily for 5 days, do all the values agree with one another or bounce around randomly? Matthew Giese and his collaborators were able to address all of these questions satisfactorily and to show that their methods yielded stable and repeatable answers. Once you have settled on a solid method that replicates over time, then you can address a whole series of new questions. These include tracking what happens when you profile different samples of purportedly the same variety from one growery; how similar are they from one another? Or even how do values vary from different parts of the same individual plant? How about samples of a particular variety from a dispensary in Denver compared to one in San Francisco? Or Girl Scout Cookies buds sold at the exact same dispensary 6 months apart? The chemovar approach to assaying terpenes is also important because it will ultimately allow researchers to address the frustratingly under-examined claims regarding entourage effects. Once you are able to quantify all of the ingredients in the recipe of a particular variety, then some of the elusive claims made about synergism of different cannabinoids and terpenes will be fully testable.

This is the promise of the chemovar approach to key ingredients. It does seem to offer a reliable method to compare apples to apples, or at least buds to buds, across different laboratories. But until there is some sort of state or federally mandated approach to cannabis fingerprinting, different dispensaries are free to use any assay laboratory and method that they choose, or none at all. What is important to know but is not addressed, is the presence of contaminants within the sample. Assaying for fungi, heavy metals, pesticides, and bacteria call for a completely different set of procedures. Currently, all of these assay procedures are loosely regulated (or not regulated at all) at a local level, and crying out for standardized, uniform federal rules so that cannabis users anywhere can be confident in what they are purchasing.

If I just persuaded you that chemovars represent the final word on state-of-the-art, science-based cannabis classification, a group of researchers at Washington State University just moved this process a significant step further down the road [6]. They accomplished this by looking not only at the chemical composition of cannabis, but by linking this information to measures of expression of the genes that synthesize the plant's key chemicals. The group, led by B. Markus Lange, did not come up with a catchy title for their approach, so as a default let's term it "Geno-Chemovars." Previous analyses have shown that as one would expect, drug marijuana and non-drug hemp differ in their ability to synthesize cannabinoids. Marijuana has a specialized form of the gene that makes THC, (known technically as the tetrahydrocannabinolic acid synthase allele) and hemp typically has instead a different variant of the gene specialized in making CBD (the cannabidiolic acid synthase allele). When other scientists performed RNA analyses of female flowers to determine how switched on these forms of the gene are, (transcriptome analysis) in marijuana compared to hemp, then, cannabinoid pathway genes were significantly ramped up in marijuana, as expected.

Analogously, once the genes that are responsible for the manufacture of terpenes are switched on, they make corresponding RNA copies of themselves, which then direct the cell in manufacturing those molecules. With his collaborators, Lange extracted RNA from different marijuana strains by abrading trichomes with tiny glass beads, perhaps inspired by Herman Hesse. This released the plant cell contents, the resulting RNA was sequenced and the related switched-on genes specifically identified. Then the various networks of genes contributing to individual patterns of cannabinoid and terpene production could be pinpointed in different cannabis varieties. This sort of approach leads to a complex but highly specific fingerprint of a cannabis variety, defined both by its genetic and chemical profile. I would compare it to the difference between the FBI having your fingerprints and photo on file (chemovar) versus also having a sample of your DNA (geno-chemovar).

Having powerful genetic tools at our disposal, enables scientists to ask interesting and hitherto unaddressable cannabis-related questions. A pioneering 2015 genetic study set out to address a number of fundamental cannabis issues, and ended up shattering many preconceptions [7]. Sean Myles and his colleagues from Dalhousie University looked at DNA derived from over 40 hemp and more than 81 drug cannabis samples. The investigators were full of questions. What is the underlying genetic structure of all plants from the genus Cannabis? Are hemp and drug-type marijuana genetically distinct? Are indica and sativa genetically close relatives or distant ones?

SNPs in DNA are a major source of variation. These SNPs, (that we encountered earlier in the Dunedin study) show up in cannabis DNA as often as 1 in 100 base pairs, so there is plenty of genetic variation in cannabis. There was no obvious DNA fingerprint to parse C. sativa from C. indica. And correlation between the genetic structure of marijuana varieties and their reported ancestry were modest to weak. Prior to the Dalhousie study, scientists had assumed that cannabis varieties used for fiber and seed production derived from C. sativa. The genetic results were completely incompatible with that hypothesis. Hemp was genetically more similar to C. indica type marijuana than to C. sativa strains. In fact, a marijuana strain's genetic distance to hemp is negatively correlated with its reported C. sativa distance. A paragraph or two back, we chatted about cannabinoid expression in hemp versus marijuana. The Dalhousie scientists showed that the genetic differences between drug cannabis and hemp were not simply restricted to locations on genes involved in cannabinoid production. They were far more fundamental and widely distributed across the entire cannabis genome.

Another myth busted by the study was that marijuana "strain" names delineate very specific distinct chemovars. Instead, the Canadian group showed that these names do not reflect a meaningful genetic identity. In some cases, the assignment of indica versus sativa ancestry disagreed wildly with genotype data. For example, Jamaican Lamb's Bread (reported as 100% sativa) was 98% identical to a reported 100% indica strain from Afghanistan. So that budtenders may be inadvertently passing along highly inaccurate descriptive information that they have received from equally uninformed cannabis breeders. In over a third of these researchers' comparisons, "samples were more genetically similar to samples with different names than to samples with identical names."The researchers stated that "....inaccuracy of reported ancestry in marijuana likely stems from the predominantly clandestine nature of cannabis growing and breeding over the past century." Recognizing this, marijuana strains sold for medical use are often referred to as sativa or indica "dominant" to describe their morphological characteristics and therapeutic effects. Our results suggest that the reported ancestry of some of the most common marijuana strains only partially captures their true ancestry...... We conclude that the genetic identity of a marijuana strain cannot be reliably inferred by its name or by its reported ancestry."

The last, truly interesting important finding from Myles and his scientific collaborators was that levels of genetic diversity were significantly greater in hemp than in marijuana. This suggested to the researchers that hemp cultivars come from a much broader genetic base than those of marijuana strains. Of course, another possibility is that marijuana, like lineages of Egyptian pharaohs or Habsburg monarchs, is characterized by inbreeding among close relatives. These two possibilities are not mutually exclusive. The bottom line here is that cannabis sativa and indica may represent clusters of genetic diversity, where one can be parsed from the other up to a point using DNA analysis strategies, but that extensive cross-breeding has scrambled their gene pools confusingly.[a]

If cannabis varieties are not what they seem, standardized testing of cannabis samples presents an even more dismal picture. As expected, in the complete absence of federal regulation, where labeling regulations vary tremendously among different US states, and even their enforcement is inconsistently applied

---

a. In contrast to other clonally propagated crops like apples and grapes, however, strain names are assigned to marijuana plants even if grown from seed. Thus, a marijuana strain name does not necessarily represent a genetically unique variety.

within states, testing of dispensary marijuana reveals disparities and confusion. For example, without even getting into the complexities of DNA analysis, the same named variety from different dispensaries may be very different in terms of identifying the underlying chemovar as identified by its cannabinoid/terpene signature. Correspondingly, some strains with different names turn out to be almost identical from a chemovar perspective, in a parallel with what we saw earlier in the genetic realm with the Afghani/Jamaican, indica/sativa confusion. Some dispensaries quantify a restricted range of cannabinoids and terpenes, while others are more thorough. But which laboratories they use for their assays is often up to the choice of the dispensary owner, and different laboratories use different assay procedures that in turn have different levels of reliability. How the contents are reported also occurs with widely varying levels of precision; one dispensary listed flower with a range of THC content of "13%-30%." Chemical testing of marijuana plants in reliable laboratories reveals a number of unexpected findings. One is that even within a single marijuana bud derived from one plant there is a gradient of THC, with the highest concentrations being at the tip. Dr. Nirit Bernstein of the Volcani Center in Israel is an expert on cannabis cultivation. She says "If we look at a tomato, some parts will be sweeter or will contain more lycopene, but it doesn't matter because it all goes in the salad or the ketchup. With medical cannabis, if one piece has twice as much THC it's too strong a dose and if another piece has less THC the dose will not be effective. So, our main goal is to learn how to grow plants to increase standardization for medical use" [8]. Another problem is that there are surprisingly few studies on which particular agricultural growing practices produce optimal amounts of cannabinoids. One might assume that the healthiest, plumpest, best-nourished cannabis plants contain the highest concentration of desirable compounds but that's not necessarily the case. Recall that in the wild the greatest amount of cannabis resin is produced by stressed plants at high altitudes, so that perhaps plants thrive on a moderate amount of neglect. Dr. Bernstein's controlled research suggests that is the case.

Because CBD and THC are completely odorless, consumers perceive differences among cannabis varieties based primarily on aroma, and make choices regarding potency, price, and likelihood of purchasing based on this information [9]. These aromatic differences in turn are almost completely dependent on terpenes.

Meet the terpenes

So, what are terpenes, why are they important, what do they have to do with cannabis, and what if any are their major psychoactive effects and purported health benefits?

I've always been attuned to tastes and smells. Friends accuse me of having been a dog in a prior life. I'm the guy who spent an entire morning in my brother-in-law's Body Shop cosmetics franchise trying to mix the perfect blend of musk, sandalwood, and citrus. As a 6-year-old, I remember accompanying my parents to a perfume factory at Grasse in the South of France and spending an afternoon of vivid olfactory bliss among mountains of rose petals and dried lavender flowers. When the McCormick spice factory in downtown Baltimore moved to Baltimore County's Hunt Valley, I anticipated the varied daily symphony of spice and flavor chemicals that the wind would blow my way on my drive home. Subsequently, as part of a set of experiments I set up at Johns Hopkins Hospital to measure people's memory for odors, a handful of us from my laboratory journeyed to the McCormick plant. We sought their donation of a variety of pleasing and nasty smelling liquids that we could use in our research. The day we arrived at McCormicks, dozens of trained tasters were sampling and rating diligently, from small plates, what seemed to be hundreds of candidate recipes for a new variety of bacon bits. For our own initiation, my laboratory members were each given an orange-colored, orange-tasting, hard candy and asked to say what flavor we thought they were. We responded unanimously "orange of course." The correct answer was "lemon." We had all been fooled by the citrusy flavor paired with the orange color cue. And one of the chemicals that makes lemons smell "lemony" is the terpene compound limonene.

Terpenes are strong-smelling carbon-containing (organic) compounds produced by many plants, (including trees), and by some insects. They are major ingredients in so-called "essential oils." The word "terpene" sounds like "terrapin" and "turpentine," but it is from the latter substance, and not the feisty little reptile from which the name is derived. Turpentine is an oily, evergreen-scented liquid made from distilled pine resin. Turpentine is one ingredient in Vick's Vapor Rub. It has also been used in medicine for millennia to kill worms in the gut and as pine tar for skin problems. If you've ever been unlucky enough to swallow any turpentine, you will know that it tastes disgusting and burns the tongue. If you swallow too much, it can damage your kidneys and make your pee smell like spring violets. As Al Capone could tell you, it was an ingredient of bathtub gin during Prohibition, for reasons that will become apparent further.

Terpenes pop up all over the natural world. Terpene trickery is used by a variety of insects to communicate, and to deter insect predators. These chemicals can have marked effects on insect behavior [10]. Swallowtail butterflies, some species of ants and termites exploit this characteristic. Some termites squirt terpenes directly onto potential predators, for example. During a hike in the foothills of the Swiss Alps a Swiss friend had me gently poke an ants' nest. He encouraged me to sniff, as the tiny, alarmed, scurrying insects released a terpene odor redolent of lemon zest, lemon Verbena, and citronella. "Predators, get lost" was their message. The lemon smell from citronella ants is citronellol, whose odor according to textbooks, resembles that of lemongrass and Kaffir limes. Humans have copied citronellol's insect-repelling properties to make anti-mosquito citronella candles. Some species of butterflies have also gotten in on the act, terpene-wise. Merely sitting quietly on a leaf, tiger swallowtail caterpillars already look threatening to would-be predators. Some species bear black or yellow eye spots that make them resemble an angry, if tiny snake. Other, related swallowtail caterpillars, looking strikingly distinguished in stripy coats of yellow, green and black, conceal within their body folds twinned, orange-yellow colored tongue-like structures. If you gently poke or harass such a caterpillar it will unfurl both of these almost fluorescent organs that look like they were made from a melted cantaloupe jellybean. Once deployed, they emit a terpene odor redolent of orange peel and spice (from limonene and myrcene) accompanied by a funky smell of rancid butter. This combo is a pretty effective predator deterrent. Other swallowtail cousins lack the specialized organ, and repel predators by other mechanisms. Those with the structure possess an expanded gene cluster whose DNA codes for a series of terpene-synthesizing enzymes, similar to those found in cannabis trichomes. But many other plants manufacture terpenes, for similar predator-deterring purposes. Terpenes in fact, are among the most widely distributed secondary compounds (ones not essential for key functions such as photosynthesis) in plants. They are responsible for the pleasant odors given off by pine trees and herbs, and they are the principal ingredients in perfumes and essential oils. When you walk into a pine forest, that characteristic fresh green smell is mostly terpenes, particularly alpha-pinene and D-limonene. Some have been used by humans for hundreds of years, as ingredients in scents, cosmetics, and flavors. These fragrant oils are not synthesized by botanicals for our olfactory enjoyment but are important plant defenses. At high concentrations some terpenes can be toxic to predators and disease-causing organisms. Thus, in nature, they protect plants from bacteria, fungi, insects, and a variety of other environmental stressors, including would-be herbivores. Individual terpenoid compounds with high insect repellent activity include alpha-pinene, limonene, citronellol, citronellal, camphor, and thymol. In cannabis, terpenes, perhaps 200 of them, [6] are secreted for the same defensive purposes, from the resin glands that also produce cannabinoids.

As you walk into a citrus grove, pine forest, or cannabis growery, the perceptible chemical haze and many of the characteristic odors are produced by terpenes. In plants these aromatic essential oils are squirreled away in specially-designed, segregated containment structures (such things as oil tubes and glands, including trichomes). These storage areas may also protect the plant from its own toxins. Plants manufacture terpenes as the Swiss Army knives of plant defense. They repel and even kill insects, discourage grazing and browsing by animals, and are plant antibiotics in terms of reducing infection. Pinene, for example, is an anti-fungal. Other terpenes used in some plants are used to attract pollinators, but wind-pollinated cannabis obviously has no need of these. CBD and THC have no odor of their own, so that cannabis terpenes explain why skunk varieties smell "skunky" and super lemon haze smells "lemony." The characteristic olfactory signature of different cannabis strains is due entirely to the terpene compounds that the plants brew in their trichomes. The aromatic qualities that they instill within marijuana flowers, provide the bouquet that budtenders, given an ounce of encouragement, are apt to describe in flowery language. The cannabis plant is not just a one trick pony that exists to produce THC and CBD. The plant is a complex, busy chemical factory producing hundreds of compounds, and the more we probe, the more we realize that many of these chemicals affect the quality of the cannabis "high," the plant's widely varied medicinal applications, and the profile of the accompanying side effects, both desirable and undesirable.

I have to admit that I had always thought of the terpenes in cannabis as amusing little ornamental chemicals along for the ride in terms of providing the chemical basis of taste and smell, but contributing nothing substantive to the essential qualities of the "high," being merely intriguing to geeky factoid-hoarders and "cannasseurs." As an analogy I would have rated it similar to knowing that David Bowie had one blue and one brown-appearing eye; an interesting tidbit, totally irrelevant to his musical performance. But I was wrong. About the terpenes, that is. Terpenes have their own distinct story to tell, not only exhibiting mild psychoactive effects-comparable to aromatherapy-in fact many of them are used as aromatherapy oils, essential oils, and perfumes. Their independent, individual psychoactive effects and their interactions in boosting, lowering or modulating THC-related behaviors are not yet well studied, and deserve further research. If we think about it, a simple push and pull relationship between THC and CBD doesn't seem sufficient to explain the wide variety of things cannabis does to our body and mind. To give even a simple example, why do some chemovars seem to unleash the energy of Lucifer while others usher in the torpor of Rip van Winkle? Outside of the large number of cannabinoids, that are only partly explored in terms of their effects on human physiology and most of which are not tracked in standard assays, terpenes may offer an important part of the explanation. It is likely that there are synergistic effects between terpenes and cannabinoids, known as "entourage effects," a term frequently bandied about by budtenders. The underlying concept is that the entire terpene plus cannabinoid melange contributes interactively, and is responsible for the resulting high [11-14]. As we will learn, terpenes may both modify THC and CBD metabolism and have their own independent psychoactive properties, affecting mood, arousal, and pain sensitivity. Some terpenes such as beta-caryophyllene even bind directly to cannabinoid receptors in the nervous system. So to understand the whats and whys of the world of terpenes, we will take an informative detour into the chemical world.

Where and how are cannabinoids and terpenes synthesized in the cannabis plant?

Chemical synthesis; meet botany. For those of you like Sam Cooke, who don't know much biology (and would prefer not to), the bottom line of the following paragraph is that both cannabinoids and terpenes are manufactured inside the cannabis plant as different products via substantially overlapping chemical pathways [3]. Or as one source phrases it, they "share a common biosynthetic origin" [6]. The operating instructions in the plant's genes can make relatively simple (terpenes) or more complex (cannabinoids) from the same stockpile of ingredients. Furthermore many of the essential terpenes involved in this process are delicious-smelling oils and resins named for various non-cannabis plants ranging from lemon to guava, where they are also produced. In cannabis, all of this chemical transmutation takes place in the lollipop-appearing trichomes of the female flowers.

If you're more interested in the twists and turns of the molecular operations by which the plant makes the ingredients over which budtenders and "cannasseurs" wax poetic, we begin with a compound named Farnesol. Farnesol is named for a flower extract from the Farnese acacia tree that is used in perfumes and deodorants. The Italian Renaissance Farnese family, (that included Pope Paul III and numerous Dukes of Parma) constructed beautiful botanical gardens in Rome, where this particular Caribbean tree was cultivated by Cardinal Odoardo Farnese, and thus named. It is also synthesized in the human body and reacts on many receptors in cell nuclei [15]. Farnesol's chemical relative in the cannabis plant, Farnesyl diphosphate, is converted by enzymes into sesquiterpene compounds including beta-caryophyllene (a food flavoring found in black pepper that is incidentally the compound that drug-sniffing dogs zoom in on) and alpha-Humulene (named for hops). The chemical building bricks used to assemble Farnesyl diphosphate can alternatively be used to build Geranyl diphosphate, (named for the showy red or white pelargonium geraniums that decorate Swiss window boxes, whose blooms are rich in the chemical). In turn, Geranyl diphosphate is used to synthesize a whole series of monoterpenes, many named for the plants that produce them in large quantities. These include Beta-myrcene (my speech software helpfully rendered that as "murder scene"), named for the myrcia plant, that (along with cloves, guava, allspice, and eucalyptus), is a member of the myrtle family, alpha-pinene (named for pine trees), and limonene (named for lemons). Not content with its role in monoterpene synthesis, Geranyl diphosphate is also used by the cannabis plant to assemble cannabigerolic acid, usually abbreviated to CBG. Depending on which synthetic enzyme carves up CBG, it is transmuted either into tetrahydrocannabinolic acid (THCA), or canabidiolic acid (CBDA). With a little bit of heat and the addition of carbon dioxide, these magically become THC and CBD respectively.

All of this chemical synthesis occurs in the cannabis trichomes. There are various trichome types on the plant, but the important ones are glands on the end of a long stalk, with the whole apparatus resembling a mushroom or lollipop. When I say "long stalk" this is speaking relatively; an entire trichome is barely visible with the naked eye. Thousands of them are packed on the surface of the flower bud, where they participate in an active dialogue between the plant and the environment.

Trichomes (the term derives from the Greek word for hair), are hardly unique to cannabis: they exist on about one third of all plants. Under the microscope, they can resemble thorns, needles, hooks, or mushrooms. When you grab a stinging nettle, the needle-like hollow stinging hairs injecting you with itch-inducing painful nastiness are trichomes. Other plants store important secretions they have manufactured in tiny bulbs or glands on the end of hair-like structures. These mushroom-shaped structures are called glandular trichomes, for obvious reasons. Plants use them as mini-factories to manufacture chemicals they would like to keep on their surface, often to deter pests of various kinds or to protect the plant from other potentially harmful things in the environment, such as ultraviolet rays or excessive heat. Plants have an inherent drive to make these chemicals, and if attacked by insects or burned by the sun, will bump up production accordingly. From our cannabis-centered perspective, glandular trichomes are especially important because many of the specialized compounds produced in this way are terpenes. Plants including mint, tomato, tobacco, petunia, hops, and cannabis all manufacture terpenes in these specialized structures. In the case of cannabis, the glandular trichomes synthesize both terpenes and cannabinoids in the form of resin. Trichomes use both their chemical and physical powers to guard the plant from insects, browsing animals and from drying out. This is what you might call their "resin d'etre." Recall that THC and CBD have no taste or smell, but to disease-causing fungi, bacteria (e.g., killed by limonene), insect pests (e.g., repelled by geraniol or killed by caryophyllene), and herbivores such as sheep, terpenes are distasteful deterrents. For cannabis, the non-chemical, physical part of its defense exists in the plant's other set of trichomes, which are non-glandular. These guys resemble tiny needles or teardrop-shaped mini-thorns. If you rub your finger across the cannabis leaf it feels a little bit like sharkskin; this sandpapery texture is the non-glandular trichomes making you aware of their presence. These structures effectively deter munching insects, crawling caterpillars, and egg laying plant parasites. In the case of the needle trichomes found on nettles, there is extra protection against large mammals [16]. The cannabis plant is thus equipped by evolution with multiple mechanisms to deter predators. The hapless insect zooming in to munch a leafy snack or to lay its eggs may unwittingly be trapped in sticky resin, have its body pierced or its legs ripped off by spiky trichomes, or be poisoned by terpenes.

As cannabis flower buds mature, the chemical processes described earlier proceed busily and the glands gradually change their appearance from milky white, to clear and then finally to a beautiful shade of liquid amber, that matches the changing stigmas as they ripen from white to orange-yellow. As Alexandre Huchelemann and his group stated in 2017 [17] "Understanding the way glandular trichomes develop to finally turn into highly efficient biochemical factories in the epidermis of non-model plant species is of key importance and could lead to more applied outcomes."

One truly tantalizing question is why the cannabis plant bothers to make cannabinoids. We don't know the answer to that question, so just this once, let's speculate with a plausible "Just So" story. Like many plants in the botaniverse, cannabis started off making terpenes for the usual predator-deterring purposes. Due to some random evolutionary coin toss, the plant changed an existing gene's DNA, allowing it to make a new enzyme that synthesized cannabinoid precursors from the same chemical toolkit that it had always used to manufacture terpenes. This small extra step conferred the evolutionary benefits in terms of protection from UV light and from heat, or to trap insects, allowing the plant with this altered gene to expand its territory higher up the mountainside and into hotter, dryer areas. Such plants were then chosen by natural selection for these beneficial features, and when early humans came along to exploit the plant for fiber and seeds, they discovered medicinal and psychoactive properties in the cannabinoid-manufacturing cannabis plants as an added benefit. In turn, this latter property was yet another reason that people cultivated the plant and spread its seeds wherever they migrated. An imperfect analogy might be a blacksmith who makes nails for horses' hooves in his forge realizing 1 day after a random accidental hammer mis-strike, that if he copies that one extra step and bends his nails into a circle, then he can sell them for 10 times the price to tourists as decorative hand-crafted rustic finger rings.

Because we use them in other contexts as flavors, perfumes, and essential oils, the terpenes found in cannabis are legal to purchase. Many are approved by the FDA as additives to food or cosmetics. Anyone can go online and buy them individually by the bottle or in blends that mimic the odor profile, (and by extension the chemical and perhaps behavioral fingerprint) of well-known cannabis chemovars. For a little over $20 you can purchase the liquid terpene profile of Super Lemon Haze, Dogwalker OG, or Blue Dream. These little bottles of bliss contain blends of pure, food-grade botanical terpenes that more or less mimic those found in particular chemovars. For example, Lemon Skunk's terp profile blend from Trim Buddies is 97% limonene with small amounts of beta carophylline and myrcene, plus a tiny sprinkle of other terpenes. So a reasonable question is whether assembling a recipe mimicking the proportions of cannabinoids and terpenes found in a given chemovar will replicate its behavioral profile. Or, put another way will a couple of squirts of lemon skunk essence out of a bottle spritzed on to a bowl of average street weed replicate precisely the buzz obtained from high-end Lemon Skunk? The answer to that question turns out to be complicated.

The average person thinks of individual chemical compounds as having straightforward, perhaps unique effects, and logically that dosing with a mixture of compounds will produce the sum of their effects, but nothing more. Valium makes you relaxed and ecstasy lovey-dovey. Therefore Valium plus ecstasy should make you super-chilled and extra cuddly. If only life were so simple. In reality different compounds taken together frequently boost or undermine each other's effects. A well-known example is the medications given to combat HIV, where drug "cocktails" of multiple virus-killing medications proved to act synergistically in ways that were effective significantly above and beyond those of each alone. Analogously, psychological effects resulting from cannabis are often said to rely on "entourage" effects. In addition to the complicated rivalry between THC and CBD, terpene compounds also have biological effects that interact with those of phytocannabinoids in ways that are only just beginning to be studied scientifically. Budtenders use "PhytoFacts" to describe terpene content and behavioral effects, but let's dive a little deeper into the terpene pool.

Terpene chemistry

There are around 200 total terpenes in the cannabis plant [14,18-21]. Around 50 are commonly found in US cannabis and around 10 of these are commonly quantified for dispensaries. Even the best laboratories max out at around 20 terpene assays. In discussing the terpenes, I will explore a small handful in detail because of their special interest or unusual properties. In regular flower cannabis, terpenes account for 2%-5% of total weight. Many chemical extraction methods produce relatively pure samples of THC or CBD for use in dabs. Terpenes boil off at different temperatures and are collected separately, usually first in the purification process. Typically, the final THC concentrate extracts will have terpenes blended back in separately, and they typically constitute around 10% of the final wax or shatter product, adding aroma and flavor. Because terpenes are oil and alcohol-based, the more of them added back to a concentrate, the more waxy and pliable and the less brittle is the resulting product.

Terpenes are classified by chemists in terms of their molecular structure. The basic chemical skeleton on which all terpenes are built is the 5-carbon compound isoprene. If you polymerize this stuff, you end up with old-fashioned rubber, as in car tires. Among the simplest members of the terpene family are mono-terpenes, (confusingly named, since they consist of 2 isoprene units, not 1), and thus contain 10 carbon atoms. Mono-terpenes include pinene, myrcene, limonene, linalool, and geraniol. Sesquiterpenes are constructed out of 3 isoprene units. These 15-carbon compounds include caryophyllene, nerolidol, humulin, and guaiol. All of them are found in cannabis flower buds.

Along with their own molecular characteristics, terpenes are the building blocks of other, increasingly complicated chemical groups. For example, *terpenoids* are related, slightly more complicated chemical structures containing additional chemical groups to those that characterize terpenes. The label "terpenoid" signifies that extra functional chemical groups, have either been tacked on to the underlying terpene structure, or snipped out. For example, molecules doubling the complexity of sesquiterpenes, (termed tri-terpenoids) have at their core 6 isoprene units. Such compounds as squalene and friedelin, belong to this class. What's confusing is that many people, including many chemists, refer to all terpenes and their kin under the general term "terpenoid." Yet more confusingly, others use the term "terpene" as a catchall to include both terpenes and terpenoids. Let's tiptoe past this troublesome terpenoid terminology, to clarify that the tri-terpenoid compound friedelin, found in the roots of cannabis plants, is the only known member of this family found in marijuana. But the reason that I mention the terpenoids, and it's an important one, is that after further judicious molecular tweaking and even more chemical complexity, the whole family of cannabinoids is synthesized from them within the cannabis plant.

Terpenoids, like terpenes are compounds we encounter every day. They contribute to the scent of eucalyptus, the flavors of cinnamon, cloves, and ginger, the yellow color in sunflowers, and the red color in tomatoes. Well-known terpenoid molecules include citral, menthol, camphor, chemicals found in the *Ginkgo biloba* tree, the curcuminoids found in turmeric and mustard seed, and the hallucinogenic chemical salvinorin-A derived from the plant *Salvia divinorum*. And incidentally, inside our bodies, yet more complicated and usually considerably less odorous chemicals found in the living world are built from them. The tri-terpenoid molecule squalene, for example, is the main precursor that we use to bio-synthesize steroids such as testosterone as well as vitamin-A (a tetra-terpenoid), a few more steps down the synthetic chain.

Genes found in dozens of plants and some insects are able to synthesize terpenes. They achieve this by using enzymes, logically enough known as terpene synthases. The same enzymes found in our livers that help break down harmful products in our food, and metabolize many of the pharmaceuticals we ingest including terpenes, also modify the synthesis of terpenes in plants. It has been recognized in the last few years that for us (humans), terpenes are not just inert compounds, but may bind to or modify CB_1 or CB_2 receptors. For example, the terpene beta-caryophyllene binds quite tightly to CB_2. Terpenes may also modify the permeability of the blood vessel barrier that acts as a living filter between the brain and the rest of the circulation for cannabinoids such as THC and/or CBD. This alteration results in letting more or less of these cannabinoids cross into the brain. Finally, terpenes may even alter the metabolism of cannabinoid compounds. Boiling points of commonest terpenes found in cannabis are similar to that of THC at around 157°C/314°F. Pinene is a little lower, myrcene and Linalool a little higher, but these compounds will vaporize off along with THC and CBD at the same temperatures.

What are behavioral effects of terpenes in humans?

Until recently, we knew very little about this question, but terpenes may possess independent psychoactive effects in the absence of cannabinoids, and additionally interact with cannabinoids regarding their metabolism, receptor binding, and possibly alter the passage of cannabinoid compounds across the blood-brain barrier. This latter is a protective mechanism that partially excludes many compounds from crossing directly from the blood into the brain; there is a suspicion that some terpenes may open up this barrier so that cannabinoid compounds can enjoy a resulting boost in their psychoactive effects.

An open question is whether terpene effects on behavior are best studied in specimens taken from cannabis plants, or by adding precisely measured amounts of terpenes from a purified compound prepared by a chemical manufacturing company and mixed with a plant specimen. There may be no difference between the two methods but nobody has actually tested this, to my knowledge. The fate of terpenes in the body is complicated. There have been many studies in animals to trace how terpenes are broken down into new compounds, some of which have novel biological activity as reviewed [18]. The key enzymes to metabolize terpenes are the CYP family in the liver.

Eva Heuberger [18] summarizes comprehensively that effects of odors on the brain's electrical activity as assessed by EEG usually focus on the theta rhythm generated primarily by our memorable friend the hippocampus. This is a logical choice because theta activity increases as animals sniff things in their environment, and the hippocampus encodes smell data along with memory, emotional, motor, and cognitive activity associated with the odor. Humans are much less sensitive to odors than many other mammals, but our EEGs also respond differentially to different-smelling compounds. Even a cursory review of the existing EEG/smell literature leads the average person to conclude that it is full of methodological problems, contradictory findings, and a lack of clear conclusions. Important basic questions are often not covered in human odor research projects. Examples: was a subject able to recognize and label the smell? Did it trigger any particular pleasant or unpleasant memories? Was it consciously perceived? Did it have any personal relevance? Did they like it or dislike it? Many of the same problems plague research into effects of odors on behavior and cognition. Although aroma therapists may claim that some essential oils are "relaxing" and others "energizing", the results of many studies on this topic are very hard to interpret, and results frequently contradict each other.

A summary of what's known regarding terpenes and their interactions with other cannabinoids is outlined in Russo's paper [22] that explores their synergy with THC, effects that can include both facilitation and antagonism. Low concentrations of terpenes can be potent behaviorally; this would seem to be necessary since concentrations in the plant are often relatively low. Terpene effects are strongest when the compounds are inhaled, and may be absent when they are taken orally. Interestingly, cannabis samples provided to researchers by the National Institute on Drug Abuse are relatively devoid of terpenes, containing approximately 8 times less than equivalent dispensary-purchased flower.

Let us review some of the commoner terpenes and their effects. A word of warning, though; many of the claims made regarding terpenes have little experimental backing, despite being frequently touted. Where there is good evidence to support health-related effects, I have tried to provide the relevant citations, but that is the case for a minority of the compounds described, where many of the claims are at a level of folk wisdom rather than science.

Mono-terpenes

Alpha-pinene smells sharply "green" like the odor of pine trees and turpentine. It is among the commonest terpenes found in cannabis. Physicians have long believed that pinene can reduce inflammation and opens up airways in the lungs, (hence the use of turpentine and Vicks Vapo Rub for people with chest complaints). This compound is found in conifer trees, including in pine needles, as well as in the herbs rosemary and parsley. Alpha pinene is prominent in the cannabis chemovars Jack Herer, Strawberry Cough, Romulan, Blue Dream, and OG Kush. Unexpectedly, it is an acetyl cholinesterase inhibitor, (i.e., it inhibits the metabolic breakdown of acetylcholine, a key brain neurotransmitter) [23,24] that among other functions helps underpin memory. This inhibitory property is shared with medicinal compounds with similar, but much stronger activity that is prescribed by physicians to patients with early Alzheimer's disease to help slow down memory loss. Thus, theoretically pinene may help with memory decline or boost memory and alertness [25,26]. Pinene also is purported in terpene charts displayed in dispensaries, to promote alertness and to be a cognitive stimulant, although the actual evidence for all of these effects is somewhat thin, and the topic needs to be much better studied and quantified objectively. To complicate things, cannabis also contains compounds (including THC) that block cholinergic receptors and inhibit acetylcholinesterase that would therefore logically impair short-term memory [3]. Indeed such amnestic effects were demonstrated in a careful placebo-controlled study of pure THC [27]. To further muddy the waters, pinene's inhibition of acetylcholinesterase is multiplied by the effect of other terpene-like compounds including camphor [23]. Which terpene wins this brainy tug-of-war over acetylcholine may well vary from one chemovar to another.

Alpha-pinene has an interesting back story. It is found abundantly in the common Juniper shrub, whose somewhat bitter, resinous, fragrant-smelling berries add most of the flavor to gin, and in larger, upright tough-as-nails Juniper trees whose leaves, twigs and cones constitute the main diet of the wild rodent known as Stephen's wood rat. In the United States, these foot-long rats dwell in the Juniper Woodland of Arizona, New Mexico, and Utah and other dry Juniper-heavy woodlands across the globe. Stephen's wood rats are rather endearing creatures, with large dark soulful eyes, delicate feet, big rounded ears, and long feathery tails that resemble grayish, furry ears of barley. Like archetypal stoners they are more active at night, sleepy during the day, move relatively slowly, and are known for their noisy chatter. They are attracted to bright shiny objects such as sparkly rocks and soda can tabs that, given their destiny as pack rats, they cache in their complicated twig nests. They are also notoriously distractible-anyone who has witnessed one discard a juicy-looking juniper cone in favor of a sparkly button, only to trade that a minute or 2 later for a tempting berry, will be struck by the behavior. Unlike caricature stoners though, they are literally "bright-eyed and bushytailed." This wood rat, one of a dozen species found in the United States, is of particular interest to us because its diet consists of 90% Juniper, from the crunchy foliage to the fleshy, waxy cones. Thus, at any given time one might expect the Stephen's wood rat's body to be awash in terpenes, especially pinene. Given that terpenes are synthesized by trees and plants to deter animal and insect predators, it is logical to ask why these creatures would choose an exclusive terpene-enriched diet. The answer is that compared to other species of wood rats that eat more promiscuously, evolution has equipped Stephen's wood rat with plentiful liver enzymes and probably specialized gut bacteria targeted specifically at breaking down terpenes and other toxins located in Junipers.

Back to gin: like cannabis, gin at various times in its long history has been a medicine, recreational compound, fad consumption item, and social hot button issue. Love it or hate it, those who disdain gin's flavor often accuse the beverage of "tasting like a cheap Christmas tree air freshener." That's not far off the mark; the same terpenes that live in Christmas trees (and of course juniper berries) permeate this alcoholic spirit. In traditional pot-distilled gin, the still workers batch-distill neutral spirit, put in carefully measured amounts of botanical ingredients, predominantly Juniper berries, to soak and impart their flavors and fragrances, before re-distilling the mash. In case you were wondering, it's called a "pot still," not because of any distant connection to cannabis, but because of the traditional alchemists alembic-shaped copper pot in which the ingredients are heated. However, there is a weed connection. Gin contains multiple terpenes derived from juniper berries; as an informed wood rat could tell you, these are many of the same terpenes found in cannabis, and include alpha pinene, limonene, beta-myrcene, and alpha-terpeneol. The first two terpenes impart a piney, woody, citrusy bouquet while the third adds a more in-your-face herby skunky, twiggy, slightly resinous flavor. Many gin distillers add coriander as an additional flavorant. This contains terpenes linalool and geranyl acetate, also found in cannabis. These compounds add flowery, lavender-like, spicy and rose-scented notes. A few brands add dried citrus peel to dial up the limonene, or other botanicals containing yet other terpenes. German gin-distilling purists disdain flavor ingredients other than juniper berries; in fact German law forbids further additions.

Because of human environmental exposure to pinene from cypress, cedar, and pine trees used in building, furniture, and decorative materials, there is a fair amount of laboratory research on effects of pinene exposure on humans. This is primarily to assess its safety, but several of these studies offer additional clues to behavioral effects. Sawmill workers in Sweden are famously exposed to large amounts of pine-derived terpenes by breathing in the coniferous fumes as they saw and stack raw lumber. They have thus been the subject of many pinene-related health studies. A 1990 article [28] measured the uptake, distribution in the body, and elimination of alpha pinene at various concentrations in human volunteers. These laboratory subjects sat in an exposure chamber two at a time and breathed in different amounts of pinene for 2 hours while they gamely pedaled away on an exercise bicycle. Because the compound is fairly soluble in blood and fat, around 60% of the terpene was absorbed into the volunteers' bodies, at proportionately greater amounts as concentrations rose. While a proportion of the chemical was rapidly metabolized, some was sucked up into fat and stayed there. The investigators estimated that it would take more than 2 days for the body to completely eliminate it. The Scandinavian scientists focused mainly on lung-related effects of the inhalation (which turned out to be trivial), and not on behavior, so we do not know whether the subjects inhaling high doses felt especially alert (bright-eyed and bushytailed, in fact), had better memories, or perhaps were more attracted by bright shiny objects. In 2016, Japanese investigators exposed 13 young women to alpha pinene vapor and showed significant effects on heart rate (which fell significantly). They also demonstrated increased high-frequency heart rate variability. Both of these changes were consistent with physiological relaxation mediated through the parasympathetic nervous system [29]. Work by others has shown that pinene reduced activity in prefrontal cortex and induces feelings of comfort.

Beta-Myrcene: this mono-terpenoid is the most prevalent terpene in many cannabis chemovars in the United States and Europe [3]. It is also present in hops, lemongrass, citrus fruits, mango, and thyme. It smells skunky and musky, with slight lemony, fruity undertones. Beta myrcene is in fact what makes skunk cannabis smell "skunky," although actual four-legged stripy skunks defensively squirt a completely different stinky chemical. Beta-myrcene is said to have moderate sedative and (somewhat weaker) pain-relieving and anti-inflammatory properties. If any terpene is responsible for couch lock, this soporific, muscle relaxing, and mildly hypnotic compound is the likely culprit [22]. There is some evidence that it may enhance psychoactive effects of THC. This musky terpene chemical can be found relatively prominently in indica chemovars; Skunk XL, White Widow, Green Crack, and Blue Dream, are said to contain relatively high percentages. Now we'll examine briefly some miscellaneous terpenes that are worth mentioning because they are frequently listed in dispensaries. *Terpinolene* is found in parsnips, citruses, mint, sage, rosemary, juniper, parsnip, cumin, lilac, apple, tea tree, nutmeg, and conifers. This minty-tasting terpene emits an aroma described as woody and smoky, with a hint of lilacs and apple blossom. It is found in the chemovars Girl Scout Cookies, OG Kush, and Jack Herer. Traditionally, that is, as conveyed by budtenders, terpinolene is said to be sedative and partly responsible for couch-lock, consistent with experiments in mice [30]. However other reports suggest confusingly that terpinolene-dominant chemovars were consistently rated as energizing [3]. Such disparities in terpene effects are another reminder that properly controlled double-blind studies of single and multiple terpenes need to be carried out. *D-Limonene* is found in citrus fruits, (especially their rinds), in juniper trees, and peppermint. Unsurprisingly, it smells like lemon oil. As mentioned earlier, it is manufactured along with related compounds by some insects such as ants. Limonene is purported to be a stimulant. There are suggestions that it has effects on adenosine receptors in the brain that are connected to wakefulness. These receptors are where the caffeine in coffee acts to keep us awake. One report claims that this terpene helps turn unhealthy white body fat into more healthy brown fat. A major terpene in citrus fruits as well as in cannabis, limonene has been used clinically to dissolve gallstones, improve mood, and relieve heartburn and gastrointestinal reflux, as an anticonvulsant, and has been shown to destroy breast-cancer cells in laboratory experiments. Also its antimicrobial action can kill pathogenic bacteria and perhaps fungi. It is prominent in Super Lemon Haze, Sour Diesel, and OG Kush. *Ocimene* is a pleasant-smelling terpene used in perfume manufacture. Little is known concerning its behavioral properties. *D-Linolool* is a terpene found abundantly in the lavender plant. Its pleasant, characteristic floral smell contributes a fair amount to marijuana's characteristic odor. It is purported to have anticonvulsant, mildly sedative, calming and relaxing properties, and may aid in skin absorption of THC/CBD. It is a reputed anticonvulsant, and it also amplifies serotonin-receptor transmission, conferring a possible antidepressant effect. It is prominent in the chemovars Lavender, Amnesia Haze, Special Kush, and Skywalker. *Geraniol* is found in many flowers fruits and vegetables-including oil of geranium, tobacco, and rose oil. This terpene is pleasantly and sweetly floral with a faint citrusy, peach, and rose backdrop. These properties account for geraniol's use in perfume making. It is also used by flavor chemists to add fruity notes to foods, and is commonly used in shampoos, soaps, and skin lotions. The compound repels mosquitoes and is said to be an antioxidant and neuro-protectant. Geraniol is produced by bees, who use it (among other compounds) to mark their territories. It may or may not have buzz-inducing properties. It is found in chemovars Amnesia Haze, Great White Shark, Islands Sweet Skunk, and Master Kush.

Sesquiterpenes are 15-carbon compounds consisting of 3 isoprene units. Minor members of this group include *Trans-Nerolidol*, a secondary fruity/floral smelling terpene occurring in Jasmine, tea tree oil and lemongrass. Its scent resembles roses and fresh apples. It is said to be sedative, antiparasitic, and antifungal. Trans-nerolidol is found in chemovars Skywalker OG, Sweet Skunk. One of the major sesquiterpene players is *Beta-caryophyllene*. This spicy-smelling, somewhat volatile sesquiterpene compound that is generally described as wood-scented or like black pepper, is found in several thousand plants, including black pepper (unsurprisingly), and in cloves, hops, oregano, rosemary, basil, cloves, cinnamon, and many green, leafy vegetables and fruits [31]. Beta-caryophyllene, or BCP for short, is also synthesized by the Artemisia (wormwood) plant that is used to flavor absinthe and used in clove/cinnamon chewing gum. The pure compound is a pale yellow oily liquid that barely dissolves in water and vaporizes at a relatively low 266°F. BCP is one odor that sniffer dogs key in on when searching for illicit substances. Perhaps coincidentally, one of the prominent chemovars expressing notable BCP levels is in fact Chemdawg, along with Girl Scout Cookies, White Widow, OG Kush, Super Silver Haze, Skywalker, Bubba Kush, and Sour Diesel. Almost uniquely among terpenes, BCP indisputably interacts with the peripheral CB_2 cannabinoid receptor system. BCP may be mildly sedative, and there are claims that it is an antioxidant, anti-inflammatory, and pain reliever, also that it kills parasites and their larvae. There are also claims that BCP has anti-anxiety properties, and that it reduces alcohol cravings in mice. It is also claimed to be gastro-protective, good for treating certain ulcers, and shows promise as a therapeutic compound for inflammatory conditions and autoimmune disorders, perhaps because of CB_2 interactions.

The earlier experimental results mainly derived from BCP given to laboratory rats and mice [32-34]. Because the compound is so fat-soluble, it quickly crosses from the blood to the brain. BCP is especially interesting as a potential medicine for humans, because at least in animals, it very effectively reduces both acute and chronic pain sensation as well as neuropathic pain that results from long-term nerve damage [31]. A potent peripheral pain reliever and anti-inflammatory drug in mice and rats [35], it is effective within 1 hour of administration [36]. The compound also seems to help in protecting the brain and peripheral nerves from varied forms of damage [37,38]. It therefore shows promise as a therapeutic compound for painful inflammatory conditions and autoimmune disorders. As noted, BCP binds directly to peripheral CB_2 cannabinoid receptors, where it acts as an agonist [39]. It is not clear whether this effect persists in the presence of other terpenes and cannabinoids commonly found in cannabis [3]. BCP is safe to administer to humans and has "Generally Recognized As Safe" status with the US Food and Drug Administration, so that it is approved by them for use as an additive to foods and pharmaceuticals. The laboratory research to back up this decision shows that BCP is not clinically toxic [29,40] when given either acutely or over many months to rats *or* mice [41-43], nor does it damage genes [30,44]. Again when administered to pregnant animals, BCP has never been shown to have been toxic on fetuses, or to cause mutations, even in very large doses. BCP has effects when administered orally, with behavioral responses within an hour of administration [31,33,36]. Liquid preparations are currently available over the Internet. A Brazilian researcher at the University of São Paulo, Lyvia Izaura Gomes Paula-Freire carefully investigated the painkilling properties of caryophyllene at a drug receptor level. She showed that mice given an oral dose of the drug showed significantly reduced pain responses [38], but that this effect was reversed by both the mu opioid receptor blocker naloxone (the ingredient in Narcan) and by an experimental cannabinoid receptor blocker. She concluded that caryophyllene's pain-reducing effects required participation of both the opioid and endocannabinoid systems.

In conclusion, a substantial number of rat and mouse studies support BCP as an effective pain reliever [31,33,38,45-47]. However, there have been no such studies of BCP yet in humans, in part because nobody has yet determined outside of animal models how the drug is metabolized in the body, what might be an appropriate dose for people, how bioavailable the drug is, and how long its effects last. We know from the FDA that BCP is presumptively safe in humans, but no one has studied effects of very high doses in people. Some biochemists have already begun making complicated preparations of BCP with other compounds to make it more soluble and more available as an oral medicine [48]. As far as we know at present, there does not seem to be any major drawback to BCP. It may cause sensitization and perhaps dermatitis/allergies in some susceptible individuals [49]. It might also weakly inhibit their liver enzyme CYP3A4 that breaks down some medications, but again this has "negligible pharmacological and/or toxicological impact" [50]. As a result, some laboratories, including my own, have begun thinking seriously about exploring BCP as a safe pain reliever in humans.

Humulin is also found in hops, coriander, basil, and cloves; it vaporizes at a very low 222°F. Humulin smells "hoppy," earthy and woody. It is claimed to have anti-inflammatory and appetite suppressant properties and may have inhibitory effect on cancer cell growth. The compound inhibits fruit fly maturation, which may give clues to why it is produced by the plant in terms of affecting potential insect pests. The compound is relatively overrepresented in strains such as White Widow, Headband, Sour Diesel and Girl Scout Cookies. *Guaiol* smells "piney," and in fact occurs in Cypress pine trees. It has a very low boiling point of 92°C. This terpene is used in traditional medicines, and claims have been made that it has antibacterial, antifungal, and insecticidal properties. It is found in cannabis varieties including Liberty Haze and Blue Kush.

In summary, the world of terpenes is considerably more complicated than anybody thought. Terpenes are very common across the plant kingdom, serving multiple adaptive, especially protective purposes. In the cannabis plant, these chemicals are believed to be primarily defensive against pests and diseases of various sorts. From the point of view of cannabis consumers, terpenes not only interact with cannabinoids in ways that have yet to be properly elucidated, but likely also have their very own complex effects on thinking and behavior as well as drug "high". At least one of them interacts significantly with CB_2 receptors. Several terpenes possess promising medical potential. This is an area that remains wide open for scientific exploration, and it's probable that many reports on this topic will be published over the next few years.

Cannabis concentrates

For all naturally occurring drugs ranging from opium to cocaine to THC, we tend to find more efficient methods to extract and concentrate the active chemical principles on the route to a higher high. In the case of cannabis, concentrates are enriched preparations containing more of the desirable compounds then the parent plant. Depending on the extraction method, along with various cannabinoids, concentrates may also include proportions of terpenes and flavinoids. Because cannabinoids are fairly fat-soluble there is a fairly discrete set of approaches that are commonly used to harvest them from plant material.

The traditional method of enriching THC in countries including Morocco and Afghanistan is by manufacturing hashish, commonly known as hash or cannabis resin. Hashish is composed of trichomes that can be collected and concentrated in a variety of ways; the time-honored approach is pressing or rubbing female cannabis leaves or buds between one's palms and rolling the resins that stick to the skin into a small ball. The plant material can also be rubbed or sieved through screens or measures of varying diameter to yield a powder that can be molded into bricks using heat. Another method to strip off the trichomes is to use ice water; the little glandular hairs become brittle at low temperatures and can be sheared off by agitating the cold mixture and because of their density will sink to the bottom of the slush. Depending on the chemovar of origin, hashish can contain anywhere from 5% to 25% THC. The color and consistency varies widely from paste-like to bouillon cube and yellowish-brown, through burnt umber to almost black. Easily the most interesting book that I've read on the topic of hashish is Robert Connell Clark's nearly 400 page long *Hashish!* [51].

The hash that I used in my birthday cake recipe was the form by which most cannabis reached England in the 1960s and 70s. Before the invention of efficient lighting and heating systems, it was a major challenge to grow cannabis plants in the unpredictable and frequently cloudy English weather. Importing floral marijuana was not efficient because of the relatively low THC content of a few measly percent back then, and its relative bulk compared to the more concentrated, easier-to-conceal hash or khif. I remember an acquaintance showing up at a student party with a chunk of hash half the size of an average paperback book, and as thick as an average McDonald's hamburger. One surface was still covered with a layer of what appeared to be light gray paint. The most immediate origin of this object had been the cabin wall of a Pakistani freighter that had recently docked at Newcastle. An entire wall panel had been carefully substituted with an equivalent amount of hashish that had then been carefully painted over to conceal its true nature. The French connection had nothing to rival this.

More modern methods of concentrate production borrow from the same processes used to create essential oils (many of which are terpenes) and to decaffeinate coffee. Precise methods are always being tweaked in order to maximize the desirable chemicals in the concentrate and minimize undesirable ones such as chlorophyll [52]. The starting plant material needs to be carefully prepared in terms of how it's harvested, dried, ground up, and milled, akin to grinding coffee for an espresso shot. The more plant material that can be packed into the extraction vessel, the more efficient and cost-saving is the procedure, so that high-end grinders and pulverizers are much sought after to yield precise particle sizes that stop the system from clogging up. One advantage of the extraction procedure is that the starting materials do not have to contain high concentrations of THC, so that leaves can be used, for example. Outside of DIY procedures for making your own mini-quantities of rosin dabs at home using a hair straightener set around 300°F [53], the crudest scaled-up process involves using open-source illegal extraction techniques, sometimes referred to as "open blasting." Only slightly more refined is using butane solvent to create butane hash oil (BHO). Compressed butane liquid is what's found inside of cigarette lighters-because it's under high pressure and extremely flammable there's always a danger of fires and explosions, so that smoking cannabis during the preparation procedure is unwise. The essential steps in making BHO are heating the ground plant material with liquid butane solvent under increasing pressure and then sucking away the liquid from the marinade by evaporating it in a vacuum. The lowered pressure changes the solvent from a liquid to a vapor that is whisked away and recycled, leaving behind the concentrate. The process is only moderately effective and marred by the fact that residual butane usually contaminates the concentrate; the chemical is unhealthy. BHO is preferred by some producers because the extraction process is relatively cheap and low-tech. Liquid propane is sometimes used in a similar process to produce propane hash oil. Other, allied methods include using ethanol or isopropanol to obtain cannabinoids without using increased pressure, but these require finicky and exact manipulation of temperature, and a rather lengthy procedure [54]. In addition they are prone to suck out unwanted material from the plant including chlorophyll and the kind of bitter chemicals that you find in fall leaves. Some chemists use activated charcoal to remove this goop.

Super-critical carbon dioxide processing is the extraction method used inside the Denver growery that I visited. This process is used to remove caffeine from coffee beans to produce decaf, and in some dry-cleaning processes. Conceptually it is similar to the liquid extraction procedures mentioned-earlier, and works by compressing the CO_2 gas to the point where it has qualities of both a gas and a liquid, and acts as an effective solvent. At this stage, the carbon dioxide is pushed through an extraction vessel loaded with ground up cannabis. The procedure is more chemically gentle than those I've described so far, and more "tweakable" in that by subtly varying the temperature, pressure or extraction time, one can more selectively remove particular cannabis compounds. Once these are obtained, via repeated extraction cycles, then the CO_2 can be routed back into a condenser, converted back into a fluid and reused. Meanwhile neither the extract itself nor the surrounding air is contaminated by any potentially toxic chemical. The machinery required to run this procedure is rather specialized and not cheap, consisting of pricey high-pressure pumps and chambers. Part of the supercritical carbon dioxide extraction process is called "winterizing," which uses alcohol (ethanol) to remove chlorophyll and pigments from the extract. This also helps harvest terpenes. Remember earlier when we looked at individual terpene compounds, I mentioned their variable boiling points; additional winterization and separation phases that account for these temperatures are necessary to preserve terpenes for removal. If in practical terms, this leverages what is a disadvantage in the ethanol extraction methods described earlier into a plus; their tendency to soak up unwanted compounds. As a postprocessing step in supercritical CO_2 extraction, the ethanol is used as a cleanup procedure.

Concentrates have many names. Tinctures, crumble, honey oil, sap, budder, wax, shatter, rosin, live resin, dabs, BHO, and hash. We have already discussed the latter. Most of the designations just refer to texture and are thus pretty self-descriptive. The main property of concentrates is obviously that they are concentrated; while bud potency varies between 10% and 25% THC, concentrates range from 50% to 85%. But a benefit is that while high-THC cannabis flower inevitably contains low CBD, concentrates can be blended to contain any proportion that's desired. If you would prefer a 50:50 ratio of THC to CBD, that's very straightforward to make. And if you like the flavor of particular terpenes, these can be added back into the concentrate during the blending process and are available to dispensary customers in the form of what's often referred to as "terp sauce" full-spectrum hash oil, or "terp juice." Some individuals purchase high-THC concentrates and immerse them in a vial of terpenes before using them, a process known as "terp dipping." Terp juices and sauces contain very high concentrations of added terpenes, combining THC potency with flavor intensity. Perusing a Weedmaps website of concentrates such as Dabblicious reveals a wide variety of choices. In the world of live resins, there's a choice of Soul Safari, Tangie Pie, Golden Lemons, Peaches, or Super Glue to name a few examples. Bottom line: the range of key ingredient choices is much greater for concentrates than with flower. Similarly while flower can be smoked in a pipe or joint or vaporized, concentrates can be used by dabbing, ingested as an oil or tincture, or packed into a joint. Dabbing involves putting a small amount of the concentrate onto a heated nail and inhaling it (most often) through a bong-like glass apparatus; the nail can be heated with a crème brûlée-style butane torch or temperature-controlled electric coil (e-nail). Ingestible oils are by definition eaten like edibles, usually in capsule form, while tinctures are dropped under the tongue, where they are rapidly absorbed from the mucous membrane in the mouth with its many surface blood vessels. Some concentrates can be consumed using a vape pen or desktop vaporizer, for the latter by dripping them on to an absorbent pad.

Who uses them and why? Young men who are heavy cannabis users favor concentrates. They are generally regarded as being further along the user consumption path that involves consuming larger quantities more frequently. But there are other considerations. Choice is one; as mentioned, a much wider variety of concentrations and ratios is available in concentrate form. Bang for the buck is another; if you're pursuing intoxication, then what you are really after buying in a dispensary is THC, and with concentrates you're getting straight to the bottom line, even if a gram costs anywhere between $70 and $100.

Using a dab rig delivers a lungful of extremely concentrated THC vapor that gets you really high really quickly. Setting the temperature high, say around 575-600°F, is apparently a recipe for the cannabis user equivalent of being hit over the head with a sledgehammer, while setting the temperature a bit lower softens the blow. Common additional unwanted effects are coughing (especially if you scorch the product at too high a temperature), lethargy, mild nausea, and fleeting paranoia. Because the concentrations of THC are so high, choosing a precise dose is tricky, and definitely a learned skill

Inhaling such a large dose of THC, particularly if indulged in frequently, is likely a risk factor for cannabis addiction. As noted in Chapters 7 and 8, high-THC potency cannabis raises the risk for psychotic illness; use of concentrates is a concern in that regard but no research has yet been published on that topic. Inhaling unwanted leftovers from the extraction process such as butane is known to be unhealthy. Scientists, including a group at Portland State University in Oregon are concerned that the high temperatures of dabbing degrade cannabis compounds themselves into potentially harmful chemicals such as benzene (a cancer risk) and methacrolein [55]. The Portland group used our old friend gas chromatography/mass spectrometry (GC-MS) that we encountered in probing Shakesperean pipes and Chinese tombs, to analyze the vaporized products of cannabis concentrates heated on a ceramic dab nail; terpenes in the mixture such as myrcene, produced worrisome potentially toxic isoprene-related compounds at high temperatures.

In summary, when it comes to discussing cannabis don't say "variety," say "chemovar." The ever-present urge for more and better intoxication, through concentrating key plant-derived chemicals has led to new compounds, new practices such as dabbing, and perhaps new risks that are as yet poorly characterized. The cannabis plant contains multitudes of chemicals, each with its own pharmacologic effects. Its overall medicinal efficacy (as we shall see in the next chapter) may consist of more than the sum of those individual parts. Terpene compounds found in the plant represent a fascinating alternative window into both the medicinal and intoxicating properties of the plant that deserve much deeper study.

References

[1] Mechoulam R. Marihuana chemistry. Science 1970;168(3936):1159–66.

[2] Clarke RC. Marijuana Botany. Second ed. Berkeley, CA: Ronin Publishing; 1993.

[3] Lewis MA, Russo EB, Smith KM. Pharmacological foundations of cannabis chemovars. Planta Med 2018;84(4):225–33.

[4] Available from: https://phytofacts.info.

[5] Giese MW, et al. Development and validation of a reliable and robust method for the analysis of cannabinoids and terpenes in cannabis. J AOAC Int 2015;98(6):1503–22.

[6] Zager JJ, et al. Gene networks underlying cannabinoid and terpenoid accumulation in cannabis. Plant Physiol 2019. Available from: https://doi.org/10.1104pp.18.01506.

[7] Sawler J, et al. The genetic structure of marijuana and hemp. PLoS One 2015;10(8):pe0133292.

[8] Leichman A.K. The woman who's going to teach us how to grow cannabis properly, 2019. Available from: https://www.israel21c.org/the-woman-whos-going-to-teach-us-how-to-grow-cannabis-properly/.

[9] Gilbert AN, DiVerdi JA. Consumer perceptions of strain differences in cannabis aroma. PLoS One 2018;13(2):pe0192247.

[10] Aharoni A, Jongsma MA, Bouwmeester HJ. Volatile science? Metabolic engineering of terpenoids in plants. Trends Plant Sci 2005;10(12):594–602.

[11] Chen A. Some of the parts: is marijuana's "entourage effect" scientifically valid? Scientific American. 2017. Available from: https://www.scientificamerican.com/article/some-of-the-parts-is-marijuana-rsquo-s-ldquo-entourage-effect-rdquo-scientifically-valid/.

[12] Wilkinson S, et al. Educating African-American men about prostate cancer: impact on awareness and knowledge. Urology 2003;61(2):308–13.

[13] Williamson EM. Synergy and other interactions in phytomedicines. Phytomedicine 2001;8(5):401–9.

[14] McPartland J, Russo E. Non-Phytocannabinoid Constituents of Cannabis and Herbal Synergy. Handbook of Cannabis. Oxford, UK: Oxford University Press; 2014.

[15] Joo JH, Jetten AM. Molecular mechanisms involved in farnesol-induced apoptosis. Cancer Lett 2010;287(2):123–35.

[16] Levin DA. The role of trichomes in plant defense 1973;Quart Rev Biol 48(1):3–15.

[17] Huchelmann A, Boutry M, Hachez C. Plant glandular trichomes: natural cell factories of high biotechnological interest. Plant Physiol 2017;175(1):6–22.

[18] Baser KHC, Buchbauer G. Handbook of Essential Oils: Science, Technology, and Applications. 2nd ed. Boca Raton, FL: CRC Press; 2015.

[19] Russo EB, Marcu J. Cannabis pharmacology: the usual suspects and a few promising leads. Adv Pharmacol 2017;80:67–134.

[20] Russo EB. Beyond cannabis: plants and the endocannabinoid system. Trends Pharmacol Sci 2016;37(7):594–605.

[21] Walsh Z, et al. Medical cannabis and mental health: a guided systematic review. Clin Psychol Rev 2017;51:15–29.

[22] Russo EB. Taming THC: potential cannabis synergy and phytocannabinoid-terpenoid entourage effects. Br J Pharmacol 2011;163(7):1344–64.

[23] Perry NS, et al. In-vitro inhibition of human erythrocyte acetylcholinesterase by salvia lavandulaefolia essential oil and constituent terpenes. J Pharm Pharmacol 2000;52(7):895–902.

[24] Volicer L, et al. Effects of dronabinol on anorexia and disturbed behavior in patients with Alzheimer's disease. Int J Geriatr Psychiatry 1997;12(9):913–9.

[25] Hasselmo ME, Stern CE. Mechanisms underlying working memory for novel information. Trends Cogn Sci 2006;10(11):487–93.

[26] Mercier B, Prost J, Prost M. The essential oil of turpentine and its major volatile fraction (alpha- and beta-pinenes): a review. Int J Occup Med Environ Health 2009;22(4):331–42.

[27] Ranganathan M, D'Souza DC. The acute effects of cannabinoids on memory in humans: a review. Psychopharmacology (Berl) 2006;188(4):425–44.

[28] Falk AA, et al. Uptake, distribution and elimination of alpha-pinene in man after exposure by inhalation. Scand J Work Environ Health 1990;16(5):372–8.

[29] Harumi Ikei et al. Comparison of the effects of olfactory stimulation by air-dried and high-temperature-dried wood chips of hinoki cypress (Chamaecyparis obtusa) on prefrontal cortex activity, 2015. Available from: https://link.springer.com/article/10.1007/s10086-015-1495-6.

[30] Ito K, Ito M. The sedative effect of inhaled terpinolene in mice and its structure-activity relationships. J Nat Med 2013;67(4):833–7.

[31] Sharma C, et al. Polypharmacological properties and therapeutic potential of beta-caryophyllene: a dietary phytocannabinoid of pharmaceutical promise. Curr Pharm Des 2016;22(21):3237–64.

[32] Cho JY, et al. Amelioration of dextran sulfate sodium-induced colitis in mice by oral administration of beta-caryophyllene, a sesquiterpene. Life Sci 2007;80(10):932–9.

[33] Klauke AL, et al. The cannabinoid CB(2) receptor-selective phytocannabinoid beta-caryophyllene exerts analgesic effects in mouse models of inflammatory and neuropathic pain. Eur Neuropsychopharmacol 2014;24(4):608–20.

[34] Andrade-Silva M, et al. The cannabinoid 2 receptor agonist beta-caryophyllene modulates the inflammatory reaction induced by Mycobacterium bovis BCG by inhibiting neutrophil migration. Inflamm Res 2016;65(11):869–79.

[35] Chavan MJ, Wakte PS, Shinde DB. Analgesic and anti-inflammatory activity of caryophyllene oxide from Annona squamosa L. bark. Phytomedicine 2010;17(2):149–51.

[36] Gertsch J. Anti-inflammatory cannabinoids in diet: towards a better understanding of CB(2) receptor action? Commun Integr Biol 2008;1(1):26–8.

[37] Machado KDC, et al. Effects of isopentyl ferulate on oxidative stress biomarkers and a possible GABAergic anxiolytic-like trait in Swiss mice. Chem Biol Interact 2018;289:119–28.

[38] Paula-Freire LI, et al. The oral administration of trans-caryophyllene attenuates acute and chronic pain in mice. Phytomedicine 2014;21(3):356–62.

[39] Gertsch J, et al. Beta-caryophyllene is a dietary cannabinoid. Proc Natl Acad Sci USA 2008;105(26):9099–104.

[40] Oliveira G, et al. Non-clinical toxicity of beta-caryophyllene, a dietary cannabinoid: absence of adverse effects in female Swiss mice. Regul Toxicol Pharmacol 2018;92:338–46.

[41] Alvarez-Gonzalez I, Madrigal-Bujaidar E, Castro-Garcia S. Antigenotoxic capacity of beta-caryophyllene in mouse, and evaluation of its antioxidant and GST induction activities. J Toxicol Sci 2014;39(6):849–59.

[42] Seebaluck R, Gurib-Fakim A, Mahomoodally F. Medicinal plants from the genus Acalypha (Euphorbiaceae)-a review of their ethnopharmacology and phytochemistry. J Ethnopharmacol 2015;159:137–57.

[43] Bahi A, et al. beta-Caryophyllene, a CB2 receptor agonist produces multiple behavioral changes relevant to anxiety and depression in mice. Physiol Behav 2014;135:119–24.

[44] Di Sotto A, et al. Genotoxicity assessment of beta-caryophyllene oxide. Regul Toxicol Pharmacol 2013;66(3):264–8.

[45] Quintans-Junior LJ, et al. Beta-caryophyllene, a dietary cannabinoid, complexed with beta-cyclodextrin produced anti-hyperalgesic effect involving the inhibition of Fos expression in superficial dorsal horn. Life Sci 2016;149:34–41.

[46] Cheng Y, Dong Z, Liu S. beta-Caryophyllene ameliorates the Alzheimer-like phenotype in APP/PS1 mice through CB2 receptor activation and the PPARgamma pathway. Pharmacology 2014;94(1-2):1–12.

[47] Basha RH, Sankaranarayanan C. beta-Caryophyllene, a natural sesquiterpene, modulates carbohydrate metabolism in streptozotocin-induced diabetic rats. Acta Histochem 2014;116(8):1469–79.

[48] Liu H, et al. Physicochemical characterization and pharmacokinetics evaluation of beta-caryophyllene/beta-cyclodextrin inclusion complex. Int J Pharm 2013;450(1-2):304–10.

[49] Paulsen E, Andersen KE. Colophonium and compositae mix as markers of fragrance allergy: cross-reactivity between fragrance terpenes, colophonium and compositae plant extracts. Contact Dermat 2005;53(5):285–91.

[50] Nguyen LT, et al. The inhibitory effects of beta-caryophyllene, beta-caryophyllene oxide and alpha-humulene on the activities of the main drug-metabolizing enzymes in rat and human liver in vitro. Chem Biol Interact 2017;278:123–8.

[51] Clarke RC. Hashish! 2nd ed. Los Angeles, California: Redeye Press; 1998.

[52] Tilray. Cannabis extracts: the science behind cannabinoid and terpene extraction methods, 2016. Available from: https://www.leafly.com/news/science-tech/cannabis-extracts-the-science-behind-cannabinoid-and-terpene-extr.

[53] Bennett P. How to make Rosin dabs, 2016. Available from: https://www.leafly.com/news/cannabis-101/how-to-make-rosin.

[54] Damon Anderson P. The best cannabis extraction methods for marijuana concentrates, 2018. Available from: https://www.labx.com/resources/the-best-cannabis-extraction-methods-for-marijuana-concentrates/147.

[55] Bear-McGuinness L. Revealing the potential risks of cannabis dabbing, 2018. Available from: https://www.analyticalcannabis.com/articles/revealing-the-potential-risks-of-cannabis-dabbing-306016.

Chapter 10

Medical marijuana and clinical trials

"In morbid states of the system, it has been found to produce sleep, to allay spasm, to compose nervous inquietude, and to relieve pain.... The complaints to which it has been specially recommended are neuralgia, gout, tetanus, hydrophobia, epidemic cholera, convulsions, chorea, hysteria, mental depression, insanity and uterine hemorrhage."

Wood GB and Bache F. 1854, The Dispensatory
of the United States. Philadelphia: Lippincott 1854, page 339.

"There really is no such thing as medical marijuana."

U.S. Health and Human Services Secretary Alex Azar March 2, 2018

"Medical marijuana is used to treat a host of indications, a few of which have evidence to support treatment with MJ and many that do not"

Hill, 2015 [1].

You can't spell 'healthcare' without THC

Anonymous.

Cannabis is a drug that has known medical uses dating back 4000 years, extending all the way to the official US Pharmacopeia, up until 1942. In this chapter, we will review the historical use of cannabis as a pharmaceutical; examine some of the arguments for the use of medical marijuana and of CBD as well as the numerous conceptual, practical, and legal difficulties that stand in the way of the necessary, definitive clinical trials. Such investigations will tell us whether or not cannabis and/or its many constituents have the utility for particular medical conditions. We will review existing data regarding medical marijuana's effectiveness in treating a variety of illnesses and try to gauge the strength of the supporting evidence for a handful of the most studied of them. This chapter does not attempt to review the detailed evidence for cannabis compounds as potential treatments for many medical disorders where only pilot or small-scale clinical trials have been conducted. That's worthy of a completely separate, stand-alone volume. Indeed there are several such books and reports currently available. Rather, what I have tried to concentrate on more is how we should best weigh the utility of medical cannabis, and how scientific thinking and evidence can aid that process. Let's start at the beginning, historically speaking.

Weed Science. http://dx.doi.org/10.1016/B978-0-12-818174-4.00010-0

For ancient medical use we need to separate out medical from spiritual and ceremonial uses of cannabis, as these often are confounded; the origins of many medical diseases were then obscure and felt to be located in the spiritual realm. Ethan Russo [2] has written a comprehensive historical account of medical cannabis. He reviews the evidence for cannabis use as a medicament from ancient cuneiform clay tablets excavated from the Mesopotamian city of ancient Nineveh in Iraq dating to 2600 years ago. These integrate medical knowledge from prior Acadian and Sumerian cultures, dating further back to 4000 years before the present day. Although it's difficult to identify the herb that they describe with certainty, and modern medical diagnoses do not align with those of antiquity, the fact that the plant in question is referred to both as a medicine, and used in spinning and cable making, strongly suggests that cannabis is being specified. Its indications included use as a pain reliever, possibly treating spasticity, kidney stones, lung congestion, depression, anxiety, and probable nocturnal epilepsy. As well as, it was employed as a tonic and love potion for impotence. Around 2700 years ago, this near-eastern culture began to use a new term for cannabis- "kunubu," that sounds very much like the contemporary name for the plant. It was used in sacred rituals. Interestingly, kunubu was also used as a female personal name and as a term of endearment. I fully expect that such cannabis-themed girls' names will emerge in the next few years to replace the current Mary-Jane; stay tuned for Cannah Montana, Canndace, Cannabina, and Canna-Mae.

Ancient Egyptian medical papyri dating to over 3500 years ago recommended cannabis for treatment of eye diseases and gynecologic disorders. At the same time, the plant was used as a source of fiber, and hemp threads have been discovered found in Egyptian tombs dating to over 3300 years ago. In Chapter 4, we alluded to the more recent medical use of cannabis as an inhalant to facilitate the birth process [3] through the discovery of the 1600-year-old Israeli tomb containing the remains of the 14-year-old girl who had died in childbirth together with burned THC-containing cannabis.

Use of cannabis in ancient Indian medicine is claimed in the Atharva Veda around 3600 years ago, where "bhanga" is one of five herbs employed to "release us from anxiety/grief," echoing Assyrian citations. Cannabis was well-established in Ayurvedic medicine by the year 300 to 400 A.D. for treating a variety of conditions including headaches, pain during childbirth, epilepsy, and insomnia. Hemp was first used in China 12,000 years ago, initially for fiber and as a staple grain; its medical use followed later. The Chinese documented that the female plant was most appropriate for medicinal use, and various descriptions of preparation for such purposes can be traced back to around 4700 years ago, when cannabis was used to treat gout and rheumatism, conditions involving pain and inflammation. Cannabis has been used in sub-Saharan Africa for at least 2000 years based on pollen samples. Termed "Dagga," the drug was again used as a pain-reliever in tea for both headaches, and in childbirth. Around 1200 years ago cannabis appears in the Middle East in medieval Arabic medicine, where it was used to treat nausea, epilepsy, pain, inflammation, and fevers.

800 years ago, Moses Maimonides, the Jewish physician/philosopher practicing in the Islamic world, included cannabis in his materia medica as "qinnab." Cannabis arrived in Brazil nearly 500 years ago, along with the slave trade from Africa, and was used in folk medicine for relieving the pain of rheumatism and toothache. In 1758, the Frenchman Marcandier in his *"Treatise on Cannabis"* referred to the use of the plant's root as a poultice for the treatment of arthritis and gout and suggested its utility for treating "tumors." Cannabis roots harbor various terpenoids, sterols, and alkaloids but no cannabinoids.

These traditional uses of medical marijuana: to treat pain of various sorts, epilepsy, inflammation, insomnia, obstetric/gynecologic problems, and anxiety seem strikingly consistent across cultures and echo current recommendations for medical marijuana.

The modern European use of cannabis as medicine and intoxicant harks back to Napoleon's invasion of Egypt in the 1790s. His military forces were accompanied by an expedition of hundreds of historians and scientists, who excavated ancient sites, documented monuments, and discovered the Rosetta Stone. The scientific team also wrote extensively regarding the cannabis that they discovered was being used there as both a medication and recreational drug. The substance was brought back to France and used medically across continental Europe as a possible treatment for the plague, (and in some quarters in France as an intoxicant, by the cultural and literary elite). Moving forward nearly 50 years to the late 1830s, the experiments of the Irishman William Brooke O'Shaughnessy in Calcutta, India suggested the utility of cannabis extracts in medicine. This physician was a polymath and innovator who had earlier suggested the pioneering use of intravenous rehydration as a treatment for cholera. Outside of cannabis' medical use, he was also well aware of its intoxicant properties, describing how it produced "ecstatic happiness, the persuasion of high rank," (i.e., grandiosity) "the sensation of flying, voracious appetite, and intense aphrodisiac desire." This nicely captures the "happy, hungry, horny" properties of the drug. He published the results of his cannabis studies in scientific journals, and as a result of his influence, soon thereafter various cannabis extracts and tinctures began to be used for medicinal purposes in Europe and North America. On O'Shaughnessy's return to England in 1841, he brought cannabis herbal material and seeds with him and shared samples with physicians throughout the United Kingdom. He experimented with cannabis for the treatment of tetanus, and suggested that the drug had great promise in treating epilepsy and rheumatism. Interestingly, he noted that over-treatment with cannabis could lead to delirium. His tincture of Indian hemp (cannabis dissolved in alcohol) developed with the pharmacist Peter Squire, was known as "Squires Extract." A sample of this tincture from an old Victorian bottle analyzed in 1984 using GC-MS, showed traces of THC, the THC breakdown product cannabinol (that we encountered earlier) and significant amounts of cannabichromene. Many contemporary practitioners became very interested in cannabis as a medical treatment for a variety of disorders, and chemists attempted without success to extract the key curative chemical compound from it. Unlike opium, which could be easily made into a standardized tincture (Laudanum) that could then be dosed reasonably accurately, cannabis was problematic to pharmacists. It was difficult to standardize, quality control was elusive, and maintaining consistent accuracy in clinical dosing was especially challenging. All of these pharmacologic problems held back more extensive use of cannabis as a medicine, with use finally ceasing when the drug was outlawed in the mid-20th century. Physicians back then had no inkling of the identity of either THC or CBD.

In the 1990s interest in medical cannabis in the United States was reawakened for a variety of reasons. These included scientific discovery of the endocannabinoid system that helped provide a scientific context for the drug's effect in the brain and elsewhere. As we've seen, this discovery ultimately led to interest in intervening in the system not only through CB receptor agonists and antagonists but also evolved into use of drugs with more complex receptor effects such as partial agonists and allosteric modulators. Attempts began as well to try to manipulate the endocannabinoid system, via drugs influencing its key enzymes such as FAAH. A second factor in the rebirth of medical cannabis was a gradual alteration in public opinion that began to favor access to medical (as well as recreational) cannabis. Part of this was due to persistent lobbying by organizations such as NORML, as well as to the coming-of-age of the 60s generation who were unimpressed by horror stories attached to the drug by their elders. A third factor pulling the bandwagon forward was persistent anecdotal reporting of the plant's medical benefits. The sea change came with the Institute of Medicine's 1999 report [4] that recommended that cannabinoids might have a role in the treatment of pain, movement, and memory disorders. The report also emphasized that risks could be associated with use. Its major recommendations were to better evaluate physiological and psychological effects, individual health risks and the role of various delivery systems, as well as to shorten clinical trials to determine the drug's effectiveness in targeted conditions. These forces led to California being the first state to legalize medical marijuana in 1996.

Opinions on cannabis have now changed sufficiently that in a Gallup poll at the end of 2018, 66% of Americans surveyed, expressed the opinion that marijuana should be legalized [5]. However, the federal government resists moves in this direction. For example in 2018, Health and Human Services Secretary Alex Azar famously stated that there was "no such thing as medical marijuana." Unfortunately, because of medical marijuana's federal designation, public opinion has not yet translated into consistent federal or state medical regulations. Medical facilities do not have consistent policies regarding patient use of physician-certified medical cannabis. For example, when such patients are hospitalized, many facilities (even in states that have legalized medical marijuana), will not allow them to use their own cannabis during their hospital admission, but instead prescribe alternatives such as opioids. Individuals admitted to federally funded group homes have had their CBD for epilepsy discontinued, provoking a return of their seizures [6]. But the tide may be turning. In 2019, the FDA held its first public hearing on regulation of CBD products [7]. A problem worth mentioning is that there is frequent confounding regarding the legalization of medical versus recreational cannabis. Some states began legalizing medical cannabis, but the extremely wide range of medical conditions for which the drug was permitted, and the relatively cursory examination by a physician necessary to obtain a medical marijuana card, for example, in California, led many to suspect (not entirely unreasonably) that legislation represented a backdoor approach to legalizing recreational cannabis. An argument often heard in the physician community is that if legalization of medical marijuana represents a "veiled step towards allowing access to recreational marijuana" then the best solution is to decriminalize all cannabis, and to leave the medical universe out of the process.

Conceptual difficulties

I'm one member of the Board of Physicians that advises the Connecticut Department of Public Health on which conditions may be appropriate for treatment with medical marijuana. I frequently talk with medical practitioners who for a variety of reasons find it very difficult to understand the concept of cannabis as a medical treatment. I try and explain later what it is that puzzles them, but by and large the confusing issue is not that the same compound can be used both as a recreational drug and as medical treatment. All of them are familiar with the examples of opioids and cocaine in this context. The first perplexing issue for physicians is that medical cannabis generally uses the entire plant (e.g., as flower) rather than one or more specific constituents, such as THC. Several other plant-derived drugs such as digitalis from foxglove and morphine from opium poppies are purified chemicals derived from plants. But many physicians are aware that cannabis contains almost 500 distinct chemicals in the form of cannabinoids, flavonoids and terpenes. As one of them confided in me, "I keep hearing about the concept of entourage effects, but I have a really hard time wrapping my head around that." Another problem is that physicians are used to prescribing standardized medications—100 mg of this pure compound, 5 mg of that one, but the percent of THC and CBD at the same cannabis dispensary vary not only from one chemovar to another, but from batch to batch. They are not comfortable with this product variability, the relative lack of standardized assays and consistent monitoring for the drug. When patients ask them which cannabis chemovar is most effective for treating a specific medical condition they have no idea how to respond, because there are no reliable data to inform them. Even more perplexing are recommendations on dosing. As one physician told me, "what do I tell my patients, take two bong hits and call me in the morning?" Also, because of the biphasic effects of cannabis, control of dose is extremely important. Using one quantity may improve anxiety levels, but a larger quantity will likely boost them. Naturally-occurring individual compounds in cannabis occur not only in different proportions in different chemovars, but may have opposite effects, say on appetite, as discussed further. Allied with this is the fact that the drug is not being administered under medical oversight, but by the patient him or herself, in unclear quantities and at uncertain intervals. Physicians are concerned about this, not only because of their lack of control over the proceedings, but because it's unclear what exactly their patient is taking. Vaping devices that utilize standardized concentrates and deliver them in metered doses whose data can be transmitted back to both the dispensary and physician, such as the GoFire show some promise in minimizing this uncertainty. But again it's unclear in what form cannabis should best be administered for a particular condition, for example, via edibles, vaporizing, or tinctures of the sort familiar to O'Shaughnessy. Where promising findings of medical efficacy do exist, there's also no obvious way to extrapolate from studies on individual cannabinoids, for example, THC or CBD to herbal cannabis and vice versa.

Many physicians do not routinely ask their patients if they are using medical marijuana. Few consider medical marijuana or CBD when deciding to change treatments. Also, many physicians are unable to make recommendations regarding usefulness or dosage when patients solicit their advice on this matter. They are unclear whether cannabis-related compounds interact with other medications that they are already prescribing. There is already some evidence that this may be the case for general anesthesia, recovery from surgery, and response to blood thinning medication [8,9]. In large part, their lack of knowledge truly reflects the dearth of reliable research on these questions [10]. Another issue is that there has been remarkable inconsistency with implementation of medical marijuana regulations, to date, from state to state. Merely listing and comparing the medical and psychological conditions for which medical marijuana is allowed in the 33 states that have passed such laws to date is educational. In some states the list of approved medical diagnoses is as long your arm, comprising panoply of diverse disorders, for some of which there is little or no supporting research to speak of. Other states are much more restrictive in their listings, but not necessarily more evidence-based in their choices. Because there's no consistent federal approach to this issue, states are currently free to do as they wish, and indeed they have implemented medical cannabis legislation in an incredibly diverse and inconsistent manner. Many states have seized on marijuana legalization both for medical, or medical plus recreational purposes as a quick way to help solve their budget deficits, akin to legalizing gambling. Unfortunately, in the case of medical marijuana this means that more thought and planning has sometimes gone into crafting the sales and tax aspects of the legislation, with relative neglect of consideration of the medical side. This has been so particularly when it comes to weighing evidence for and against inclusion of particular medical diagnoses. In turn, this lack of thoughtful planning is apparent in the confusion expressed by practicing physicians.

The most familiar refrain that I hear from doctors interested in medical cannabis is "We never learned any of this in medical school. It's not in the textbooks. I'd be a lot more comfortable using this drug if the FDA did some decent clinical trials to provide me some direction in prescribing this stuff properly.

But where are the evidence-based data?" So what's the proper answer to that last, perfectly reasonable question? Before approving a drug for any specific indication (e.g., treatment of gout) the US Food and Drug Administration (FDA) requires evidence from two or more adequately-powered (usually construed as containing 200 or more patients), randomized clinical trials. In other words, the study needs to contain sufficiently large numbers of patients to be convincing, with doses of the drug or placebo randomly assigned to combat the powerful placebo effect that we discussed in Chapter 8. But the FDA currently officially designates cannabis as a schedule 1 compound, meaning that it has a high potential for addiction and no currently accepted medical treatment use in the United States. Other schedule 1 substances include heroin, fentanyl derivatives, LSD, ecstasy, and peyote. As has been pointed out, in a move that seems to directly undermine this categorization, the US government itself holds US patent US6630507B1 for CBD, that covers "pharmaceutical compounds and compositions that are useful as tissue protectants, such as neuroprotectants and cardioprotectants. The compounds and compositions may be used, for example, in the treatment of acute ischemic neurological insults or chronic neurodegenerative diseases." Obviously, this dual-faced approach, (essentially, "we have classified the substance as a dangerous drug with no medical use, but we've also taken out a patent because it may be useful in treating strokes, heart attacks and chronic neurologic illnesses") makes no logical sense. Because of its current federal scheduling, proposing a clinical trial of cannabis or even one of its constituent cannabinoids quickly becomes ensnarled in a welter of federal bureaucratic regulations; what investigators in the field refer to as "green tape." It's not completely impossible to conduct such studies, but jumping through multiple time-consuming hoops is certainly a requirement and therefore a significant discouragement. Thus far no federal administration has been willing to bite the bullet when it comes to re-scheduling cannabis or its derivatives, although several 2020 presidential candidates support such a move. In the interim, because of the federal law on cannabis, many companies that routinely test staff for drug use will terminate employees for a positive THC urine test, even in the face of a valid medical cannabis card. In addition, because of a lack of standardized testing, some individuals have purchased what they believe to be pure CBD oil, only to discover that it contains a sufficient amount of THC to show up as positive on their employment drug test. Meanwhile, legal or not, use is growing steadily. Quest Diagnostics recently published a study showing that marijuana use has increased by 16% since 2014 to 2.8% of workers; almost 1% of individuals in safety-sensitive jobs such as airline pilots and train operators screened positive for THC [11].

Another issue holding back medical marijuana research is that it's not possible to patent a plant. So that while pharmaceutical companies may be interested in trials of a more exotic synthetic drug that they have designed, such as an FAAH inhibitor, there is absolutely no incentive for them to conduct a medical marijuana trial. And until recently the National Institutes of Health were more focused on demonstrating problematic effects of cannabis, and therefore much more likely to fund studies on psychosis, or driving impairment, for example, than investigations aimed at demonstrating possible beneficial effects. This stance may well be appropriate for the National Institute on Drug Abuse, whose mission is to focus (of course) on drug abuse. But recently the NIH's National Center for Complementary and Integrative Health began funding grants to explore trials of cannabinoid compounds for pain. In 2017, the National Academy of Sciences and Medicine (NAS) published *The Health Effects of Cannabis and Cannabinoids; The Current State of Evidence and Recommendations for Research"* [12]. In the section *"Challenges and barriers in conducting cannabis research,"* this report made the point that despite the changes in both US state policy and cannabis use, the federal government "has not legalized cannabis, and continues to enforce restrictive policies and regulations on research into the health harms or benefits of cannabis products that are available to consumers in a majority of states." It is true that the federal government has not significantly updated its restrictive policies regarding cannabis research in over 40 years, since the Nixon administration. The National Academy committee identified multiple barriers to research on cannabis and cannabinoid research. These included the necessity for any application to be reviewed by multiple federal and local agencies, and the rule that any substances must be obtained from the National Institute on Drug Abuse (NIDA). In practical terms that means that cannabis for research purposes must be obtained from a single source at NIDA, sourced from the University of Mississippi, which grows smallish quantities of the plant material. This cannabis has been criticized for being both less potent than and unrepresentative of the drug that can be obtained from a local dispensary. Research carried out with this cannabis is therefore likely lacking desirable validity and generalizability, although this situation may change both for quantity and diversity of plant material. The issue of the low THC content of NIDA-supplied cannabis is easily compensated for experimentally by using a vaporizer loaded with sufficient weight of material to mimic a typical street dose. Researchers at the University of Northern Colorado performed genetic analyses on 49 different cannabis samples including NIDA marijuana, non-THC containing hemp, as well as various indica, sativa, and hybrid chemovars. In a research report published on the online site bioRxiv [13], the authors demonstrated that non-drug hemp and THC-containing cannabis were genetically distinct strains. However, the NIDA drug samples shared a closer "genetic affinity with hemp samples in most analyses," compared to commercially available cannabis. In response to criticisms regarding its low output, the University of Mississippi recently (2019) agreed to boost its annual cannabis production for researchers. Several years ago, the federal government also agreed to make cannabis for experimental purposes available from a larger variety of suppliers, as opposed to the current single source. But in practical terms they have made little progress yet to solve that issue.

Non-flower formulations as edibles and concentrates are also not yet available through NIDA, although this federal agency is moving slowly to become more responsive to such requests from clinical investigators. In 2015 NIH spending on all cannabinoid research totaled just over $111 million. The 2017 NAS report concluded that "a diverse network of funders is needed to support cannabis and cannabinoid research that explores the harmful and beneficial health effects of cannabis use." Orrin Devinski, director of NYU's Epilepsy Center, stated "We have the federal government and the state governments driving 100 miles an hour in the opposite direction when they should be coming together to obtain more scientific data...... It's like saying in 1960, "we're not going to the moon because no one agrees how to get there"[6]. Access to CBD for experimental purposes is likely to improve, since 2018 legislation removed hemp-derived products such as CBD from schedule 1 status, and allowed cultivation of hemp, officially defined as having less than 0.3% THC content. In 2018, the Senate Agriculture Committee passed the 2018 Farm Bill that legalized the cultivation of non-THC-containing hemp, and allowed its cultivation by universities. These provisions were backed by Senate Majority Leader Mitch McConnell of Kentucky, in part to help support farmers in his home state concerned about lost revenue from falling demand for tobacco, and seeking alternative crops.

Overall, from the small number of clinical trials of medical cannabis that successfully negotiated all of the hurdles I outlined earlier, what can we learn? After all, we need to be informed by properly designed trials in order to move beyond the stage of anecdotes ("medical marijuana helps my wife's neuralgia") toward rational medical prescribing of cannabis-based drugs. So, what do we know right now, and what do we still need to discover before understanding marijuana's medical risks and benefits? To be honest, one of the more obvious lessons is that, in part because of all of the restrictions and lack of funding, sadly there are predictable design problems with many of these existing studies. Let's enumerate some of them. One difficulty is that many of the clinical trials are not properly randomized, (where, who, gets which active medicine or placebo should be totally random), and double-blind (to deal with expectation effects; neither the prescribing physician nor the patient knows exactly what they're receiving). Placebo-controlled investigations with adequate sample sizes, say in the realm of 200 patients, are required. Patient samples of that magnitude are sufficiently powered to reach confident conclusions about a drug's usefulness, and also to gauge how powerful its effects actually are, (what statisticians refer to as its "effect size"). Another design issue is the need for subjects to be fully blind as to whether they are receiving active drug or not. It is pretty obvious to people participating in a cannabis trial when they are receiving placebo, because they will experience no psychoactive THC effect, (unless they are part of a CBD-only study). Ideally, clinical trial designers should insist on some form of "active placebo" that is, a non-cannabinoid drug that had some form of psychoactive effect, so that it is not clear to the participant whether he or she was actually receiving active cannabis or not. In a practical sense this is difficult to set up. For example, what substance should be used for the active placebo? Are we sure that whatever drug we choose for that purpose does not have its own medicinal effects? A related problem is that many studies do not adequately document other substance use, both legal (such as prescribed opioids), and illegal, in their subjects.

To be properly informative, a well-designed trial also needs to report all relevant outcomes. For example, if cannabis is being prescribed for painful spasms resulting from multiple sclerosis, it's important to know not only whether pain was relieved, but also whether there was worsening of pre-existing cognitive impairment. Pain is an inherently subjective and private experience. It's not possible for researchers to stick a theoretical "painometer" into a patient and obtain an objective, reliable readout of the sufferer's pain level. These difficulties have not deterred a number of states including California and Iowa from designating chronic pain alone as a qualifying condition for their medical cannabis laws. Thus, for cannabis studies focused on pain relief, it's most informative where there are adequate and comprehensive outcome measures, including, for example, a reduction in opioid use, or a return to work, to accompany the patients' subjective pain reports. Knowing the underlying medical condition provoking the pain is also useful. Another frequent design problem in trials is the lack of long-term follow-up of study subjects, to assess both efficacy and the possible development of later-emerging problematic/adverse side effects such as tolerance, or the development of psychosis symptoms.

Physicians and nurses who conduct clinical trials for a living usually raise a number of other related questions. Some of these include asking whether the sample studied was representative of the target population. For example, did it capture the typical age range of individuals affected by a particular disorder, and the usual proportion of men and women affected by the disease? These clinicians are interested in whether the duration of the clinical trial was sufficiently long, and if the outcome measures were adequate to test the hypothesis. For example, if the experiment was testing effectiveness in treating ADHD, were subjects asked only if they felt that they could concentrate better, or were college grades tracked to validate the outcome? In some of the published cannabis clinical trials, it's hard to determine what the primary outcome actually was, because it's not clearly stated. Another issue is what standard the cannabis should be tested against. Rather than just active drug versus placebo, there may be alternative existing FDA-approved treatments available against which the cannabis can be tested. Cannabis doesn't always meet this challenge. For example, in the case of glaucoma, it's less effective and has more side effects than available existing medical treatments. Other issues that have complicated the head-to-head comparison of medical cannabis trials are that some studies have employed whole plant marijuana, while others have used individual constituents such as THC or CBG. Other trials have used a wide variety of administration routes and substances (e.g., smoked vs. oral versus vaporized THC, or oral mucosal sprays). The NAS report concluded that proper study of the long and short-term health outcomes of cannabis necessitated developing improved and better standardized study methods. For those readers interested in finding out more information regarding issues involved in designing quality clinical trials, they are listed at the following websites; GRADE working group [14] and BMJ Clinical Evidence [15].

Given these almost inevitable problems with many existing studies of medicinal cannabis, how can we weigh the evidence in the absence of proper clinical trials that frequently fail to meet FDA standards? The honest answer is that it's very difficult to come to any conclusion because of the conundrum of few conclusive clinical trials being feasible with an FDA schedule 1 substance that by definition has no accepted medical use. The only effective way forward will be to have the federal government reschedule cannabis and its constituents, and to encourage the flourishing of well-conducted clinical trials. Given all of these caveats, and not forgetting the many current barriers impeding current quality research, let's look at what pharmacologists and clinicians have discovered to date.

Whether cannabis use helps or hinders long-term symptoms in anxiety and mood disorders is hotly debated and complicated by the fact that there are non-straightforward relationships between dose and subjective effects. George Mammen and colleagues at Toronto's Center for Addiction and Mental Health carried out a systematic literature review that sifted through over 10,000 citations to yield 12 relevant studies. These latter had followed a total of almost 12,000 individuals, diagnosed with conditions including depression, panic disorder, bipolar disorder and PTSD for variable lengths of time, up to 5 years [16]. The bottom line was that across 11 of the 12 studies, cannabis use over the last 6 months was associated with the patients with these disorders having higher symptom levels over time, compared to a similar patient group that used little or no cannabis. In addition 10 of the studies also suggested that cannabis use was associated with less symptom improvement resulting from prescribed treatments such as medication or psychotherapy. Since the cannabis use in the subjects was self-reported and the cannabis itself was neither tested nor standardized in any way, these results are hardly definitive. But they do not suggest that cannabis has beneficial effects in these conditions. They also are consistent with prior studies reporting that reducing cannabis use is associated with improved mood in young women with depression [17]. Of course the best way to clarify these treatment issues is to put together a properly controlled trial of cannabis compared to placebo. Sue Sisley of the Scottsdale Research Institute has recently done just that, using NIDA-supplied marijuana flower samples in a sample of US veterans suffering from PTSD. She completed the study recently, but has not yet published her results.

Studies with cannabidiol

CBD has been widely covered in the news recently [18,19]. The most dramatic result was that well-conducted clinical trials demonstrated the effectiveness of CBD in treating two uncommon diagnoses leading to previously untreatable chronic, intractable and disabling pediatric epilepsy (Lennox-Gastaut and Dravet syndromes). In 2018, the FDA approved a CBD oral solution for the treatment of these two neurologic conditions and their accompanying symptoms of repeated seizures. Sometimes there are many of these events in the course of the day that are both frequently disabling to the affected epileptic children and exceptionally hard to treat with other approaches. The drug approved for use in these conditions in individuals aged 2 years or older, was Epidiolex (CBD), which had passed a standard FDA clinical trial. The medication caused side effects of sedation in some individuals, and appetite decreases, diarrhea, and sleep disturbance in others. The most serious unwanted effect noted was an increase in liver enzymes when the medication was administered at high doses. A fascinating account of CBD's effectiveness in treating these seizure conditions was the cover story in a 2019 New York Times Sunday Magazine article by Moises Velasquez-Manoff [19].

However, outside of this obviously effective application CBD is also marketed in multiple forms: tinctures, oils, sprays, lotions, gummies, etc. to treat multiple medical conditions and symptoms, including chronic pain, low libido, arthritis, inflammatory bowel disease, anxiety, peripheral neuropathy, and psoriasis. "It's a new kind of snake oil in the sense that there are a lot of claims and not so much evidence," said Dustin Lee, a Johns Hopkins psychiatrist [18]. Because the FDA classifies CBD as a drug (although no longer Schedule 1) they do not allow it to be sold in foods, drinks, or dietary supplements. Although many of us will have encountered establishments in multiple cities selling CBD-infused sodas, coffees, smoothies and candies, some of which provide doses, others not so much. CBD currently has a cachet because of its association with cannabis and its reputation as a cure-all for people and their pets that is essentially devoid of side effects. Industry analysts predict the market in the United States will reach 1 billion a year by 2020 [20]. These opinions have some basis in fact; CBD is less psychoactive/intoxicating than THC (but recall its measurable psychoactive effects in the Solowij experiments in Chapter 8). Unlike THC, CBD does not seem associated with dependence or withdrawal syndromes. And while it possesses fewer side effects than THC, that does not imply that it has none, as published in the results of the Epidiolex trial, above. Problematically, quality control is lax, since the substance is not federally regulated by the FDA, other than as a drug for one specific indication. Routine testing of samples of commercially available CBD oil has revealed widely different CBD concentrations, different levels of accompanying THC, and various contaminants, few of which were reported accurately on content labels. Also, despite Internet marketing implying that CBD products are legal to purchase, they are not, although currently there is little state or federal enforcement.

In addition to THC, CBD may hold promise as a pain-relieving and anti-inflammatory drug [21–24]. There is now considerable support from animal studies that CBD has pain-relieving properties, and strongly suggestive evidence of the same effects from the smallish number of human clinical trials that have measured CBD effects separately from those of THC. While both central and peripheral cannabinoid receptors appear to be implicated in THC's effects in animal models, repeated low-dose CBD treatment in rats seems to relieve pain mainly through activating $TRPV_1$ receptors [25].

We previously discussed the utility of CBD as a potential add-on treatment for schizophrenia in Chapter 8. Here conclusions as to its efficacy are mixed, with two well-conducted recent studies from different groups coming to opposite conclusions [26,27]. It's fair to say that this issue of CBD's usefulness in treating schizophrenia remains currently undecided. Overall, the many claims made for CBD as being useful in a wide variety of medical and psychiatric conditions run way ahead of the evidence. Hopefully that will change over the next few years as properly designed clinical trials clarify what conditions, under what circumstances, and at what doses, the drug is helpful.

What's the evidence for utility of medical cannabis in other conditions?

It's beyond the scope of this book to provide an exhaustive review of the massive, if generally preliminary literature concerning various conditions in which various cannabinoid compounds seem to show promise. The reader is referred to two recent excellent comprehensive reviews [12,28]. As I remarked earlier, it's striking that the conditions for which medical marijuana or its constituents are most frequently prescribed (insomnia, anxiety, pain, inflammation, nausea, and epilepsy) are almost precisely those to be found in historical pharmacopeias from Assyria to O'Shaughnessy. Even the use of medical marijuana to increase appetite (in recent years used to help address weight loss associated with HIV/AIDS) is hinted at in a 16th century book. This first mention of the "munchies" derives from Garcia da Orta, a Spaniard working for the Portuguese crown in 1563, who published the first European account of Indian hemp originating in Portugal's contemporary Indian colony of Goa. "Those of my servants who took it, unknown to me, said that it made them so as not to feel the work, to be very happy, and to have a craving for food." As is common in plant-derived medicines, cannabis also contains cannabinoids with opposite, appetite-diminishing effects, such as THCV.

A very brief summary of conclusions from the comprehensive reviews referred to at the beginning of this section, are that in addition to the medical illnesses that we have reviewed earlier, there is moderate or promising evidence for the efficacy of different forms of cannabis or its derivatives in several diagnoses. Conditions repeatedly implicated are, chemotherapy-related nausea and vomiting, weight loss associated with HIV/AIDS, neuropathic and inflammatory pain associated with multiple disorders, particularly the painful spasticity and neuropathic pain associated with multiple sclerosis. Evidence for efficacy in Tourette's syndrome, Crohn's disease, and ulcerative colitis is less, but suggestive. In many other conditions, evidence is either more flimsy, or encouraging but very preliminary, or confusing/contradictory. These include Parkinson's disease, PTSD, agitation in Alzheimer's disease, addictive states including opioid addiction, and anxiety, various cancers, and schizophrenia.

Currently available legal cannabinoid drugs and their indications

Some of these compounds are synthetic copies of THC, while others contain CBD or THC derived from cannabis plant extracts.

Nabiximols, marketed as Sativex, is a THC plus CBD oral spray, consisting of a mixture extracted from cannabis sativa plant material, marketed legally in 15 countries including Canada, Mexico and parts of Europe, for treatment of painful spasticity and neuropathic pain in multiple sclerosis. Pure CBD gel (Zynerba) is currently in FDA Phase I and II trials for several disorders. There are now several FDA-approved, THC-based medications. Dronabinol (marketed as Marinol) is a synthetic form of THC in an oily base, administered as capsules approved by the FDA as an appetite stimulant in AIDS wasting syndrome. It's also used for combating cancer chemotherapy-induced nausea and vomiting. Another form of the same drug is approved in the United States in the form of an oral solution, marketed as Syndros. Dronabinol is the drug that we administered to research subjects at Johns Hopkins in our time estimation study, referred to in Chapter 6. Nabilone (marketed as Cesamet and Canames) is another THC analog in capsule form for treatment of nausea in cancer chemotherapy patients, licensed in the United States. Epidiolex, that we discussed earlier is a purified, plant-derived, orally-administered CBD solution, currently FDA-approved in the United States for treatment of two forms of severe childhood epilepsy (Dravet syndrome and Lennox-Gastault syndrome). Finally, Bedrocan is a form of dried cannabis flower containing THC and CBD in different ratios for oral administration, marketed in Europe.

What are the arguments against using medical marijuana?

Outside of the ill-informed "there is no such thing as medical marijuana" argument, more reasonable concerns and objections subsume a variety of topics. Let us examine these one at a time. Lack of clear-cut data on cannabis' medical benefits is the issue most frequently raised by the medical community. As stated earlier this can obviously be solved by boosting the amount of good-quality medical cannabis research. We need to examine not only efficacy, but long-term usefulness, and the emergence of potential side effects that might accompany chronic use. Safety data still need to be collected to answer many questions, such as whether particular cannabis-based medications are safe to administer to children, or pregnant and lactating women. Second-hand smoke is an issue with tobacco smoking, but is little-studied in the realm of cannabis. If a pregnant woman is in a room, is it safe for her and her fetus if somebody is smoking medical cannabis in the vicinity? Diversion of the drug to non-medical users, especially to vulnerable teenagers whose brains may be at particular risk of psychiatric and cognitive morbidity such as psychosis or drops in IQ, is an issue frequently brought up by concerned legislators and physicians. The remedies here seem to lie as much with legislation as with medical practice. Physicians worry that significant potential health risks will accompany the use of therapeutic cannabis. These include the concern that tolerance will occur to the drug's beneficial effects, leading patients to escalate their doses, with consequent development of adverse effects such as delirium, anxiety, psychosis, amotivation as well as cognitive compromise, and increased potential for abuse and addiction. Since the dependence rate for individuals using marijuana daily is somewhere around 10%, the increased availability of medical marijuana would potentially create an unwanted and growing substance abuse problem. We know from PET brain imaging studies that repeated cannabis exposure is accompanied by down-regulation/desensitization of brain CB1 receptors in human subjects, similar to that seen for certain types of brain nicotine receptors in smokers that relates to tobacco craving.

Right now, there is no proper quality control in terms of standardized methods for analyzing and reporting percentages of cannabinoids and other content in medical marijuana, or ensuring an absence of pesticides, heavy metals etc. FDA based-federal legislation could effectively deal with this issue. Another set of objections comes from the public health sphere. There are worries that increased use of medical (as well as recreational) cannabis will result inevitably in increased numbers of marijuana-related motor vehicle accidents and fatalities.

Yet another issue involves therapeutic trade-off. For example, in multiple sclerosis, relief of painful spasms may be obtained at the cost of worsened dementia. This is certainly not an issue unique to medical treatment with cannabis, but merits consideration for a number of disorders. A related issue is one of harm reduction. If cannabis can help reduce opioid consumption (an issue discussed in Chapter 7), then the equation has to be weighed carefully, if we are potentially trading one addiction with a high morbidity and mortality rate, for a lesser, but still non-trivial second addiction. Part of what has to be balanced here is whether there are alternative, less harmful, alternative but effective non-cannabis treatments. Understanding the risk-to-benefit ratio is important here, so that well-informed medical care professionals can provide evidence-based guidance on these questions. A potential area for future research is to study interactions of medical marijuana with opioid analgesics, to examine issues of cross-tolerance, safety and the trade-off between benefits and harms.

A final issue is that improperly labeled or packaged medical cannabis products (e.g., THC-containing chocolate or gummies) will lead to inadvertent poisoning of children and household pets. An effective way to accomplish this would be to institute standardized warnings regarding common health risks of THC, for example, as part of standardized safety rules for packaging.

Recommendations

No US insurance companies currently cover medical cannabis except for limited numbers of synthetic cannabinoids and plant-derivatives listed earlier. The handful of currently available FDA-approved medicines are in general expensive, and covered by some insurance companies only for their listed FDA indications, usually on a case-by-case basis. This situation needs to change, which will likely happen only when the FDA approves more cannabis products for medical use. In turn, that will only occur when federal and state governments both encourage and fund medical cannabis research conducted under FDA clinical trials standards. The FDA will hopefully standardize analysis and formulation of medical cannabis through regulation and licensing of drug production and distribution. Deepak Cyril D' Souza at Yale has made the point that we would benefit from better understanding the physiological mechanisms underlying potential beneficial effects of marijuana and its constituents in particular medical conditions [29]. Better-quality standardized clinical trials may able to demonstrate that there are indeed specific physiologic pathways or mechanisms through which cannabinoids are acting, (e.g., via endocannabinoid-related mechanisms related to specific disorders) rather than merely providing non-specific subjective relief, in a manner similar to Valium, for example. In the United States there are no prescribed medicines administered through smoking. In the case of cannabis we need to improve drug delivery systems other than through the gut (the absorption of dronabinol is notoriously variable) or lungs (which are potentially irritated even by vaping). Sublingual and nasal sprays or tinctures seem to be an obvious alternative to explore. Harm minimization in general is an essential issue to study in more depth not only from the point of view of drug administration, but also from making determined attempts to reduce known cannabis side effects and complications. As has occurred with certain medications such as amphetamines, THC-containing medical marijuana may need to be accompanied by a list of patients for whom the drug is unsuitable, for example, those suffering from substance dependence, schizophrenia, or bipolar disorder. Currently, prescription monitoring databases exist for such drugs as opioids, to ensure that doctors are not over-prescribing in "pill mills" and that patients are not obtaining similar potentially abusable medicines from multiple physicians. There have been suggestions that similar provisions should be made for medical marijuana.

Ultimately, as has happened with many plant-derived medicines, it's likely that in the long-term, new, powerful and specific synthetic compounds that target the endocannabinoid system will replace those found in cannabis. Since such compounds can be patented by pharmaceutical companies, and are therefore potentially profitable, this is an area of active investigation by such manufacturers.

References

[1] Hill KP. Medical marijuana for treatment of chronic pain and other medical and psychiatric problems: a clinical review. JAMA 2015;313(24):2474–83.

[2] Russo EB. History of cannabis and its preparations in saga, science, and sobriquet. Chem Biodivers 2007;4(8):1614–48.

[3] Zias J, et al. Early medical use of cannabis. Nature 1993;363(6426):215.

[4] Clarke RC. Marijuana and medicine. Washington, D.C: National Academy of Sciences; 1999.

[5] Hawkins S. Pot never should have been illegal in the first place. USA Today. 2018. Available from: https://www.usatoday.com/story/opinion/2018/10/31/pot-never-should-have-been-illegal-editorials-debates/1838195002/.

[6] Devsinki O, Taylor M, Bailey M. Medical marijuana's 'Catch-22': limits on research hinder patient relief, 2018. Available from: https://www.npr.org/sections/health-shots/2018/04/07/600209754/medical-marijuanas-catch-22-limits-on-research-hinders-patient-relief.

[7] US Food and Drug Administration. Scientific data and information about products containing cannabis or cannabis-derived compounds; public hearing; request for comments; 2019. Available from: https://www.federalregister.gov/documents/2019/04/03/2019-06436/scientific-data-and-information-about-products-containing-cannabis-or-cannabis-derived-compounds.

[8] Ruder K. If you smoke pot, your anesthesiologist needs to know. NBC News. 2019. Available from: https://www.nbcnews.com/health/health-news/if-you-smoke-pot-your-anesthesiologist-needs-know-n1038191.

[9] Pardini C. Cannabis use leads to INR elevation in warfarin-treated patient; 2019. Available from: https://www.empr.com/home/news/cannabis-use-leads-to-inr-elevation-in-warfarin-treated-patient/.

[10] Ginsberg S. Doctors and patients are flying blind as medical marijuana use rises, research lags. USA Today. 2019. Available from: https://www.usatoday.com/story/opinion/2019/07/23/medical-marijuana-use-rises-so-should-its-research-column/1794459001/.

[11] Workforce Drug Testing Positivity Climbs to Highest Rate Since 2004, According to New Quest Diagnostics Analysis; 2019. Available from: https://www.questdiagnostics.com/home/physicians/health-trends/drug-testing.

[12] National Academies of Sciences, Engineering, and Medicine. The health effects of cannabis and cannabinoids: the current state of evidence and recommendations for research. Washington, DC: The National Academies Press; 2017.

[13] Schwabe AL, et al. Research grade marijuana supplied by the National Institute on Drug Abuse is genetically divergent from commercially available Cannabis. BioRxiv 2019.

[14] Schünemann H, et al., editors. GRADE handbook 2013. Available from: https://gdt.gradepro.org/app/handbook/handbook.html.

[15] Evidence at the Point of Care. British Medical Journal. 2019. Available from: https://bestpractice.bmj.com/info/evidence-information/.

[16] Mammen G, et al. Association of cannabis with long-term clinical symptoms in anxiety and mood disorders: a systematic review of prospective studies. J Clin Psychiatry 2018;79(4).

[17] Moitra E, Anderson BJ, Stein MD. Reductions in cannabis use are associated with mood improvement in female emerging adults. Depress Anxiety 2016;33(4):332–8.

[18] Rabin RC. CBD is everywhere, but scientists still don't know much about it NY Times. 2019. Available from: https://www.nytimes.com/2019/02/25/well/live/cbd-cannabidiol-marijuana-medical-treatment-therapy.html.

[19] Velasquez-Manoff M, Can CBD really do all that? The New York Times Magazine; 2019.

[20] Doheny K. As CBD oil flirts with mainstream, questions mount. Web MD. 2018. Available from: https://www.webmd.com/pain-management/news/20180605/as-cbd-oil-flirts-with-mainstream-questions-mount.

[21] Aviram J, Samuelly-Leichtag G. Efficacy of cannabis-based medicines for pain management: a systematic review and meta-analysis of randomized controlled trials. Pain Physician 2017;20(6):E755–96.

[22] Hill KP, et al. Cannabis and pain: a clinical review. Cannabis Cannabinoid Res 2017;2(1): 96–104.

[23] Serpell M, et al. A double-blind, randomized, placebo-controlled, parallel group study of THC/CBD spray in peripheral neuropathic pain treatment. Eur J Pain 2014;18(7):999–1012.

[24] Russo EB, Guy GW, Robson PJ. Cannabis, pain, and sleep: lessons from therapeutic clinical trials of Sativex, a cannabis-based medicine. Chem Biodivers 2007;4(8):1729–43.

[25] De Gregorio D, et al. Cannabidiol modulates serotonergic transmission and reverses both allodynia and anxiety-like behavior in a model of neuropathic pain. Pain 2019;160(1):136–50.

[26] McGuire P, et al. Cannabidiol (CBD) as an adjunctive therapy in schizophrenia: a multicenter randomized controlled trial. Am J Psychiatry 2018;175(3):225–31.

[27] Boggs DL, et al. The effects of cannabidiol (CBD) on cognition and symptoms in outpatients with chronic schizophrenia a randomized placebo controlled trial. Psychopharmacology (Berl) 2018;235(7):1923–32.

[28] Fraguas-Sanchez AI, Torres-Suarez AI. Medical use of cannabinoids. Drugs 2018;78(16): 1665–703.

[29] D'Souza DC, Ranganathan M. Medical marijuana: is the cart before the horse? JAMA 2015;313(24):2431–2.

Chapter 11

Economics

"Marijuana: a $75 billion market by 2030?"

Sean Williams. April 15, 2018, Motley Fool Investing. [3]

"Legalization should be limited to nonprofit production."

Jonathan Caulkins, Against a weed industry March 15, 2018, National Review. [17]

This chapter poses many inter-related questions.Who stands to benefit from sales following marijuana legalization? Tobacco and vaporizer companies? Nonalcoholic beverage retailers? How are they already getting into the game? Who loses? Drug dealers, cartels, owners of prison businesses, possibly liquor distillers, and retailers all risk being disadvantaged. Should national or local governments be in the business of selling? We will also take a brief trip into another aspect of cannabis capitalism; accoutrements for the well-off cannabis consumer, including "high-end" bongs, vaping systems, glassware, and electronics. One strong thread in this chapter will be serious concerns regarding teenage consumption of cannabis and efforts to encourage their use.

Beneath the ongoing debates regarding marijuana legalization, the real story revolves around cannabis commercialization. They call it the "dot-bong boom." At least according to Havocscope's global black market information website, worldwide, marijuana occupies fourth place in the list of the world's biggest illicit businesses, at almost $142 billion, outranked only by counterfeit drugs, prostitution, and counterfeit electronics [1]. From a financial perspective then, federal marijuana legalization in the United States is a big deal. How big is the United States cannabis market? Cannabis is a consumer product transitioning to a hip, trendy purchase that's moved from the counterculture to the culture counter. Free-market capitalism has cannabis as a commodity firmly in its sights. For North America as a whole, in 2017 legal sales jumped by one-third to almost 10 billion, with further annualized growth of nearly 30% through 2021, on track to reach nearly $25 billion by then [2]. Even more bullish investors, looking forward to the year 2030, predict a $75 billion market [3]. However, many believe that these estimates are bloated and overly-optimistic, blithely ignoring risks of oversupply and a likely significant fall in price per gram, as well as assuming that current patterns of consumption (e.g., flower versus edibles) will likely remain stable. According to a recent article in Forbes magazine, the US marijuana industry in May 2017 was expanding so quickly that were it legalized, it would

Weed Science. http://dx.doi.org/10.1016/B978-0-12-818174-4.00011-2

have total sales close to $45-$50 billion annually, greater than that of ice cream, or movie tickets and ironically an order of magnitude greater than US sales of snack items such as Doritos and Cheetos [4]. In 2016 legal cannabis sales alone in the United States totaled $4-$4.5 billion, still catching up on those for ice cream at $5.1 billion. Logically, Ben & Jerry's has proposed coming out with cannabis-infused ice cream, that would appear to be the ultimate self-selling product [5]. At our weekly laboratory meeting, suggested candidate flavor names included banana spliff, chunky junkie, cow-a-bonga, case the joint, and toasted Bud. In addition to over the counter sales, legal cannabis brings jobs. A report in Marijuana Business Daily [6] estimated that the cannabis sector employs between 165,000 and 230,000 full- and part-time workers, outnumbering bakers, massage therapists, and dental hygienists. The knock-on economic effect is to boost real estate prices, as cannabis businesses move into formerly vacant properties, attract tourism, and expand the local economy.

Where are things headed?

So, many businesses and would-be entrepreneurs, seeing the chance to make a profit from cannabis legalization want in on the act. Peter Bourne, Jimmy Carter's drug Czar, stated in an interview that as far back as the late 1970s, tobacco companies were already exploring the market in the event that marijuana was legalized. These large-scale commercial ventures will likely compete with small, local gourmet blends and so-called "micro-potteries," paralleling the phenomenon of craft beers going head-to-head with the commercial brewery industry. The major concern here is that commercialization will greatly increase cannabis use, as has happened with tobacco and alcohol. We will return to these issues a little later in this chapter.

Who may benefit financially from marijuana legalization, and who may potentially lose out?

State and federal entities are a major driver in terms of hoping that the tax revenue generated from legal marijuana sales will bail out their state budgetary deficits. New York State, for example, estimated that it would garner somewhere between $248 million and $678 million annually, depending on where they set the retail price of cannabis and the retail tax rate. They correctly noted that these estimates were very preliminary, based on uncertainties regarding pricing, consumption, effects of legalization on the unregulated market, and whether prices would push consumers in the direction of the legal or illegal marketplace. They also stated that "some states overestimated revenue initially, as they did not account for the length of time it takes for a recreational market to become established, leading to fewer than expected sales." Tobacco and soft drink companies are also pushing ahead with plans to invest in the cannabis industry.

One potential loser is the US alcohol industry. According to the International Wines and Spirits Record, a British-based alcohol beverage market analysis research group, "Consumers will continue to look to cannabis products over alcohol for occasions when they are feeling creative, need to get motivated, or seeking health, medical or wellness benefits." The report notes that "not every dollar spent on legal cannabis is a dollar taken from alcohol-it is much more complex than that ...but that there is a risk to alcohol due to legal cannabis, and the risk will be bigger as cannabis acceptance and consumption grows" [7]. Similarly, the Molson Coors 2018 annual shareholder report worried that "the emergence of legal cannabis in certain US states and Canada may result in a shift of discretionary income away from our products or change in consumer preferences away from beer." Daniel Rees, an economics professor at the University of Colorado, Denver, forecasted confidently that consumers would substitute marijuana for alcohol. But an examination of that state's tax revenue suggests that following legalization, alcohol purchases have remained pretty rock-steady [8].

Naturally, alcohol companies are taking preemptive action by embracing the cannabis sector. They would far rather be potential winners than losers. For example, Constellation Brands, owner of Corona beer and Robert Mondavi wine, invested $4 billion in Canopy Growth, the $10 billion publicly traded Canadian cannabis producer in 2018, as announced in the New York Times under the irresistible headline "This Bud's for You" [9]. Heineken has already begun marketing nonalcoholic, THC-containing Lagunitas brand sparkling water. And Aphria, the Canadian marijuana grow complex, recently recruited their new chief operating officer from Diageo Canada, the UK-based liquor conglomerate that produces Guinness beer and several Seagrams products. The top 12 Canadian cannabis companies are worth around $42 billion and reportedly pulling in investors hand-over-fist. One of the better known of these outfits is Tilray, based in British Columbia [10]. According to the Economist, "the industry is not particularly lucrative yet"; the 100 cannabis-related firms followed by Bloomberg lost $1.2 billion last year, (although some did make profits). Their overall revenues were a paltry $2.5 billion from a combined market value of $76 billion. The US Securities and Exchange Commission in 2018 warned investors about potential "investment fraud and market manipulation" with listed cannabis firms [11]. So while eager investors sniff around for the coming "green rush," profits are still relatively elusive. In the meantime, Diageo, Coca-Cola, and Altria (the former tobacco behemoth Philip Morris) were apparently expressing interest in potential deals with Canadian cannabis companies [11]. These investment shifts from big tobacco and big alcohol to cannabis are occurring on a global scale. Imperial Brands, marketer of Kool and Winston cigarettes and the world's fourth biggest cigarette producer, invested a rumored $10 million in a British cannabis research startup. Interestingly Snoop Dogg's investment firm also bought into the same enterprise, Oxford Cannabinoid Technologies. Imperial Brands had previously evinced interest by bringing on board a medical cannabis company chairman [12]. At the same time, alcohol companies are hedging their bets. The Intercept unearthed a campaign financing report indicating that the Beer Distributors PAC donated $25,000 to various anti-marijuana legislation initiatives in Massachusetts [8,13].

One set of potential losers is the class of individuals who have been sucked into the legal system for possession of small amounts of marijuana. In the future, such persons will no longer garner criminal records, suffer from family disruption, job loss, or deprivation of voting rights. But there is a strong movement to push for persons who've already suffered one or more of these losses to gain some sort of restitution or restorative justice in the form of employment in the expanding cannabis industry. Because of their past involvement with the substance, there is a certain argument for such people to be given a form of employment priority in expanding cannabis ventures. But currently because of their ethnic minority and low socioeconomic status, it is precisely these individuals and their communities who are most likely to be left behind. Massachusetts, for example, recently passed a law to train those jailed in the past for non-violent marijuana offenses for jobs in the now-legal cannabis industry, and to help them qualify for assistance in raising capital [14]. Several cities in California have also started similar so-called "equity programs."

Another potential loser is the illicit cannabis market extending from low-level growers and street dealers, all the way up the pyramid to drug barons and cartels. If cannabis is legalized, then the black market will shrink, but only to the extent that marijuana is not over-taxed in a way that encourages illicit purchases, for example, as happens with cigarettes in high tobacco tax states such as New York. In Colorado, a number of users continue to purchase from their street dealers because these are individuals with whom they have built up trusting, quasi-social relationships over the years. New York State's July 2018 assessment of marijuana regulation/legalization emphasizes the possibility of "a reduction in violent crime due to the substantial reduction in the unregulated market, which would lead to a decline in home invasions associated with marijuana and the associated violence."

A third obvious loser is the so-called prison industrial complex. Tens of thousands of individuals have been arrested for low-level marijuana possession offences that disproportionately affect poor minorities and people of color. Enforcing these laws costs taxpayers money, and individuals who are locked up often feed into for-profit private prison companies who contract out to government penal agencies. These private companies can profit from multiple aspects of incarceration, including construction, food supply, provision of medical care to inmates, hiring prison guards, and even providing probation services once inmates are released. Inmates themselves may constitute a cheap source of exploitable labor. Business models dictate that to maintain profitability, prisons need around 1000 beds, running at least a 90% occupancy rate [15]. Because of their nature, it is in the interest of these profit-driven enterprises to maintain or increase incarceration rates, and the companies have been often accused of lobbying legislators to pass stringent laws, and judges to hand down longer sentences in order to accomplish this goal. More than any other factor, the so-called "war on drugs" over the past 40 years has contributed to the expansion of the prison-industrial complex. Drug-offense convictions, including those for marijuana transgressions, have led to a majority of the US inmate population in federal prisons [16]. A majority of those inmates are black or Latino; in Massachusetts in 2013, 75% of inmates in prison for mandatory drug sentences fitted that description, but constituted a mere 22% of the state's population [14].

What is likely to happen to cannabis prices over time?

Because of economies of scale related to industrial-scale production, costs to the consumer will likely come down. As more large-scale producers get into the act, cannabis will be cheaper to grow, process, and distribute. This trend already occurred in Washington state, where after-tax cannabis prices fell by 70% between August 2014 and August 2017 [17]. Meanwhile, in Ontario Canada, FSD Pharma has taken over an extensive former Kraft plant, with the aim of building the world's largest hydroponic indoor cannabis production and processing facility. By the year 2025, the company aims to produce 400,000 kg of dried cannabis annually and to turn it into cannabis products. In 2019, it acquired a US-based research and development pharmaceutical company that will focus on developing synthetic compounds for therapeutic use that influence the endocannabinoid system. Jonathan Caulkins and his co-authors examine the pricing question in detail in their comprehensive book *Marijuana Legalization* [18].

Is youth marketing of cannabis in our future?

We've seen that the alcohol and tobacco industries have an avid interest in moving into cannabis sales, and if there's one thing in which both tobacco and alcohol companies excel, it's targeting young people. The story of youth cigarette marketing through such cartoon characters as Joe Camel is well known. Perhaps less recognized is the alcohol beverage industries targeting of young people. The alcohol industry spends more than $2 billion annually promoting their products. Research carried out by the Center on Alcohol Marketing and Youth, at Johns Hopkins School of Public Health, found that magazine ads in publications with substantial youth readership tended to feature ads encouraging alcohol overconsumption, with almost 25% of them containing sexual connotations or sexual objectification [19]. Youth are targeted by alcohol ads aimed at their age demographic, and as multiple studies have determined, the more young people are exposed to alcohol advertising and marketing, the more likely they are to drink. Youth exposure to alcohol advertising is also associated with starting underage drinking, drinking more alcohol and adverse health and social problems. Between 2017 and 2018, adolescents under the legal drinking age were exposed 28.5 billion times to alcohol ads on cable TV [20]. Sweet, fizzy alcoholic beverages, such as Alcopops are generally based around a malt beverage-fruit juice flavored mixture. They are often sold in relatively inexpensive, large, brightly-colored cans. Consuming one supersized Alcopop meets the criterion for binge drinking in terms of the alcohol content. Although originally designed to attract millennials who disdain traditional alcohol tastes, popular brands such as Twisted Tea, Sparks, Seagram's Escapes, and Strawberry Rum Job, are youth magnets.

Given the earlier, what might happen in the worst of all possible worlds? Teens, at least in trend-setting Colorado, are already moving toward greater use of vaporizing, edible use and dabbing [21]. If you were to ask me about my biggest concern regarding cannabis sales in the future, it's this scenario. Recreational cannabis is legalized at a federal level. Soon thereafter, high-THC, low-CBD extracts with attractive flavorings aimed at teens, are marketed for use in small, pocket-sized, inconspicuous e-cigarettes, ostensibly to adult consumers. These products then rapidly disseminate into an enormous teenage marketplace. Teen cannabis use multiplies quickly, with short- and long-term adverse health consequences. But before leaping headlong into alarmism over this issue, let's take a brief but instructive diversion into seeing what has played out in the world of e-cigarettes, officially known as electronic nicotine delivery systems (ENDS), since these vaping devices were first imported into the United States in 2006.

Alongside the battery and heating element, e-cigarettes contain a replaceable pod of liquids (e-liquids) that are a mixture of nicotine, flavorings, and moisture-preserving chemicals such as vegetable glycerin. Although these devices may help some adult tobacco cigarette smokers cut their consumption or quit smoking, for youth and young adult users, ENDS encourage both nicotine dependence and the risk of transitioning to regular combustible tobacco cigarettes. When the latter happens, prior e-cigarette use is associated with greater frequency and intensity of subsequent cigarette smoking, according to the National Academies of Sciences, Engineering and Medicine 2018 report on the public health consequences of e-cigarettes [22]. What's indisputable is that the number of teens vaping nicotine skyrocketed, and e-cigarettes became wildly popular among young people. Juul, resembling a flash drive, is the most well-known brand. Because it is relatively cheap, inconspicuous, and effective, it is way out in front in terms of brand popularity among ENDS, enjoying a 70% plus market share. Juul use increased nearly 80% among high school students, and nearly 50% among middle school students, with 3.6 million youths reporting that they used e-cigarettes in 2018, according to the 2018 National Tobacco Youth Survey. Although in that year, the US FDA banned the gas station and retail store-based sales of most flavored e-cigarette cartridges, they are still available for online purchase. There are plans to raise the minimum age for buying these and other tobacco products to 21 years [23]. The FDA has warned device manufacturers to stop marketing to teens or be banned. "Tobacco companies have fought cutting flavors from e-cigarettes, saying they are not aimed at youths but at adults who need them as a way to transition from tobacco cigarettes. But health advocates point to the packaging and youth appeal of a variety of flavors, including chicken and waffles, rocket Popsicle and unicorn milk as well as fruity tastes like mango creme." In the world of nicotine, e-cigarettes are seen as "different" than traditional tobacco cigarettes. Users are less likely to equate inhaling nicotine from a personal vaporizer with smoking, and the equivalent amount of the substance from cigarettes. Smoking a pack a day is intuitive; inhaling from a device is harder to quantify and keep track of. Major tobacco companies view e-cigarettes "as critical to their survival now that smoking rates have declined to their lowest levels in the United States" [24,25]. In 2019 Altria, maker of Marlboro, and at $100 billion market value, the US's biggest tobacco company, purchased a "35% stake in Juul, whose valuation soared on the investment to $38 billion" Altria spent over $10 million on lobbying in 2018, while Juul spent $1.6 million [25].

As time goes by, cannabis will likely be available for consumption in more forms that are cheaper and more convenient to use. Cannabinoid-containing oils for use in personal vaporizers are likely to become extremely popular, copying nicotine e-cigarettes. Such products are already available at dispensaries in many states that have legalized marijuana. Since they are highly effective as a means of delivering cannabinoids, easy to conceal and convenient to carry around, they are likely to be adopted increasingly. It's the logical next step that I believe should definitely be avoided, as mentioned a little earlier. If the pods for "e-joints" also contain teen-attracting flavors such as mango creme, even if officially marketed to adult users, they will predictably become immensely popular among teenagers. If the price is kept low, this will be a super-desirable product for teens. Other related, possible nasty combinations to be equally shunned would include "e-spliffs" containing super-addictive nicotine/THC combinations and teen-targeted "e-joint" marketing campaigns featuring cartoon characters. Joe Camel, meet Marley the addictive and harm-maximizing Rastafarian magpie. But recent (2019) outbreaks of serious lung illnesses and deaths related to vaping products threaten the burgeoning vaping/e-cigarette market. This story is evolving rapidly even as I write this chapter.

Smart Approaches to Marijuana (SAM)

In the context of all the earlier-mentioned developments, (SAM Inc/Project SAM) [6] is worth a mention. The organization was co-founded in 2013 by former Congressman Patrick Kennedy (D-Rhode Island) and Kevin Sabet. Kennedy, the youngest son of the late Sen. Teddy Kennedy, has been frank regarding both his prior abuse/addiction to various substances (cocaine, alcohol, prescription opioids) and his diagnosis of bipolar disorder. He has been a consistent supporter of federal legislation designed to improve addiction and mental health care delivery and helped found the One Mind research initiative to promote neuroscience discovery impacting mental health issues [26]. (Patrick is a different person than his cousin Robert F Kennedy Jr. the environmental law specialist, also a health activist, but one who has spread misinformation about childhood vaccines and their association with autism through the organization Children's Health Defense, which he chairs). Patrick Kennedy's SAM co-founder is Kevin Sabet, a long-standing anti-drug crusader with a particular and long-standing focus on marijuana. Sabet is a former 3-time advisor to the White House Office of National Drug Control Policy, and like myself on Yale's faculty. When it comes to cannabis, he is effectively a prohibitionist. (See Joe Dolce's interview with Sabet in the former's book *Brave New Weed* [27]). SAM's abiding concern when it comes to cannabis legalization is that America is in the process of creating a new big tobacco scenario. SAM now encompasses a multi-state lobbying group with a fair amount of clout that inserts itself into many statewide cannabis legalization initiatives. The group is well-funded, and uses effectively-organized public information campaigns that bring in national figures to testify before legislatures in any state considering legalizing recreational marijuana. Given the complex interrelationships between corporate lobbyists, politicians, political action committees, and one-off state initiatives, it's important to know where SAM's funding derives from, but this information is hard to ferret out.

SAM generally supports cannabis decriminalization, but not legalization, and is concerned (some would say overly-concerned) about public health repercussions of marijuana legalization. It generally encourages both medical marijuana research and the development of FDA-approved cannabis-based medicines. One of SAM's main aims is to counter the "predatory" pot industry [28], which as we've learned is indeed becoming increasingly entangled with big tobacco and big alcohol. I share their concern on this major topic. Clear lessons have emerged from the marketing strategies of alcohol and tobacco conglomerates. Commercial marketers for profitable consumable products such as cigarettes and booze push hard to set prices low in order to boost sales, and rely on continued consumption by the minority of frequent, (i.e., daily) users who purchase the overwhelming majority of the product. Or as Mark A R Kleiman phrases it, the cannabis "industry will do everything in its power to create and sustain the biggest possible population of chronic stoners" [29].

SAM ostensibly promotes the view that our national and international drug policies should be founded on science, a principle that chimes with the intent of this book. But the organization concludes from their own examination of the available scientific evidence that marijuana should remain illegal. This is not my judgment. I find their overall position on marijuana logically untenable. Their policy reminds me of the old joke about your uncle the cop, who after sharing a joint with you then promptly arrests you for drug use. Karen O'Keefe, director of Connecticut state policies for the Marijuana Policy Project believes that SAM plays fast and loose with facts. For example, she states that their testimony during a recent (June 2019) failed legislative initiative to legalize recreational marijuana in Connecticut was misleading. "Including inaccurately claiming that marijuana use went up among teens in Colorado (after legalization in 2014), when it actually went down, and falsely claiming the graduation rates went down, when they went up" [30]. So let's take a look at SAM's policy positions. I would summarize these as anti-legalization, partial decriminalization, opposition to harm reduction policies, and skepticism toward initiatives to tax and regulate marijuana. A careful reading of SAM's policy and press statements, website and numerous related publications, leads me to think that implementation of their main agenda would translate into only minimal change to the status quo. SAM states (I believe quite reasonably) that they oppose enriching a small number of pro-cannabis business people. But they have little to say regarding the immense amounts of money already being poured into the war on drugs by federal and local governments, including policing and incarceration expenses that are ultimately taxpayer-financed. SAM also states that they are extremely concerned that cannabis legalization will lead to a rapid wave of high schoolers and teenagers becoming cannabis dependent. But as we saw in Chapter 7, such rates have stayed rock-steady in those states that legalize recreational cannabis. Outside of science-based issues, Sabet relies on cultural arguments. For example, he states that "Alcohol, unlike illegal drugs, has a long history of widespread, accepted use in our society, dating back to before biblical times. Illegal drugs cannot claim such pervasive use by a large part of the planet's population over such a long period of time" [31,32]. But that view is contrary to the historical facts I summarize elsewhere, that illustrate that societies in India and the Islamic worlds learned to accommodate successfully large-scale public consumption of cannabis in various forms. Besides, Americans are ever-eager to adopt a panoply of desirable foreign customs ranging from eating quinoa and goji berries to driving compact vehicles.

It's all about marketing

Speaking of those gaining from legal cannabis sales, in the world of marijuana-themed product marketing, the Higher Standards store in New York City is an example of a non-dispensary purveyor of literally high-end accessories for well-heeled, stylish cannabis users. This establishment is several steps up from the usual head shop in terms of atmosphere, decor and price. As Katherine Bernard stated in her New York Times article [33] "You can buy cannabis accessories in a room with exposed brick and industrial-inspired shelving, just like at a place where you can buy all of your other class signifiers.... If high is highbrow, then cannabis can be consumed tastefully and stylishly, like cupcakes or wine". Elsewhere in the United States, skilled glassblowers have produced high-end cannabis-related objects embodying beauty and fine craftsmanship that can cost you $100,000 or more. To check out some examples of super-expensive bubblers, dab-rigs, and bongs, visit this website [34]. Meanwhile, at the inexpensive and illicit end of the market, the US DEA in Oakland and Emeryville California succeeded in inducing several individuals to plead guilty for their role in manufacturing and distributing marijuana-infused products that mimicked familiar candy bars in slightly disguised form, with titles such as "Puff-a Mint Patty," "Twixed," "Pot Tarts," and "KeefKat".

What are some of the procedural hurdles to be overcome?

How do we tax legalized cannabis? Any chosen strategy evokes a likely consumer counter-move. If tax rates are set too high then this pushes consumers back to the illegal market. This has likely occurred in California, where almost three-quarters of cannabis purchased is still sold illicitly [35]. But if rates are set too low then this encourages over-use with its concomitant risks. Although states that have legalized marijuana had originally hoped for very significant revenue boosts, there was a general failure to predict just how steeply prices would fall. One proposed strategy is to tax marijuana by weight [36]. California does this; Maine began this policy in 2018. The obvious consumer counter-strategy is to purchase concentrates, to get more bang for their buck. Thus states will need to either set a limit on THC potency, or to figure out a complex formula to tax by the amount of THC being sold. For example, back in 2013 the Netherlands reclassified hyper-potent cannabis products with a THC of 15% or more under a legally restricted drug schedule. Taxing in this manner correlates better with level of intoxication, similar to the system used with alcohol. The consumer counter-reaction to similar restrictions in the United States, would likely to encourage illicit production and sale of butane hash oil and similar concentrates. A related issue is where all types of cannabis products are available for legal sale, how can taxation be somehow made both equivalent and fair across edibles, concentrates, flower, and tinctures. One possibility is to have a simple retail price-based tax, although states are wary of this because it is potentially a more unstable revenue source. States seeking income through taxation also worry about where to set the limits on cannabis plants grown by individuals, which by definition remain untaxed.

Banking and money-laundering is another thorny issue. Under the 1986 Money Laundering Control Act (MLCA), banks are prohibited from providing financial services to any business engaged in illicit activities. Because at a federal level, marijuana is still illegal, cannabis sales even in individual states that have legalized it, are by definition illicit under the Controlled Substances Act. In July 2019, the US Senate Committee on Banking, Housing and Urban Affairs held a hearing to review a bill, the Secure and Fair Enforcement (SAFE) Banking Act. If enacted, this legislation would allow federally supervised financial institutions such as banks to have cannabis-related businesses as customers. According to Ben Curren's article in Forbes magazine, SAFE would protect banks from federal prosecution as long as the cannabis businesses they work with comply with the laws in the states where they operate [37]. SAFE was proposed in order to "create protections for depository institutions that provide financial services to cannabis-related legitimate businesses, and service providers for such businesses." The major purpose of this Act is to increase public safety by ensuring access to financial services to cannabis-related legitimate businesses and service providers, and reducing the amount of cash at such businesses. According to the Washington Post, cannabis dispensaries have dropped off duffel bags and suitcases full of cash to pay their taxes, sometimes driving hundreds of miles to do so [38]. Currently, the cash-only nature of dispensary transactions makes them potential robbery magnets. The worlds of legal cannabis retailing and those of blockchains/cryptocurrencies are thus fated currently to buddy up to one another. Another issue is the following. Because cannabis is currently a Schedule 1 drug, it falls under section 280E of the US tax code. Businesses that sell such substances are not allowed to take the normal corporate income tax deductions, except for the cost of the goods. This results in extremely high tax rates for profitable businesses. Those dollars accrue to the federal government, and this revenue source would shrivel up, were cannabis to be legalized federally. The government may be disinclined to lose this income stream unless it can find an alternative. How the current tax revenue would be offset by federal marijuana legalization leading to a larger market, and thus more sales at a lower federal tax rate, is not altogether clear.

What are some alternative possibilities?

Several cannabis-concerned economic thinkers have pushed out-of-the-box ideas when it comes to marijuana production and sales. Nonprofit production is one. This is not just a pipe dream of ageing hippies, but has been mooted as a serious consideration by Jonathan Caulkins of the RAND corporation [17]. He argues that the debates over cannabis legalization immediately narrow the focus to bypass numerous viable but usually unconsidered options contained within the legalization umbrella. These range along a wide spectrum between small-scale personal cultivation and large-scale commercial sales. Caulkins proposes several novel ideas that would gradually phase out for-profit businesses. These could be replaced by nonprofit organizations whose boards would aim to protect public health and constrain demand to current levels. Optionally, cooperatives could supply only their own registered members. Another of his considerations revolves around the possibilities that might occur if we treated marijuana like alcohol [18]. This chapter attempted to provide a 20,000-foot view of the snarled and tangled universe of commercial cannabis. Resolving these issues will be one essential part of the future of federally legalized cannabis. How this future may shape up is the theme of the next chapter.

References

[1] Marijuana facts. Available from: https://www.havocscope.com/tag/marijuana/.
[2] Williams S. Nearly 37 million Americans used marijuana in 2017; Motley Fool. 2018. Available from: https://www.fool.com/investing/2018/09/01/nearly-37-million-americans-used-marijuana-in-2017.aspx.
[3] Williams S. Marijuana: a $75 billion market by 2030?; Motley Fool. 2018. Available from: https://www.fool.com/investing/2018/04/15/marijuana-a-75-billion-market-by-2030.aspx.
[4] Borchardt D. Report: total marijuana demand tops ice cream in U.S; Forbes. 2017. Available from: https://www.forbes.com/sites/debraborchardt/2017/05/17/new-report-says-total-marijuana-demand-tops-ice-cream/.

[5] Ben & Jerry's founders open to MJ infused ice cream; MJ Biz Daily. 2015. Available from: https://mjbizdaily.com/ben-jerrys-founders-open-to-mj-infused-ice-cream/.

[6] Available from: https://learnaboutsam.org/.

[7] Waterworth S. IWSR cannabis strategic study hones in on beverage alcohol, cannabis and the changing US consumer. This is the first report to look at the impact of cannabis on alcohol through the lens of beer, wine and spirits consumers, in collaboration with BDS Analytics; 2019. Available from: https://www.theiwsr.com/cannabis-strategic-study/.

[8] Fang L. Alcohol industry bankrolls fight against legal pot in battle of the buzz. The Intercept. 2016; Available from: https://theintercept.com/2016/09/14/beer-pot-ballot/.

[9] Merced MJ. This bud's for you: What Corona owner's $4 billion bet on a marijuana firm says about pot's future. NY Times. 2018. Available from: https://www.nytimes.com/2018/08/16/business/dealbook/constellation-canopy-cannabis.html.

[10] Austen I. Marijuana legalization in Canada has companies chasing a green rush; 2018. Available from: https://www.nytimes.com/2018/10/16/world/canada/cannabis-legalization-industry.html.

[11] Cannabis stocks go ever higher. The Economist. 2018. Available from: https://www.economist.com/business/2018/10/20/cannabis-stocks-go-ever-higher.

[12] Tomoski M. Big tobacco company imperial brands gets into cannabis. Imperial Brands. 2018. Available from: https://herb.co/news/industry/imperial-brands-tobacco-cannabis-pharmacie-lo-oxford/.

[13] Hines N. These are the alcohol companies that want to keep weed illegal; 2016. Available from: https://vinepair.com/articles/alcohol-companies-that-dont-like-weed/.

[14] Wood J. Second chances: how ex-convicts are lighting up the cannabis industry; 2018. Available from: Second chances: how ex-convicts are lighting up the cannabis industry.

[15] Taylor R. War on drugs: how private prisons are using the drug war to generate more inmates; 2012. Available from: https://www.mic.com/articles/20186/war-on-drugs-how-private-prisons-are-using-the-drug-war-to-generate-more-inmates.

[16] Whitehead JW. Jailing Americans for profit: the rise of the prison industrial complex. 2012 Huffington Post. Available from: https://www.huffpost.com/entry/prison-privatization_b_1414467.

[17] Caulkins J. Against a weed industry. National Review; 2018.

[18] Caulkins JP, Kilmer B, Kleiman MAR. Marijuana Legalization: What Everyone Needs to Know. 2nd ed. Oxford University Press, Oxford, United Kingdom; 2016.

[19] Mulvey J. How alcohol ads target kids. Business News Daily. 2012. Available from: https://www.businessnewsdaily.com/2989-alcohol-companies-target-youth.html.

[20] Henehan ER, et al. Alcohol advertising compliance on cable television, July-December (Q3-Q4), 2018. Baltimore, MD: Johns Hopkins Bloomberg School of Public Health; 2019.

[21] Tormohlen KN, et al. Changes in prevalence of marijuana consumption modes among colorado high school students from 2015 to 2017. JAMA Pediatr 2019;173(10):988–9.

[22] Prochaska JJ. The public health consequences of e-cigarettes: a review by the National Academies of Sciences. A call for more research, a need for regulatory action. Addiction 2019;114(4):587–9.

[23] E-cigarettes Threaten the "tobacco end game". Available from: https://www.heart.org/en/healthy-living/healthy-lifestyle/quit-smoking-tobacco/tobacco-endgame.

[24] Kaplan S. F.D.A. Plans to ban most flavored e-cigarette sales in stores. NY Times. 2018; Available from: https://www.nytimes.com/2018/11/08/health/vaping-ecigarettes-fda.html.

[25] Kaplan S, Richtel M. Tobacco and e-cigarette lobbyists circle as F.D.A. Chief Exits. NY Times. 2019. Available from: https://www.nytimes.com/2019/03/15/health/tobacco-e-cigarettes-lobbying-fda.html.

[26] Schulzke E. Bipolar and addicted, Patrick Kennedy embodies mental health challenges; 2013. Available from: https://www.deseret.com/2013/2/18/20448423/bipolar-and-addicted-patrick-kennedy-embodies-mental-health-challenges#former-rep-patrick-kennedy-on-capitol-hill-in-washington-before-leaving-office-when-his-term-ended-in-2011.

[27] Dolce J. Brave New Weed. New York, NY: Harper Collins; 2016.

[28] Vlahos KB. Cannabis goes corporate. The American Conservative. 2014. Available from: https://www.theamericanconservative.com/articles/fear-the-rise-of-big-pot/.

[29] Kleiman MAR. How to prevent casual pot smokers from slipping into abuse and dependence. Vox. 2018. Available from: https://www.vox.com/the-big-idea/2018/4/20/17259032/marijuana-abuse-dependency-risk-policy-420-drug-addiction.

[30] Hardman R. Why did the recreational marijuana effort fail in the conn. general assembly? 2019. Available from: https://www.wnpr.org/post/why-did-recreational-marijuana-effort-fail-conn-general-assembly.

[31] Seitz-Wald A. Meet the quarterback of the new anti-drug movement. Salon. 2013. Available from: https://www.salon.com/2013/02/13/meet_the_quarterback_of_the_new_anti_drug_movement/.

[32] Gwynne K. Legalization's biggest enemies. Rolling Stone. 2013. Available from: https://www.rollingstone.com/politics/politics-news/legalizations-biggest-enemies-109821/.

[33] Bernard K. At an updated head shop, high meets highbrow. NY Times. 2018. Available from: https://www.nytimes.com/2018/08/15/style/cannabis-accessories-higher-standards.html.

[34] Benjamin D. The most expensive glass pieces in the world. Wikileaf. 2018. Available from: https://www.wikileaf.com/thestash/the-most-expensive-glass-pieces/.

[35] Williams S. California's cannabis black market is insanely larger than its legal market. Motley Fool. 2019. Available from: https://www.fool.com/investing/2019/09/14/californias-cannabis-black-market-is-insanely-larg.aspx.

[36] Humphreys K. Marijuana is getting cheaper. For some states, that's a problem. Washington Post. 2018. Available from: https://www.washingtonpost.com/business/2018/11/16/marijuana-is-getting-cheaper-some-states-thats-problem/.

[37] Curren B. SAFE act passage would mean safe cannabis banking for all. Forbes. 2019. Available from: https://www.forbes.com/sites/bencurren/2019/08/01/safe-act-passage-would-mean-safe-cannabis-banking-for-all/#1160f9e54010.

[38] Merle R. Banks want a hit of the marijuana business. Will they get to partake? Washington Post. 2019. Available from: https://www.washingtonpost.com/business/2019/02/13/banks-want-hit-marijuana-business-will-they-get-partake/.

Chapter 12

Summary

*"Very few drugs, if any, have such a tangled history as a medicine. In fact, preju-
dice, superstition, emotionalism, and even ideology have managed to lead cannabis to ups and downs concerning both its therapeutic properties and its toxicologic
and dependence-inducing effects."*

Carlini, E.A., 2004. [1].

It should be clear from the prior chapters, that legalization of recreational cannabis is likely on the horizon in the United States. Exactly what such a future
will look like is speculative, but given the scientific information that we have
reviewed, some of the key problems that will follow, as well as benefits, are
predictable. Science raises other important questions as well as suggesting appropriate answers. Informed by available research, how can we best formulate
plans and policies to minimize these potential harms and to make cannabis use
as safe as possible? If education and drug treatment are included in these policies, who will pay for them? What is the future of medical marijuana? Where
will our cannabis come from down the road? Will it be grown in greenhouses,
hydroponic facilities, agricultural fields, or tanks of yeast? This chapter enter-
tains some of these questions and ways to address them.

Cannabis is the most commonly used illicit drug in the world. As a recreational substance it is particularly favored among adolescents. If you have read
this far, you will understand that the drug offers unique subjective experiences
to its users from a recreational point of view, and that it may also have important
medicinal properties. However, like all drugs it is associated with known undesirable side-effects and harms. It has evolved from being a pariah drug to one that
two-thirds of the US adult population thinks should be legalized for recreational
use. Support for marijuana legalization currently outranks that for gay marriage.
Talking with legislators, physicians and business people, the message I hear is
the same. Federal legalization is a "runaway freight train," the "horse has already
left the barn," it's "too late to turn back the clock," and "it's inevitable now,"

So the United States is more than likely to embark on a significant social
experiment akin to the repeal of alcohol prohibition. One consequence of legalization is inescapable. There will be more users and that translates inevitably to
more problems. Simple statistics tell us that even if the percentage of individuals experiencing harm from cannabis is small, if their overall number is very
large, then many people are affected. So, for the sake of argument if 10% of

Weed Science. http://dx.doi.org/10.1016/B978-0-12-818174-4.00012-4

cannabis users become dependent, then 10% of 120 million people (i.e., 50% of the number of the US population over the age of 21), is a huge number, if each one of these individuals chose to get high regularly. Similarly the more people who use cannabis, then the more cases of psychosis and more intoxicated driving incidents will occur. The problems though, will occur most often not in the vast majority of moderate, or occasional cannabis users, but in the approximately 20% who will use frequently, if not daily. This minority consumes large amounts of the drug, and will tend to favor high-THC-containing forms of cannabis, because these products get you higher, faster, and cheaper. Jonathan Caulkins estimates that the 20% of users who consume cannabis products several times every day, actually account for about 80% of the marijuana consumed [2]. Alcohol beverage consumption follows a very similar pattern. That is one of numerous moderate users, but with around 20% of really heavy drinkers who consume a similar, significant proportion of alcohol sold. One little-asked question is what the quality of life is like for individuals who smoke cannabis every day on multiple occasions. Currently, there are over 8 million of them. In a statistical profile, up to 31% of the adult habitual users are more likely to have a high school education or less, and to have some level of impairment in work, school, and relationships [3]. There is also an association of heavy daily use with lower income, unemployment, and decreased life satisfaction [4,5].

In planning for this future, to paraphrase JG Ballard, what the legislators of today decide, society will live with tomorrow. Science, with its dependence on facts over emotional appeals should help guide this process. But science-based reports can be easily buried, as was the fate of the La Guardia Committee's advice in 1944 (by Harry Anslinger, Commissioner of the US Treasury Department's Federal Bureau of Narcotics), and the Shafer commission's recommendations in 1972 (by Richard Nixon). When the scientific conclusions are unacceptable to those in power, then the messenger is, if not literally shot, then consigned to some form of limbo along with their report. Scientists need to plan ahead to avoid being similarly silenced.

According to Susan Weiss of the National Institute on Drug Abuse, scientific knowledge and discovery can influence public policy regarding cannabis in several important ways [6]. To get this advice across to decision makers most effectively, a few key considerations always need to be kept in mind, she states.

- Public health interests must be primary when making policy decisions, not monetary ones.
- The scientific knowledge that we have available right now needs to be better disseminated; there are too many myths and too much misinformation about cannabis that need countering by available facts.
- We need to build on our existing scientific knowledge base through further research.
- Since marijuana legalization is such a polarized issue, scientists need to be clear to the public, both about what we know, and equally what we don't know.

- The public also needs to hear what's understood about real-world questions regarding cannabis, such as addiction, teen exposure, and use during pregnancy. In other words, not a general "pot is bad" message.
- In speaking about marijuana, when dealing with issues such as limiting drug potency or prohibiting certain products, it's super-important to use the correct terminology, for example, distinguishing plant marijuana from synthetics such as K2, and from concentrates such as dabs.

So, in speaking with decision-makers what are some of the salient facts requiring emphasis? To reprise, cannabis is both a medication and an intoxicant, along with a spectrum of such dual-purpose substances, which also includes opioids and cocaine. Cannabis though is associated with more moderate harms than those drugs when used recreationally, and it is gaining in popularity, such that some 37 million Americans report using it in the past year. Additionally, a majority of the US population now favors legalization. We know that cannabis, although definitely a less harmful recreational substance than either tobacco or alcohol, is now stronger in terms of THC content than the drug familiar to baby boomers, and thus carries more potential risk. The point of recreational marijuana legalization, though, is not to replace alcohol or tobacco with cannabis, but to introduce an additional intoxicant that requires judging on its own terms. In legalizing any new recreational substance, an essential consideration, which we will discuss in this chapter, is how to most effectively minimize the drug's associated harms. Many legislators do not like discussing harm reduction, but such a conversation is essential. Substance legalization always represents the devil's bargain. Recall that the US elected (after a brief, failed experiment at Prohibition) to legalize beverage alcohol, a much more harmful drug than cannabis, as I reviewed in Chapter 8. This proved to be a popular choice. According to the 2016 National Survey on Drug Use and Health, 51% of those aged 12 and older drank alcohol in the prior 30 days. We do not automatically call such moderate drinkers "alcoholics." As with alcohol, the vast majority of cannabis users indulge moderately and suffer no ill effects. Nevertheless, for cannabis there are several major concerns, and because of the ongoing opioid crisis, US policymakers are concerned to avoid unleashing new drug related problems. The first concern is that exactly as has happened with alcohol, underage use will become very common, especially on college campuses. Diversion of cannabis from dispensaries to teenagers, with their vulnerable brains, is an especially concerning related issue that will prove challenging (but not impossible) to prevent.

Marijuana is now available in multiple dosing forms, including the increasingly used edibles and vaporizing devices that were either virtually unknown or much less common in the United States even a decade ago. Although, as we've seen, edibles have been used in Indian festivals such as Holi for thousands of years. Irrespective of how it's administered, the drug has risks, and yet another concern is that of cannabis abuse and dependence. Around 10% of continuing users will develop dependence, (although physical withdrawal is relatively milder than with many other recreational drugs). A major worry is the drug's association with psychotic illnesses, including schizophrenia (a risk that rises with how often cannabis is used, how much THC it contains, and probably how young the user is when they begin). Finally, the drug impairs academic performance, (as we saw in the college students of the BARCS and the Dutch research studies), and is associated with intoxicated driving-related accidents, yet likely fewer than with alcohol. To a greater or lesser extent, all of these concerns are justified, as detailed in prior chapters.

Before we dive into harm reduction in detail, I think it's useful to discuss some general issues regarding cannabis legalization. First, it's important to realize that impending federal legalization is not absolutely certain, despite the many indicators that suggest that this is in the cards. Despite the growing number of state-licensed marijuana businesses in existence, and ever-increasing pro-legalization public sentiment, a restrictive federal government could, should it choose to do so, decide to enforce existing laws, and close down every single dispensary. If legalization proceeds though, how might things (so to speak) roll out? Harking back to the last chapter, the most likely emerging model, although by no means the only one, is the alcohol paradigm. Under this scenario, for-profit corporations produce cannabis, sell it on the open market and heavily advertise its use, according to Jonathan Caulkins of Carnegie Mellon University, and co-director of RAND's Drug Policy Research Center in Santa Monica, who has written clearly and intelligently on this issue [7]. An alternative model is legalizing cannabis analogously to tobacco. This implies focusing on short-term reduction of the product's use and its long-term extinction. Powerful commercial lobbying efforts make this very unlikely. Legalizing cannabis like alcohol focuses on harm reduction but not restricting sales, other than to vulnerable individuals such as teens. Or as Mark A R Kleiman termed it "weakly-regulated commercialized legalization" [8]. As we saw in the prior chapter, the negative side of this model is that big tobacco and big alcohol will try to sell as much product on a large scale, as cheaply as possible and to market it to youth, in order to build lifelong consumers. We've learned from the targeting of Juul fruit-flavored nicotine vaping products to teens, and from Joe Camel advertisements, how this scenario is likely to unfold [9]. So part of harm reduction is to anticipate such moves and to think about how to counter them.

To better anticipate the harms associated with widespread use, the federal government would be wise to change cannabis' Schedule 1 status, and to allow more research on these problems. As one aspect of this, they will also need to give researchers access to multiple sources of marijuana that more closely resemble what's available to consumers, for their experiments. This menu will include concentrates, edibles, tinctures, and flower marijuana from multiple sources that reflect the range of THC and CBD available in the real world. The relevant entity, likely NIDA, will definitely need to source this material from more than a single, relatively small-scale supplier at the University of Mississippi. This effort may be underway as the result of a recent physician-scientist initiated lawsuit. Another aspect to recognize is that while we have been discussing cannabis use as an isolated phenomenon, it can of course occur alongside consumption of alcohol, tobacco, and other drugs. Due to the possible synergism between alcohol and cannabis-related impairment, this mix-and-match consumption is a special concern for, among other outcomes, driving while intoxicated, and work-related accidents. Similarly for co-use with tobacco, health concerns are important.

We've learned a lot from epidemiology in terms of what outcomes to expect from boosted cannabis consumption, and we have our neighbor to the north that is already beginning to offer us practical observations. But it will likely take decades to figure out legalization's longer-term consequences. Some policymakers are already grappling with these issues based on the experience of Colorado and other states that have legalized recreational marijuana. In July 2018, New York State published a lengthy and fairly comprehensive document outlining their assessment of the potential impact, risks, and benefits of regulated marijuana for the region's health, criminal justice, public safety, economics and education [10]. Some of their conclusions are provocative, although many could be classified as overly optimistic. For example, "changes in overall patterns of use are not likely to be significant... There is no conclusive evidence about whether legalizing marijuana increases use... brief increases in use in Colorado and Washington leveled out... (but)... a regulated marijuana program should monitor and document patients' abuse to evaluate the impact of legalization." The New York subject matter experts also doubted that legalizing marijuana would increase use among youth and noted that there would likely be a reduction in the use of synthetic cannabinoids. They recommended that an appropriate regulatory framework "could support a more appropriate level of treatment for marijuana use that focuses on harm reduction."

Therefore, what might sensible, science-guided harm-reduction strategies look like? Many possibilities exist to minimize harm and maximize benefits. Some of these are more practical to enforce, and more desirable than others. Deepak Cyril D'Souza and others at Yale have devoted a fair amount of time to enumerating public policy proposals that have influenced some of the initiatives I explore next.

Set age limits on cannabis sales

Teens are uniquely vulnerable to cannabis effects, from increased psychosis risk to (disputed) effects on IQ, and associations with reduced academic grades and (perhaps reversibly) impaired educational and occupational attainment. Restricting sales to minors, with rigorous proof of identity and significant penalties for underage sales to both purchasers and vendors would all seem to be necessary and sensible policies. Indeed, these are the national measures pursued by Canada during its recent federal legalization process. One obvious question is what should the minimum age be for cannabis sales. Remembering Susan

Weiss' admonition that scientific consideration should take priority over commercial ones, and having learned about frontal lobe development and neural pruning trajectories in the human brain, we can make a few recommendations here. The lower age limit could be either 18 (as in most of Canada), age 21 (as in Québec), or possibly age 25 (the probable endpoint of the neural pruning process). If we are guided more by the neurobiology of pruning, and the psychosis risk data from the Dunedin study, the safest age would be 25. If we are led more by practicality, then age 21 is where legal limits would most likely settle. In the real world, neurobiology is likely to compromise with feasibility to settle on age 21. As in Canada, one's age at a dispensary would need to be confirmed by use of a rigorously checked federal or state official ID, and there would be severe penalties for underage purchases for both the buyer and the dispensary. Purchasing cannabis on behalf of an underage customer is also strongly sanctioned in Canada. Again, I believe that these are sensible policies.

Limit drug potency

Numerous studies, most convincingly the recent Di Forti investigation [11], demonstrate that cannabis potency is a major driver of psychosis-associated risk. So that control of THC strength, for example by restricting sale of concentrates and dabs, or perhaps mandating THC to CBD ratios, would seem to be indicated. However, passing such legislation might have unintended consequences, for example, encouraging an illicit market in the banned high-potency substances. The resulting rash of home-brewing burn victims and butane-contaminated concentrates that would accompany uncontrolled home-made BHO manufacture with its untested product would have to be weighed into any decision. The Netherlands did ban THC concentrations that exceeded a certain arbitrary limit, but such enforcement is more practicable in Holland than in the United States.

An important related issue is how cannabis users have changed their drug use behavior in response to the high-THC cannabis now available. The assumption of many anti-cannabis advocates is that high-THC chemovars necessarily have adverse effects because of their potency, but such an argument isn't true when individuals accustomed to drinking low-alcohol content "near beer" gain access to distilled spirits such as whiskey. Such persons don't imbibe equivalent volumes of whiskey, but generally modulate their alcohol intake to reach the same degree of intoxication. This question hasn't been intensely studied in the cannabis world, despite its relevance, but Korf and colleagues surveyed Dutch "coffee shop" cannabis users and found a tendency for them to adjust for stronger cannabis by smoking smaller amounts and inhaling less deeply [12]. Nevertheless, the Di Forti data are compelling in linking THC potency to psychosis risk.

The bottom line is that limiting drug potency is a terrific idea in theory, but hard to implement in practice and likely accompanied by negative outcomes.

The unknown but crucial fact is whether the number of cannabis users protected from legal high-THC cannabis by legislation is outweighed by the size of the population who will be harmed by purchasing or home-brewing illegal high-potency forms of the drug. There is no easy, a-priori method to predict the relative percentages of each.

Reduce cannabis consumption rates

Various strategies for curtailing sales have been proposed. Here's how to do it. We can limit the amount of product sold to one individual during a single sale, and restrict the locations where cannabis products are sold, (perhaps to premises controlled by states, as is the case with liquor in some regions). Drug sales to individuals can be tracked across dispensaries using the types of applications currently used to tally sales of abusable prescription drugs such as opioids, by pharmacies. One important legislative initiative would be to limit cannabis advertising, as has been mandated for cigarettes and alcohol at a national level. We saw previously that adolescents are especially influenced by prominent local dispensary ads. This seems a strong rationale for curbing such advertising. In discouraging cannabis users from over-consumption, some could be tempted to use smaller amounts of more desirable marijuana. Encouraging the growth of niche markets for unique/high quality, more costly cannabis products would have the aim of cultivating more discriminating consumers who ingest lower overall quantities of THC. Together, these standards and restrictions would have a particular focus on reducing adolescent use, and avoid some of the truly undesirable outcomes that I contemplated in the prior chapter.

Improve product quality and ensure uniformity and safety

Enforcing uniform national standards on quality testing for cannabis products will be essential to implement. This effort will entail standard testing of contents in registered laboratories. These constituents would include at a minimum the percentages and amounts of multiple cannabinoids and terpenes, similar to that used for pharmaceuticals. To keep everybody honest, disguised specimens of previously known content would be submitted to the laboratories on a regular basis by a federal agency to maintain accuracy. This type of procedure is already used as part of the approval process for medical laboratories. As the Food and Drug Administration mandates for foods, rigorous safety testing to ensure that cannabis products are free of pesticides, fungi, and heavy metals will help protect consumers from harmful chemicals. This safety information should be clearly posted on package labeling. As in Canada, standard health warning labeling should be mandated, as exists currently for alcohol and tobacco products. Ideally these should be displayed in large legible type, with content such as "unsafe for pregnant or lactating women," and "do not operate a motor vehicle or heavy machinery for two hours after using this product." As has been legislated with our neighbor to the north, there should be clear federal rules on packaging and labeling for all forms of cannabis edibles, such as standardizing their form and color, selling them individually portion-wrapped, within clearly marked childproof containers that bear standard warnings. Sales of edibles such as gummy bears, "Krondike Bars" and THC-infused chocolate, that could easily be mistaken for candy by young children, should be limited.

Dispensaries and point of sale procedures. License budtenders and cap retail cannabis outlets

Early in the book, we reviewed the sociology research where dispensaries were telephoned, ostensibly by pregnant women experiencing morning sickness who were advised by budtenders that cannabis products were a safe and effective means for treating their condition. Worrisome results like this have led many to suggest that legislation may be necessary barring budtenders from making medical claims or providing medical advice. One extension of this theme would be to have appropriate age restrictions for budtenders, and to train and license them in a standardized manner, as is already required in many states for barbers, cosmeticians, etc.

In an effort to control both inappropriate mass-prescribing by physicians in "pill mills" and to track drug-seeking patients who were receiving drug prescriptions from multiple physicians, many states oversaw the creation of a database of controlled substances for tracking purposes. Tracking marijuana purchases in a similar manner would help reduce the incidence of abuses, as I mentioned earlier.

Legalization is often followed by opening of multiple dispensaries. In Boulder Colorado, there are now more dispensaries than both Starbucks and regular pharmacies. Both living close to a dispensary and viewing its advertising increases cannabis use among youth. Thus some legislators have proposed capping the allowable number of cannabis retail outlets in a particular area, and limiting or prohibiting the use of dispensary billboards.

Appropriate product pricing

Economic research has shown that one of the primary drivers of substance purchase for relevant comparator consumables such as alcohol and tobacco is pricing [13]. For example, in 2018 Scotland set a minimum price on sales of a standard alcohol unit. Thus the bottle equivalents of a pint of beer, a shot of spirits, and a glass of wine in supermarkets and liquor stores could not be sold below a set unit price. Within the following year alcohol sales fell significantly. This is the economic equivalent of pricing cannabis based on THC content. Accordingly, decisions on drug taxation and pricing will play an important role in how much cannabis is purchased, and by whom. Price too low, and you will encourage cannabis over-consumption resulting in, for example, higher dependence rates. Price too high and you will drive consumers back to the black-market, with its links to crime and its unregulated, untested product. This is a good example of Susan Weiss' dictum that scientific considerations should override commercial ones; the temptation for states is always to push for maximum tax revenues. How should we adjust pricing to help prevent occasional cannabis consumers from becoming frequent or dependent binge users? The late Mark Kleiman has written cogently on this issue [8]. There may be a distinct price point of relevance; according to one view, if the US-wide cost of an hour of cannabis intoxication falls below a dollar, then the type of dysfunctional uses that we discussed in Chapter 6 (misguided self-prescribed therapy) will begin to spike among vulnerable individuals [14]. This should be prevented by appropriate pricing, that can be adjusted based on using dispensary sales data to track consumption rates.

Teaching teens using facts and avoiding inappropriate scare tactics; providing education that works

Credible and effective public education about the potential harms of marijuana use is important, given the increases in those who see marijuana use as harmless. Such education needs to start appropriately early. Teens, with their still-maturing brains, their tendencies toward impulsivity and living in the moment, disinclination to think about future consequences of their actions, and feelings of relative invulnerability, are predisposed to try recreational drugs of all types. Use of alcohol and other drugs is relatively common in high schools. Certainly, we know that adolescent brains are still developing, a fact that puts teens at increased risk for particular harmful effects of cannabis, including dependence and psychosis. One way of curtailing cannabis use by vulnerable teens is to educate them about these potential harms, but, this is easier said than done. Existing research data can provide useful guidance on drug education that works. "Just say no" policies clearly don't work. The best-known drug education campaign, D.A.R.E., (Drug Abuse Resistance Education) preaches abstinence and is dramatically ineffective. Former first lady Nancy Reagan supported government-funded youth anti-drug programs, including the infamous "Just say no" initiative. DARE began in 1983 as the brainchild of the LA Police Department. Almost $1 billion was spent on this campaign, and it is generally recognized to have been a failure due to its tendency to make inaccurate and exaggerated anti-drug assertions easily debunked by savvy kids, and its short-term focus. For example, a 2009 meta-analysis of 20 controlled studies revealed that teens enrolled in the program were no less likely to use drugs than those who had not been exposed. [15–17]. One illustration of DARE's general cluelessness is that it mistakenly published on its website an entire satirical report that made numerous outrageous claims regarding cannabis. These included: "Four teens become pregnant for every joint smoked," "Marijuana candies, sold on the street as 'Uncle Tweety's Chewy Flipper' and 'Gummy Satans' are taking the country by storm," and "Children are being addicted to marijuana. I knew this day would come, when a liberal president allowed a state to legally sell Marijuana Flintstone vitamins to children." Nor do exaggerated Reefer Madness scare tactics have the intended impact. Most teens are acutely aware when someone is trying to sell them a bill of goods. When in 1986 President Reagan's drug advisor Carlton Turner suggested that marijuana could make you both gay and more susceptible to AIDS, few adolescents paid attention except to roll their eyes.

So what sort of drug education does work on the school-age individuals who most need to hear science-based information? Also, what is the most appropriate place to receive this guidance? According to the 2012 National Survey on Drug Use and Health by the Substance Abuse and Mental Health Services Administration, the home front is a good starting point. Teens who consistently learn about the risks of drugs from their parents are up to 50% less likely to use them than those who don't [18]. Middle and high school is the next obvious place. The Drug Policy Alliance has what I believe is a sensible policy on high school drug education; reaching high school students who already have exposure to drugs is a particular priority. A large-scale review suggests that successful programs involve many aspects missing from the DARE approach [19]. These strategies include leveraging social influence, (e.g., through the use of peer leaders), focusing on sustained interaction between students and instructors over a long time period, and providing accurate knowledge regarding substance use norms. (For example, it is important to convey that the majority of students do not use drugs, and that student beliefs regarding frequency and quantity of use are significant overestimates). Using role-playing exercises regarding acceptable drug refusal behavior is helpful in building effective interpersonal skills. Programs with the best track records tried to elicit a commitment not to use drugs from students, and explored their intentions not to indulge. Other features that boosted success included additional community interventions, and teaching life skills to the students.

As well as dealing with cannabis-associated risks in a factual manner, important educational information needs to be embedded in a more general context that might apply to all recreational substances. Teachable skills such as how to avoid social pressures to use drugs, not using them to cope with unpleasant emotions, and stress reduction techniques, for example, can help individuals move out of an existing substance use disorder or to avoid initial drug entanglement.

We need to design effective courses that will restrict teen cannabis use. What I proposed earlier is what we've learned works. But funding for effective drug education in schools is not cheap. Mandating drug education in schools paid for by local taxes on cannabis is one obvious revenue source. In Colorado, this has not played out in the way that taxpayers imagined. The initial good intent was clearly present. For example, a pre-legalization TV ad stated "Let's have marijuana tax money go to our schools rather than criminals in Mexico."

For 2017–18, the total marijuana tax revenue for Colorado's public schools was $90.3 million, or 1.6% of the entire $5.6 billion K-12 public school budget. Of this revenue, $11.9 million, or 13% went into substance abuse and health-related programs [20]. This level of funding sounds respectable, although admittedly we would need more details of the programs themselves, and how those dollars were spent to know that for sure. But few states invest in such programs.

In addition to education programs in schools, there is a need to teach the workforce more generally in terms of cannabis-related job safety, to create policy for child protective services regarding how to respond to positive marijuana screens in newborns and pregnant mothers, post-legalization. Another obvious place to find funding for drug education is to divert federal money from the expensive and generally ineffective war on drugs. According to the drug policy Alliance, the United States spends $51 billion annually on these initiatives. Alan Leshner, a former director of the National Institute on Drug Abuse pushed consistently for better public drug education, stating that people need to be informed that addiction is "a brain disease expressed in behavioral ways that occurs in a social context" [21]. Leshner also talked about the "great disconnect," i.e. "the large gap between the public's perception of drug abuse and addiction and the scientific facts" [22]. Addressing this gap by spreading scientific information more effectively was one factor that helped motivate me to write this book.

Some of the needed adult educational content is obvious: consumers need acquaintance with facts regarding the implications of high THC and THC/CBD ratios, and authorities should be alerting individuals at high risk of psychiatric illness in terms of avoiding use. This last priority leads to my next question.

Can we identify individuals at high-risk for psychosis, and prevent them from using cannabis?

Part of the answer to this question is a practical scientific one. Even if we could come up with an ideal test, for example, a composite psychosis genetic risk score, or a brain scan to identify individuals at high risk from cannabis-related psychosis, there is no ethical way to mass-screen young people to identify who is at high-risk. Neither is screening practical from a budgetary point of view, when the tests may be expensive, the rate of schizophrenia in the population is only 1% to begin with, and not everybody who's identified as being at risk will develop the illness even if they are exposed to cannabis. Additionally, even if all of this testing were able to identify individuals who are at substantially increased risk of psychosis, those pinpointed may choose to ignore advice not to use cannabis. A more practical, if less specific approach is to educate teenagers about the risks of using cannabis, as proposed earlier. So, teach them, don't screen them. Part of that advice could be more focused. That is, if an individual has a family history of schizophrenia or another psychotic illness or if using marijuana tends to make them feel paranoid, or they have previously experienced a brief psychotic episode when using the drug, then their risk for developing a more serious psychotic illness, is likely increased significantly by further cannabis use. Furthermore, if despite the risk they wish to continue using cannabis, they can reduce their odds of psychosis by using it as infrequently as possible, avoiding high-THC-containing formulations, and possibly by picking compounds with more equal THC:CBD ratios. These are all practical and teachable harm-reduction skills.

In addition, teaching high-risk individuals stress-management techniques may offer a degree of protection. The evidence for this comes from a recent Toronto study that measured brain changes in individuals in their early 20s who were at high-risk for developing schizophrenia. These young people did not have a clinically diagnosable psychotic illness, but had experienced several months of symptoms such as believing that objects in the environment were changed in some way, were suspicious that people might be following them, and experienced declining school performance without an obvious explanation. Such people have about a one in three chance of ultimately developing a psychotic illness. The Toronto investigators used a PET scanner to measure the release of the neurotransmitter dopamine from parts of the prefrontal cortex in these individuals while they performed a stressful task that involved solving challenging mental arithmetic problems. Nine of the volunteers used cannabis regularly, while 23 had never used. The key finding was that the cannabis-smoking group released much less dopamine from this part of their brain during the task and showed more psychosis-like symptoms following the stressful math stimulus [23]. The study, (despite the relatively small number of subjects tested), is telling us something about a possible biochemical mechanism related to cannabis smoking that might help explain the biology behind their increased risk for schizophrenia, and its relationship to stress. As well, it suggests a practical strategy. Teaching stress reduction techniques to all pupils (as happens already in some schools) may reduce risk while offering general benefits and boosting life skills.

If we can anticipate a somewhat greater number of cases of cannabis-related psychosis emerging, then we can also plan ahead in terms of harm-reduction. There is evidence that early preemptive identification and prompt treatment of psychotic illnesses in young people improves long-term outcome. Part of the tax dollars derived from marijuana sales can be devoted to addiction treatment programs and also to community early psychosis detection and intervention initiatives. Patrick McGorry is an Australian psychiatrist who successfully lobbied his government for funds to create and implement a national network of early psychosis intervention and treatment centers across the country that have been widely copied elsewhere. Such steps are cost-effective. In addition to this, educational campaigns and public service announcements aimed at increasing awareness and reducing stigma will help individuals obtain speedy referral to coordinated specialty care programs. For example, simple cognitive behavioral therapy as an add-on treatment seems fairly effective in preventing first-episode psychosis [24,25].

Reducing marijuana-impaired driving

In Chapter 8, we discussed at length considerable difficulties in detecting specific cannabis-related impairment at the roadside, and the complexities of establishing recent use through blood and saliva testing. Since there is reasonable evidence that recent cannabis use does impair driving, then finding reliable, straightforward, research-based methods for detecting it, is on everybody's agenda. Hound Labs' THC "breathalyzer" device would represent a considerable step forward, if it proves as effective as advertised. Other technology to detect impairment at the roadside is under development. If effective and foolproof roadside testing for cannabis-impaired driving has yet to be realized, effective preventative measures can be implemented now. In New Zealand, controlling driving under the influence of marijuana relies in part on widespread public safety advertisements. Sample slogans include "Hits lead to hits" and "Grinding one (next to a picture of a cannabis bud) can crash the other" (next to a picture of a wrecked automobile).

Changing drug laws and drug policies

Policymakers commonly assume that tougher drug policies reduce adolescent cannabis use. Alex Stevens is a sociologist and Professor in Criminal Justice at the University of Kent in the United Kingdom who has evidence to dispute this view. He recently surveyed over 100,000 teens in 38 countries across Europe, Russia, and Canada about their cannabis use. Consistent with many prior studies, he demonstrated no link between tougher cannabis penalties and lower use, even after carefully controlling for differences among both the countries and the participants, for example, in terms of socioeconomic status and national income [26]. So, laws need to be smart, not tough.

In the prior Chapter 11 I had much to say regarding the need to change current policies on cannabis-related incarceration and to limit the prison industrial complex that informs a continuing need for drug policy change. One key issue being debated in many states is whether past records for minor cannabis-related conviction should be expunged, (restorative justice).

Testing for drugs in the workplace, following legalization

This complex public safety issue has a vague Brave New World feel to it. We have seen that acute cannabis use is accompanied by short lasting cognitive impairment. Therefore, companies may be rightfully concerned that workers will use the now-legal substance during breaks, and may be injured on the job, or perhaps damage equipment. Unlike alcohol, cannabis would not be detected on the user's breath, or necessarily by obvious intoxicated behavior. In context then, marijuana would be just one drug used recreationally among several that have entered the legal sphere, along with alcohol and abused prescription drugs. So a first step where cannabis intoxication is suspected should not be to determine whether an employee's drug screen shows up positive. As we have seen in the case of driving, toxicology is not informative in distinguishing between recent versus days-old marijuana use. The bottom line is to determine, "whatever the cause, is this person impaired right now in a way that will interfere with his or her job performance?" For individuals with key roles in the work environment where safety is paramount, such as blast furnace operators or long-distance truck drivers, some businesses have already journeyed along this path of screening employees' performance. Portable test devices exist that can be used to test a worker cognitively at baseline, and to establish their normal, (negative urine tested) sober function scores. Examples are handheld digital devices that display patterns rapidly and rely on focused attention to provide correct answers. Some of these have communication capabilities that allow for sending information via PC or smart phone from distant sites, for example, truckers out on the road. The employee can be texted on their smart phone at any point and asked to test themselves on the device, or they can be screened at the beginning or end of a work shift. If their current performance on the device fails to match their sober baseline, then they are required to report for mandatory, more detailed appropriate assessment. It would be prudent to consider such screening routinely for individuals in skilled jobs with a high public safety impact such as train engineers and airline pilots. If this seems harsh, recall that around 1% of airline pilots test positive for cannabis, although admittedly were not sure how recently they used. You should always ask yourself "does the person flying my plane right now belong to that 1%?"

What are the research issues of the future?

Outside of the issue of cannabis-associated risk reduction, there are numerous other interesting questions that research can help address, that are fodder for future research. A few of these are listed further. But it is worth noting that funding for marijuana-based research in general is disproportionately low, access to experimental substances is unnecessarily difficult, and existing research has had a hard time keeping pace with emerging public policy issues.

Prolonged use of concentrates

For individuals who use dabs and other concentrates in large amounts, the long-term consequences of such super-use are unknown at this point. Due to their relatively recent availability, users haven't consumed them for long enough yet to reveal anything about their possible long-term side effects on either lungs or brains. Neither have research laboratories challenged subjects with the high

THC doses provided by these compounds. This type of research needs to be done. But THC provided by the National Institute on Drug Abuse, for example, does not yet include these types of formulations. While concentrate users are increasing in number, it's also likely that future cannabis consumers will also incorporate upscale professionals and hipsters extolling expensive prime craft bud, following the example of current day wine snobs.

Risks of illicit vaping

In mid-2019, a spate of reports emerged nationwide in the United States of hundreds of cases of serious lung-related illnesses related to vaping and some deaths. Evidence implicated the role of illicit vaping supplies, especially "home-brewed" cartridges used for consuming nicotine and THC, manufactured in underground "pen factories." It's relatively straightforward to purchase large quantities of vape pens, empty cartridges, and realistic-looking fake packaging online either through legitimate or dark web sources. THC or nicotine liquid are both readily obtainable, and given a little time and effort, it's straightforward to mix the ingredients together, perhaps "cut" them with dubious adulterants to increase profitability, and inject them into the individual vape cartridges [27]. All of this assembly is several orders of magnitude easier than say procedures in *Breaking Bad,* as no new chemicals have to be synthesized in the pen factory. The problem is that some of the carrier material mixed with the drug (particularly the THC), may contain vitamin E acetate which has been fingered as a potential lung irritant and other potentially harmful substances. This is an evolving story that may ultimately affect the status of all portable vaping devices in the United States.

New sources of cannabis and cannabinoids

How can we grow better cannabis plants faster and cheaper? Producing larger quantities of cannabinoids at industrial scale to bring down prices will help fuel the discovery of new ways of growing both cannabis and its essential biochemical contents. Significantly scaled-up hydroponic cannabis groweries seem to be one such emerging trend, as we encountered in the previous chapter. Crop scientists are already exploring ways of growing hardy, disease-resistant hemp on a large scale. Geneticists may aid them in a quest to produce robust, easy-to-grow cannabis plants that thrive when cultivated en masse outdoors. Traits such as resistance to powdery mildew, auto-flowering/day-neutral, and disease-resistant properties need combining in plants engineered for predictable THC and CBD ratios. If security measures can be figured out for field agriculture, then outdoor growth of drug cannabis crops may soon follow. Initially such desirable traits will be engineered by plain old-fashioned plant breeding. But genetics will soon follow to accomplish this more precisely and predictably.

In 2011, Canadian biologists published the first cannabis genome by sequencing Purple Kush DNA. Companies such as NRGene are focused on using single nucleotide polymorphism markers (SNPs) to inform them about genetic variation within cannabis plants. Detailed knowledge of this information can be used to help build more desirable chemovars without relying on guesswork. Another approach is to use genetic techniques directly to custom-design marijuana plants that display particular features. For example, CRISPR-cas9 is an enzyme that can be used to cut and edit repeating DNA sequences, and to introduce new snippets of DNA into the genome that code for desired features. In the world of botany, CRISPR-based techniques allow highly specific and fine-tuned alteration of a plant without the trial and error associated with most attempts at genetic modification of cannabis. The usual stages of producing novel chemovars include selectively crossbreeding, then growing plants to maturity over 3–4 months and screening them. This is a slow, laborious trial and error process. The precision of CRISPR allows introduction of new, beneficial features (such as disease resistance or increased terpene production), or removal of undesirable ones (e.g., small plant size). Yoav Giladi is a plant breeder from the Volcani Center in Jerusalem and an advisor for the Israeli Industrial Hemp Pilot Program who has begun to focus on using these techniques [28]. Finally, geneticists are rapidly discovering the key genes that underlie the synthesis of cannabinoids, terpenes, and other compounds within the cannabis plant. With that knowledge, they can both use techniques such as CRISPR to genetically engineer superior plants, or intriguingly to introduce those same genes into yeast or into bacteria, that can then be grown on a massive scale in factory vats, bypassing botany altogether. Stay tuned for further developments in these and other areas over the next few years.

The future of medical marijuana

Medical marijuana is at a crossroads. Nearly 10% of cannabis users in the United States report using the drug for medicinal purposes [5]. Strong claims are made for its utility in treating a variety of medical conditions but solid evidence of its efficacy is mostly lacking, for a variety of reasons I've discussed in prior chapters. How can we move the field forward?

Cannabis, as we've learned, is a complicated plant, containing so many different classes of chemical compounds, with numerous compounds within each class, although we tend to oversimplify things by referring to medical marijuana as if it were a single entity. Research studies have already shown us that CBD is effective in significantly reducing the frequency and severity of certain types of infantile and childhood epileptic seizure syndromes. And there is preliminary evidence that various cannabinoids and terpenes, including THC, have neuroprotective, anti-inflammatory, pain and spasm reducing, appetite-increasing, and tumor-suppressing properties. But as I've described, much of the needed evidence for medical marijuana's real-world effectiveness, based on properly powered and adequately controlled clinical trials is still lacking. That situation needs to change. Current circumstances don't imply that constituents of cannabis don't work to treat various illnesses. We don't know that yet. What they do suggest strongly is that we need to begin carrying out such studies as soon as possible. Clinical trials need to be designed as rigorously as is practicable to begin providing definitive answers. And in order to do that, the first crucial step is for the federal government to reclassify cannabis from its current Schedule 1 status, and make it significantly easier for researchers to access these compounds. One of the first issues that needs addressing is how real and how significant entourage effects are.

If recreational marijuana becomes federally legalized, then individuals who wish to purchase cannabis to treat medical complaints can do so as they wish. Presumably the distinction between medical and recreational forms of the drug will diminish. Medical insurance is unlikely to ever cover medicinal cannabis in the form of flower or edibles, but will likely pay ultimately for standardized tinctures or pure compounds extracted from the herb. It's those pure compounds that I want to talk about next.

The pharmaceutical deployment of plant-based medicines, such as the heart drug digitalis, derived from the familiar garden foxglove, the 1950s blood pressure drug reserpine, extracted from an East Asian herb and the anti-malaria drug artemisinin from the traditional Chinese wormwood plant, tend to follow a similar pattern. First, traditional herbal medicine or initial experiments may suggest that a particular plant is useful in helping treat a medical condition. Then pharmacologists and chemists extract multiple candidate compounds from the plant and home in on the relevant one, often via animal experiments. These scientists then purify this chemical, and turn it into a potential medication with known properties that can be dosed precisely when prescribed to patients. Next, they figure out how to synthesize it, then run the compound in clinical trials to prove real-world effectiveness. If those trials show that the drug meets the FDA's criteria as both safe and effective, then they will market it. With cannabis, in most cases we're at the very start of that long process. The plant contains multiple compounds including dozens of cannabinoids, plus terpenes and other relevant compounds many of which have potential medical properties, and may also rely on entourage effects to produce optimal medical results when used in combination. The relevant chemicals may well differ for treating different disorders. So we need scientists to examine the vast library of compounds locked up inside the cannabis plant, to figure out which compounds or combinations are most effective in treating which illnesses, as the first step in the process I just described. We know that the cannabis chemical library is huge, but it is hardly infinite, and modern drug screening protocols can efficiently pick it apart and establish which compounds work best in which situations. On the other hand, as outlined elsewhere in the book, pharmaceutical companies may skip the cannabis plant altogether and decide that their best strategy is to proceed with drugs that act on the endocannabinoid system directly, such as FAAH inhibitors and CB receptor modulators. Another reason why the medical and recreational drug paths are destined to diverge is that smoking and vaping are probably not the most effective delivery mechanisms for any medicinal drug.

What can we learn from Canada's experience?

This is a question that is already occupying social policy researchers and epidemiologists and will ultimately offer lessons for consequences of legalization in the United States. In June 2018, Canada fully legalized recreational marijuana, the only country other than Uruguay to make this transition. Adult Canadians and people visiting the country will be able to purchase marijuana at stores throughout the nation, and to purchase it online. The launch began by allowing sales of cannabis flower, seeds, and tinctures; it will follow-up with later sales of concentrates and edibles. Canada has very strict rules regarding cannabis sales; some people have compared these more to the marketing of tobacco than that of alcohol. Canada has a number of legal provisions to constrain use. These laws have sharp teeth. For example, they prohibit production or sale of products that appeal to youth, particularly through packaging or labeling, and do not allow cannabis sales through vending machines or self-service displays. Violation of these prohibitions includes fines of up to $5 million or 3 years in jail. Giving or selling cannabis to a person under 18 is punishable by up to 14 years in jail, as is "using a youth to commit a cannabis-related offense." Possession of small amounts of cannabis over the statutory limit is merely ticketed, but possession of large amounts can land you up to 5 years in jail. Cannabis-impaired driving is taken seriously and punished appropriately. Cannabis products are packaged analogously to European cigarettes, in plain, unenticingly colored, industrial-style boxes, plastered with large-font health warnings. Containers for edibles are childproof and bear highly visible warning labels. Advertising is limited and kept away from children.

This legislation varies to some extent from province to province. For example, in Ontario, consumers are free to smoke or vape marijuana any place they can legally use tobacco, but in other provinces such as Manitoba there are hefty fines for the same behavior [29]. Ontario initially planned to sell cannabis only at government-run stores, but then changed its mind and switched to private licenses. The minimum age for purchase varies from state to state; federally it is 18 years but Québec has set the limit at 21 years. There have been calls for expunging the legal record of individuals who had been previously criminalized for cannabis possession; so far there's been no progress on this front. And while employers are still debating how to handle positive THC drug tests, the police are wondering how to deal with the persisting black and gray markets. Meanwhile, in Ottawa swarms of lobbyists are appealing to the government to loosen the rules [30].

Many of these Canadian initiatives address issues I discussed earlier in this chapter. For example, an important question that public policy experts are currently debating is whether these rules and regulations are sufficiently stringent to reduce harms, or overly harsh and thus encouraging an illicit use. For these reasons, keeping a very careful eye on drug use trends over the border will provide us with useful information on how future cannabis legislation should be handled in the United States. Each time a society sanctions the use of a recreational drug, it is implicitly making a series of calculations. From time to time this calculus needs to be revised and decisions revisited. So it is currently with marijuana. Hopefully emerging policies will benefit from consideration of existing science.

In conclusion, while many individuals can use cannabis without problems, cannabis use is not risk-free. Existing research can already help mitigate potential harms associated with use of the drug, and particularly impact those that will grow following likely federal legalization of recreational marijuana. Credible public education can play an important role in harm reduction, particularly given the increases in public perception of cannabis as a harmless substance. This education will enable informed decisions about safe and sensible use. Policies to protect youth from a burgeoning cannabis industry that is allied with big tobacco and big alcohol will be essential. Research-based information about potential problems associated with cannabis use should be conveyed to educators, and to medical and psychiatric providers. This will help deal with genuine risks, while avoiding panic and alarmism. State and national policy-makers will need access to relevant research data, to help them formulate laws and regulations. Still, it is wrongheaded to think that the only choices we have in drug policy are a punitive approach centered exclusively on enforcement, or one based on careless or ill-considered legalization. Neither has ever worked well. Finally, continued research is necessary to explore the still large number of unknowns surrounding this ancient and fascinating plant.

References

[1] Carlini EA. The good and the bad effects of (-) trans-delta-9-tetrahydrocannabinol (Delta 9-THC) on humans. Toxicon 2004;44(4):461–7.

[2] Caulkins J. Against a weed industry. National Review; 2018.

[3] Hasin DS, et al. Prevalence of marijuana use disorders in the United States between 2001-2002 and 2012-2013. JAMA Psychiatry 2015;72(12):1235–42.

[4] Volkow ND, et al. Adverse health effects of marijuana use. N Engl J Med 2014;370(23): 2219–27.

[5] Hill KP. Medical use of cannabis in 2019. JAMA 322, 2019, 974–975;.

[6] Weiss S. Research. Society of Marijuana, 2nd Annual Scientific Mtg. How Cannabis Research and Cannabis Policy Can Inform one Another. 2018, Colorado State University: Fort Collins, CO.

[7] Jonathan P, Caulkins B, Kilmer MAR, Kleiman. Marijuana legalization: what everyone needs to know. Second ed. Oxford University Press, Oxford, United Kingdom; 2016.

[8] Kleiman M.A.R. How to prevent casual pot smokers from slipping into abuse and dependence. Vox.2018. Available from: https://www.vox.com/the-big-idea/2018/4/20/17259032/marijuana-abuse-dependency-risk-policy-420-drug-addiction.

[9] Henehan ER, et al. Alcohol advertising compliance on cable television, July–December (Q3-Q4), 2018. 2019, Johns Hopkins Bloomberg School of Public Health Baltimore, MD.

[10] Assessment of the Potential Impact of Regulated Marijuana in New York State 2018.

[11] Di Forti M, et al. High-potency cannabis and incident psychosis: correcting the causal assumption—Authors' reply. Lancet Psychiatry 2019;6(6):466–7.

[12] Korf DJ, Benschop A, Wouters M. Differential responses to cannabis potency: a typology of users based on self-reported consumption behaviour. Int J Drug Policy 2007;18(3):168–76.

[13] Hyland A, et al. Higher cigarette prices influence cigarette purchase patterns. Tob Control 2005;14(2):86–92.

[14] Salam R. Is it too late to stop the rise of marijuana, Inc.? The Atlantic. 2018, Available from: https://www.theatlantic.com/politics/archive/2018/04/legal-marijuana-gardner/558416/.

[15] Lynam DR, et al. Project DARE: no effects at 10-year follow-up. J Consult Clin Psychol 1999;67(4):590–3.

[16] Lilienfeld SO, Arkowitz H. Why "Just Say No" doesn't work. Scientific American. 2014. Available from: https://www.scientificamerican.com/article/why-just-say-no-doesnt-work/.

[17] Hartley M. Do anti-drug campaigns actually work? Leafly. 2019. Available from: https://www.leafly.com/news/health/do-anti-drug-campaigns-actually-work.

[18] Hedden SL. et al. Behavioral health trends in the United States: Results from the 2014 National Survey on Drug Use and Health, 2015. Available from: https://www.samhsa.gov/data/sites/default/files/NSDUH-FRR1-2014/NSDUH-FRR1-2014.pdf.

[19] Cuijpers P. Effective ingredients of school-based drug prevention programs. A systematic review. Addict Behav 2002;27(6):1009–23.

[20] Brundin J. Do marijuana taxes go to schools? Yes, but probably not in the way you think they do, 2018. Available from: https://www.cpr.org/2018/10/22/do-marijuana-taxes-go-to-schools-yes-but-probably-not-in-the-way-you-think-they-do/.

[21] Grady M. Constituent groups join forces with NIDA to bridge the "Great Disconnect," 1996. Available from: https://archives.drugabuse.gov/news-events/nida-notes/1996/04/constituent-groups-join-forces-nida-to-bridge-great-disconnect.

[22] Leshner AI. Taking drug abuse research to the community, 1997. Available from: https://archives.drugabuse.gov/news-events/nida-notes/1997/02/taking-drug-abuse-research-to-community.

[23] Schifani C, et al. Stress-induced cortical dopamine response is altered in subjects at clinical high risk for psychosis using cannabis. Addict Biol 2019;e12812. Available from: https://doi.org/10.1111/adb.12812.

[24] McGorry PD. Early intervention in psychosis: obvious, effective, overdue. J Nerv Ment Dis 2015;203(5):310–8.

[25] van der Gaag M, et al. Preventing a first episode of psychosis: meta-analysis of randomized controlled prevention trials of 12 month and longer-term follow-ups. Schizophr Res 2013;149(1–3):56–62.

[26] Stevens A. No evidence tougher policies deter adolescent cannabis use. Science Daily. 2019. Available from: https://www.sciencedaily.com/releases/2019/02/190219111730.htm.

[27] Richtel M, Bosman J. Vaping bad: were 2 Wisconsin Brothers the Walter Whites of THC oils?. NY Times. 2019. Available from: https://www.nytimes.com/2019/09/15/health/vaping-thc-wisconsin.html.

[28] Kuhl L. Selective breeding: how the CRISPR-Cas9 could potentially change the way cultivators look at cannabis Pot Network. 2019. Available from: https://www.potnetwork.com/news/selective-breeding-how-crispr-cas9-could-potentially-change-way-cultivators-look-cannabis.

[29] Cecco L. Dazed and confused: Canada cannabis legalization brings complex new laws. The Guardian. 2018. Available from: https://www.theguardian.com/world/2018/oct/16/canada-legalizes-recreational-marijuana-law-problems.

[30] Austen I. Marijuana legalization in Canada has companies chasing a green rush. NY Times. 2018. Available from: https://www.nytimes.com/2018/10/16/world/canada/cannabis-legalization-industry.html.

Index

Made in the USA
Las Vegas, NV
16 September 2021

30423607R00179